D0074592

Applied Functional Analysis

APPLIED FUNCTIONAL ANALYSIS

D.H. GRIFFEL
SCHOOL OF MATHEMATICS
UNIVERSITY OF BRISTOL

DOVER PUBLICATIONS, INC.
Mineola, New York

Bibliographical Note

This Dover edition, first published in 2002, is an unabridged republication, with a few minor corrections, of the 1985 revised edition published by Ellis Horwood Limited, Chichester, West Sussex, UK. The original edition was published in 1981, also by Ellis Horwood.

Library of Congress Cataloging-in-Publication Data

Griffel, D. H.
Applied functional analysis / D.H. Griffel.—Dover ed.
p. cm.
Originally published: Chichester, W. Sussex : E. Horwood ; New York: Halsted Press, 1981.
Includes bibliographical references and index.
ISBN 0-486-42258-5 (pbk.)
1. Functional analysis. I. Title.

QA320 .G684 2002
515'.7—dc21

2002017456

Manufactured in the United States of America
Dover Publications, Inc., 31 East 2nd Street, Mineola, N.Y. 11501

Contents

PART II BANACH SPACES AND FIXED POINT THEOREMS

Chapter 4 Normed Spaces

Chapter 5 The Contraction Mapping Theorem

Chapter 6 Compactness and Schauder's Theorem

PART III OPERATORS IN HILBERT SPACE

Chapter 7 Hilbert Space

Preface

Aims. This book is intended to be a simple and easy introduction to the subject. It is aimed at undergraduate students of mathematics, and mathematical physics and engineering, though I hope to interest other readers too. I have tried to avoid difficult ideas, as far as possible; the spectral theorem, for example, is discussed for operators with discrete eigenvalues only. But within this limitation I have tried to tell a coherent story, and to give both proofs and motivation for the theory, at least in the central Parts II and III; in Parts I and IV some proofs are omitted.

To illustrate the uses of the theory, a few problems of theoretical mechanics are discussed in some detail, and a chapter is devoted to variational approximation methods. I have tried to give some sort of application of the theoretical ideas as soon as possible after they are introduced; I have found that students are more willing to grapple with compactness, for example, after they have seen the idea at work in fluid mechanics.

Prerequisites. Readers should be familiar with the ϵ-δ theory of limits and continuity, simple ordinary and partial differential equations, Fourier series, vectors and matrices. Little else is assumed, apart from a general understanding of what is meant by a mathematical proof and a mathematical definition, such as should be acquired in a course on ϵ and δ. An acquaintance with uniform convergence and with Sturm-Liouville systems would be helpful, but brief outlines of these subjects are given in the Appendices. Complex contour integration is used in one section, for the calculation of retarded waves, but this material is not used in the rest of the book, and can be omitted.

Conventions. The definitions, theorems, etc. appearing in each chapter are numbered in a single sequence; numbers such as 4.21 refer to item 21 of Chapter 4, which may be a theorem or an example, etc. Numbers in parentheses refer to equations. The symbol □ marks the end of a proof, example, etc. When □ appears after the statement of a theorem, it means that no proof will be given. Terms being defined are printed in bold type. The following rough definitions may be useful: a **theorem** is an important result; a **proposition** is a result less significant than a theorem, but still of some interest; a **lemma** is a result of little intrinsic interest which is stated because it is needed in the proof of a theorem; a **corollary** is a result which is a (more or less) obvious and immediate consequence of the preceding theorem.

Other Remarks. For an outline of the structure of the book, see the introductions to the four parts into which it is divided.

I have not used a consistent scheme of notation throughout the text. This is only partly laziness; it seems to me that students brought up on a uniform notation sometimes become addicted, and have difficulty coping with a different notation. Hence there is some virtue in inconsistency. Of course, where there is a generally accepted standard notation, I have used it.

A related matter is the question of notation for functions. I assume that my readers have learnt calculus thoroughly enough to understand clearly the distinction between a *function* $f: \mathbb{R} \to \mathbb{R}$ and the *number* $f(x)$, the value of f. I therefore feel free to use convenient expressions like "the function $f(x) = \int K(x,y)g(y)\, dy$" instead of the more correct but clumsy "the function f defined by $f(x) = \int K(x,y)g(y)\, dy$".

There are plenty of exercises at the ends of the chapters. Hints and answers to some of them are given at the back of the book. Complete solutions are also available; see p. 367.

My thanks are due to many people who have read drafts of some of the text, and to the typists in Bristol,and the staff of Ellis Horwood who have turned it into print. I am afraid that despite all their efforts, some of my mistakes will have survived; I will be grateful to any readers kind enough to point them out.

D. H. Griffel,
School of Mathematics,
University Walk,
Bristol, BS8 1TW.
England.

PREFACE TO SECOND EDITION

Many small amendments have been made in this second edition, and some new problems have been added. I am grateful to many people who have pointed out errors in the first edition, and urge readers to follow their example. I wish also to acknowledge the helpful and friendly collaboration of Ellis Horwood Ltd in the production of this book.

PART I

DISTRIBUTION THEORY AND GREEN'S FUNCTIONS

The central ideas of our subject, the theories of Banach and Hilbert spaces, are contained in Parts II and III. Part IV can be regarded as a coda, and Part I as a prelude. The theory of Parts II and III is independent of the distribution theory developed in Part I, which can be omitted if desired (though an acquaintance with Green's functions would be helpful). In Part IV, however, the two theories will be unified.

The theory of distributions is essentially a new foundation for mathematical analysis; that is, a new structure, replacing functions of a real variable by new objects, defined quite differently, but having many of the same properties and usable for many of the same purposes as ordinary functions. This theory gives a useful technique for analysing linear problems in applied mathematics, but is less useful for nonlinear problems (which is why it is not used in Parts II and III). It leads naturally to the introduction of Green's functions, which are essential to the application of functional analysis to differential equations. And it forms a useful prelude to Parts II and III because it introduces fairly abstract ideas in a fairly concrete setting.

Our treatment of distribution theory is intended to be detailed enough to give a good general understanding of the subject, but it does not aim at completeness. Many details and many worthwhile topics are omitted for the sake of digestibility. A full account would take a full-length book; see the references given in the last section of each chapter.

The plan of Part I is as follows. In Chapter 1 we set out the basic theory of distributions. Chapter 2 discusses ordinary differential equations from the distributional point of view, and introduces Green's functions. Chapter 3 begins with a discussion of the Fourier transform from both the classical and distributional points of view, and then uses it to obtain Green's functions for Laplace's equation and for the wave equation.

The following sections form a short account of Green's functions, which may be useful to readers who do not wish to learn distribution theory: 1.1; 2.3 as far as Example 2.9; 2.4; 2.5 omitting the proof of 2.15; 3.1; the second half of 3.4; 3.5; 3.6.

Chapter 1

Generalised Functions

In this chapter we lay the theoretical foundations for the treatment of differential equations in Chapters 2 and 3. We begin in section 1.1 by discussing the physical background of the delta function, which was the beginning of distribution theory. In section 1.2 we set out the basic theory of generalised functions or distributions (we do not distinguish between these terms), and in sections 1.3 and 1.4 we define the operations of algebra and calculus on generalised functions. The ideas and definitions of the theory are more elaborate than those of ordinary calculus; this is the price paid for developing a theory which is in many ways simpler as well as more comprehensive. In particular, the theorems about convergence and differentiation of series of generalised functions are simpler than in ordinary analysis. This is illustrated by examples in sections 1.5 and 1.6. References to other accounts of this subject are given in section 1.7.

1.1 The Delta Function

The theory of generalised functions was invented in order to give a solid theoretical foundation to the delta function, which had been introduced by Dirac (1930) as a technical device in the mathematical formulation of quantum mechanics. But the idea of Dirac's delta function can easily be understood in classical terms, as follows.

Consider a rod of nonuniform thickness. In order to describe how its mass is distributed along its length, one introduces a 'mass-density function' $\rho(x)$; this is defined physically as the mass per unit length of the rod at the point x, and defined mathematically as a function such that the total mass of the section of the rod from a to b (distances measured from the centre of the rod, say) is $\int_a^b \rho(x)\,dx$. This is a satisfactory description of continuous mass-distributions; dynamical properties such as its centre of mass and moment of inertia can be expressed in terms of the function ρ.

But if the mass is concentrated at a finite number of points instead of being distributed continuously, then the above description breaks down. Consider, for instance, a wire of negligible mass, with a small but heavy bead attached to its mid-point, $x = 0$. Suppose that the bead has unit mass and is so small that it is reasonable to represent it mathematically as a point. Then the total mass in the interval (a,b) is zero if 0 is outside the interval, and is one if zero is inside the interval. There is no function ρ that can represent this mass-distribution. If there were, then we would have $\rho(x) = 0$ for all $x \neq 0$, since the mass per unit length is zero except at $x = 0$. But if a function vanishes everywhere except at a single point, it is easy to prove that its integral over any interval must be zero, so that integrating it over an interval including the origin cannot give the correct value, 1. From the physical point of view, the mass-density is zero everywhere except at $x = 0$, where it is infinite because a finite mass is concentrated in zero length; and it is so infinitely large there that the integral is non-zero even though the integrand is positive over an 'infinitesimally small' region only. This makes good physical sense, though it is mathematically absurd. Dirac therefore introduced a function $\delta(x)$ having just those properties:

$$\delta(x) = 0 \quad \text{for} \quad x \neq 0, \ \int_a^b \delta(x)\,\mathrm{d}x = 1 \quad \text{if} \quad a < 0 < b \quad . \tag{1.1}$$

If one uses the function δ as the mass-density function in any calculation or theoretical work involving continuous distributions on a line, one is led to the corresponding result for a point particle, and thus the two cases of continuous and discrete distributions can be included in a single formalism if the delta function is used. It can be considered as a technical trick, or short cut, for obtaining results for discrete point particles from the continuous theory, results which can always be verified if desired by working out the discrete case from first principles.

A point particle can be considered as the limit of a sequence of continuous distributions which become more and more concentrated. The delta function can similarly be considered as the limit of a sequence of ordinary functions. Consider, for example,

$$d_n(x) = n/[\pi(1 + n^2x^2)] \quad . \tag{1.2}$$

Then $d_n(x) \to 0$ as $n \to \infty$ for any $x \neq 0$, and $d_n(0) \to \infty$ as $n \to \infty$ (see Fig. 1). Also $\int_{-\infty}^{\infty} d_n(x)\,\mathrm{d}x = 1$, and if $a < 0 < b$, then $\int_a^b d_n(x)\,\mathrm{d}x \to 1$ as $n \to \infty$. Thus we might say that 'in the limit as $n \to \infty$', d_n has the properties of $\delta(x)$. The delta function is sometimes defined in this way, but again this is not a proper mathematical definition, since $d_n(x)$ does not have a limit for all x as $n \to \infty$. However, when one uses the delta function in practice it usually appears inside an integral at some stage, and it is only integrals of $\delta(x)$, multiplied by ordinary functions, that have direct physical significance. If one replaces δ by d_n, and then lets $n \to \infty$ at the end of the calculation, the integrals involving d_n will generally be well behaved and tend to finite limits as $n \to \infty$, and the mathematical inconsistency is removed. The delta function can be considered as a kind of mathematical shorthand representing that procedure, and results obtained by its use can always be verified if desired by working with d_n and then evaluating the limit.

The point of view described above is that of many physicists, engineers, and applied mathematicians who use the delta function. To the pure mathematicians of Dirac's generation it presented a challenge: an idea which is mathematically absurd, but still works, and gives useful and correct results, must be somehow essentially right. There must be a theory in which $\delta(x)$ has a rightful place, instead of being sneaked in by the back door as a mathematical shorthand, to be justified by doing the calculation again without using it. The situation is reminiscent of the use of complex numbers in the 16th century for solving algebraic equations. It proved useful to pretend that -1 has a square root, even though it clearly has not, since one could then use an algorithm involving imaginary numbers for obtaining real roots of cubic equations; any result obtained this way could be verified by directly substituting it in the equation and showing that it really was a root. It was only much later that complex numbers were given a solid mathematical foundation, and then with the development of the theory of functions of a complex variable their applications far transcended the simple algebra which led to their introduction. In the same way, Dirac found it useful to pretend that there exists a function δ satisfying (1.1), even though there does not. The solid foundation was developed by Sobolev (in 1936) and Schwartz (in the 1950s), and again goes far beyond merely propping up the delta function. The theory of generalised functions that they developed can be used to replace ordinary analysis, and is in many ways simpler. Every generalised function is differentiable, for example; and one can differentiate and integrate series term by term without worrying about uniform

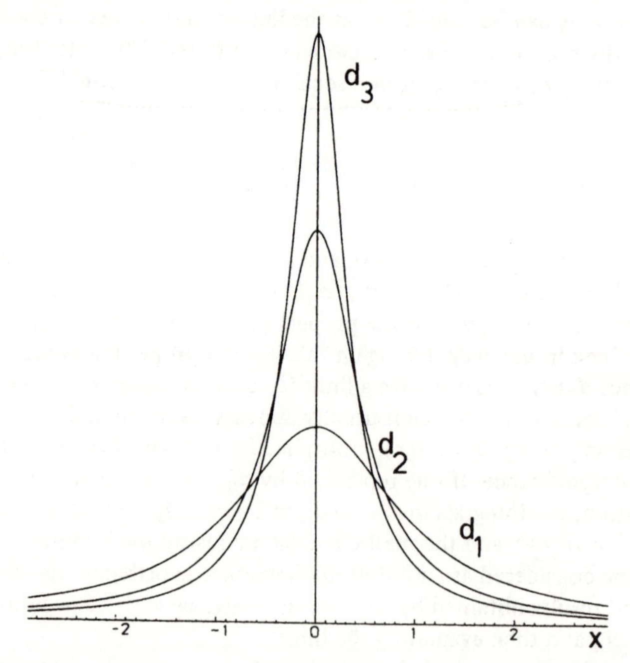

Fig. 1. The functions $d_n(x) = \dfrac{n}{\pi(1 + n^2x^2)}$ which approximate to the delta function.

convergence. The theory also has limitations: that is, it shows clearly what you cannot do with the delta function as well as what you can – namely, you cannot multiply it by itself, or by a discontinuous function. The other disadvantage of the theory is that it involves a certain amount of formal machinery.

1.2 Basic Distribution Theory

There are several ways of generalising the idea of a function in order to include Dirac's δ. We shall follow the method of Schwartz, who called his generalised functions 'distributions'; the idea is to formalise Dirac's idea of the delta function as something which makes sense only under an integral sign, possibly multiplied by some other function ϕ. Given any interval, a mass-density function allows one to calculate the mass of that interval; more generally, given any weighting function ϕ, it allows one to calculate a weighted average of the mass, such as is needed for calculating centres of gravity, etc. We shall define a distribution as a rule which, given any function ϕ, provides a number; the number may be thought of as a weighted average, with weight function ϕ, of a corresponding mass-distribution. However, we must be careful about what functions we allow as weighting functions. The definitions below may at first seem arbitrary and needlessly complicated; but they are carefully framed, as you will see, to make the resulting theory as simple as possible. The reader unfamiliar with the notation of set theory should consult Appendix A.

Definition 1.1 The **support** of a function $f: \mathbb{R} \to \mathbb{C}$ is $\{x : f(x) \neq 0\}$, written supp(f)†. A function has **bounded support** if there are numbers a,b such that $\text{supp}(f) \subset [a,b]$.

Definition 1.2 A function $f: \mathbb{R} \to \mathbb{C}$ is said to be n **times continuously differentiable** if its first n derivatives exist and are continuous. f is said to be **smooth** or **infinitely differentiable** if its derivatives of all orders exist and are continuous.

Definition 1.3 **A test function** is a smooth $\mathbb{R} \to \mathbb{C}$ function with bounded support. The set of all test functions is called $\mathscr{D}$. □

Example 1.4

$$\phi(x) = \begin{cases} \exp\left[\dfrac{1}{(x-a)(x-b)}\right] & \text{for } a < x < b \\ 0 \quad \text{for } x \leqslant a \text{ and for } x \geqslant b \end{cases}$$

is a test function with support (a,b).

† The support is usually defined as the closure of this set, but this subtlety is not needed for the simple account of the theory given in this chapter.

This is probably the simplest example of a test function. They are bound to have somewhat complicated forms, for the following reason. If $\mathrm{supp}(\phi) = (a, b)$, then $\phi(x) = 0$ for $x \leqslant a$, so all derivatives of ϕ vanish at a. Hence the Taylor series of ϕ about a is identically zero. But $\phi(x) \neq 0$ for $a < x < b$, so ϕ does not equal its Taylor series. Test functions are thus peculiar functions; they are smooth, yet Taylor expansions are not valid. The above example clearly shows singular behaviour at $x = a$ and $x = b$. Fortunately, we never really need explicit formulas for test functions. They are used for theoretical purposes only, and are well-behaved (smooth etc.) even though their functional forms may be complicated. The following result gives another nice mathematical property.

Proposition 1.5 The sum of two test functions is a test function; the product of a test function with any number is a test function.

Proof. Obvious. □

A set of functions with this property is often called a 'space', for reasons that will become clear in Chapter 4.

Definition 1.6 **A linear functional** on the space $\mathscr{D}$ is a map $f: \mathscr{D} \to \mathbb{C}$ such that $f(a\phi + b\psi) = af(\phi) + bf(\psi)$ for all $a, b \in \mathbb{C}$ and $\phi, \psi \in \mathscr{D}$. □

A map $\mathscr{D} \to \mathbb{C}$ means a rule which, given any $\phi \in \mathscr{D}$, produces a corresponding $z \in \mathbb{C}$; we write $z = f(\phi)$. We also speak of the 'action' of the functional f on ϕ producing the number $f(\phi)$. The notation $\phi \mapsto f(\phi)$ stands for the phrase 'ϕ is mapped into the number $f(\phi)$'; see Appendix A.

Examples 1.7 (a) $\phi \mapsto \phi(0)$ is a functional, easily seen to be linear. (b) $\phi \mapsto \int_{-\infty}^{\infty} |\phi(x)|^2 \, dx$ is a functional; the integral converges because ϕ has bounded support. It is not linear. (c) $\phi \mapsto \int_{-\infty}^{\infty} f(x)\phi(x) \, dx$ is a linear functional for any function f sufficiently well behaved to ensure convergence of the integral. □

We must now define convergence in the space $\mathscr{D}$. The reader will know that more than one meaning can be attached to the phrase 'a sequence of functions is convergent'. For some purposes pointwise convergence is suitable; for other purposes uniform convergence is needed (an outline of the theory of uniform convergence is given in Appendix B). One of the characteristics of functional analysis is its use of many different kinds of convergence, as demanded by different problems. The most useful for our present purpose is the following.

Definition 1.8 (Convergence) If (ϕ_n) is a sequence of test functions and Φ another test function, we say $\phi_n \to \Phi$ in $\mathscr{D}$ if (i) there is an interval $[a,b]$ containing supp(Φ) and supp(ϕ_n) for all n; (ii) $\phi_n(x) \to \Phi(x)$ as $n \to \infty$, uniformly for $x \in [a,b]$; and (iii) for each k, $\phi_n^{(k)}(x) \to \Phi^{(k)}(x)$ as $n \to \infty$, uniformly for $x \in [a,b]$, where $\phi^{(k)}$ denotes the k-th derivative of ϕ. □

This is a stringent definition, much stronger than ordinary convergence. We do not offer an example because specific examples are never needed: test functions are only the scaffolding upon which the main part of the theory is built.

Definition 1.9 A functional f on $\mathscr{D}$ is **continuous** if it maps every convergent sequence in $\mathscr{D}$ into a convergent sequence in $\mathbb{C}$, that is, if $f(\phi_n) \to f(\Phi)$ whenever $\phi_n \to \Phi$ in $\mathscr{D}$. A continuous linear functional on $\mathscr{D}$ is called a **distribution**, or **generalised function.** □

This definition of continuity is modelled on one of two alternative definitions of continuity of an ordinary $\mathbb{R} \to \mathbb{R}$ function. The other common definition for $\mathbb{R} \to \mathbb{R}$ functions is: f is continuous if for any $\epsilon > 0$ there is a $\delta > 0$ such that $|f(x) - f(y)| < \epsilon$ whenever $|x - y| < \delta$. This can be shown to be equivalent to the condition that f map every convergent sequence of numbers into a convergent sequence. We adopt the latter as our definition in $\mathscr{D}$, because there is no analogue in $\mathscr{D}$ of the modulus of a number which appears in the other definition (in Chapter 4 we shall consider this question further).

Notation 1.10 We shall use bold type to signify a distribution, and $\langle f,\phi \rangle$ to denote the 'action' of the distribution f on the test function ϕ; in other words, $\langle f,\phi \rangle$ is the number into which f maps ϕ. The reason for using this odd-looking notation, rather than $f(\phi)$, will appear shortly.

Example 1.11 The delta distribution δ is defined by

$$\langle \delta, \phi \rangle = \phi(0)$$

for all ϕ in $\mathscr{D}$. This is the functional defined in Example 1.7(a), using different notation. To justify calling it a distribution, we must show that the functional is continuous, i.e. that $\langle \delta, \phi_n \rangle$ is a convergent sequence of numbers whenever (ϕ_n) is a convergent sequence in $\mathscr{D}$. In fact it follows immediately from Definition 1.8 that $\phi_n(0) \to \Phi(0)$ if $\phi_n \to \Phi$ in $\mathscr{D}$, so δ is indeed a distribution. □

There is a class of 'regular' distributions, corresponding to ordinary functions of the following kind.

Definition 1.12 A function f: $\mathbb{R} \to \mathbb{C}$ is **locally integrable** if $\int_a^b |f(x)|\, dx$ exists for all numbers a,b. □

Any continuous function is locally integrable; so is any piecewise continuous function, as defined below.

Definition 1.13 A function $f: \mathbb{R} \to \mathbb{C}$ is called **piecewise continuous** if it is continuous except at a set of points x_i such that any finite interval contains only finitely many x_i, and if the left and right hand limits of the function exist at each x_i. □

Piecewise continuous functions may have finite jump discontinuities; they are locally integrable. Functions with mild singularities, such as $\log|x|$ and $|x|^a$ for $-1 < a \leqslant 0$ are locally integrable, but functions with stronger singularities, such as $|x|^a$ for $a \leqslant -1$, are not. Note that the behaviour of $f(x)$ as $x \to \infty$ is irrelevant to whether it is locally integrable.

Theorem 1.14 (Regular Distributions) To every locally integrable function f there corresponds a distribution f defined by

$$\langle f,\phi\rangle = \int_{-\infty}^{\infty} f(x)\phi(x)\,\mathrm{d}x \quad .$$

The distribution f is said to be **generated** by the function f.

Proof We must first show that the integral exists. Since any ϕ in $\mathscr{D}$ has bounded support, contained in $[a,b]$ say, the integral is $\int_a^b f(x)\phi(x)\,\mathrm{d}x$, which exists since the product of an integrable function and a continuous function is an integrable function. Hence f is a functional on $\mathscr{D}$. Linearity is obvious. To prove that f is a distribution, suppose $\phi_n \to \Phi$ in $\mathscr{D}$. Then there is an interval $[a,b]$ containing the supports of ϕ_n and Φ, and using the standard results $|\int p(x)q(x)\,\mathrm{d}x| \leqslant \int |p(x)q(x)|\,\mathrm{d}x \leqslant (\max|p(x)|)\int|q(x)|\,\mathrm{d}x$, we have

$$\begin{aligned} |\langle f,\phi_n\rangle - \langle f,\Phi\rangle| &= \left|\int_a^b f(x)[\phi_n(x) - \Phi(x)]\,\mathrm{d}x\right| \\ &\leqslant (\max|\phi_n - \Phi|)\int_a^b |f(x)|\,\mathrm{d}x \\ &\to 0 \quad \text{as} \quad n \to \infty \end{aligned}$$

because $\phi_n \to \Phi$ uniformly (by Definition 1.8). Hence $\langle f,\phi_n\rangle \to \langle f,\Phi\rangle$, so f is a continuous functional, that is, a distribution. □

The reason for the notation $\langle f,\phi\rangle$ is that in other branches of analysis (f,g) is a common notation for $\int f(x)\overline{g(x)}\,\mathrm{d}x$ (see section 7.1), and this theorem identifies $\langle f,\phi\rangle$ with (f,ϕ), at least when f is locally integrable and ϕ is real.

Definition 1.15 A distribution which is generated by a locally integrable function is called **regular**. All other distributions are called **singular**. □

Thus the class of distributions contains objects which correspond to ordinary functions as well as singular distributions which do not. In the next section we shall define operations on distributions analogous to the operations of ordinary algebra and calculus applied to functions. This will justify calling distributions 'generalised functions', and will allow us to use distributions for most of the purposes for which ordinary functions are used.

1.3 Operations on Distributions

Definition 1.16 (Addition) If f and g are distributions and a and b are complex numbers, we define the distribution $af + bg$ to be the functional $\phi \mapsto a\langle f,\phi\rangle + b\langle g,\phi\rangle$ for all ϕ in $\mathscr{D}$. □

Strictly speaking we should verify that this defines a linear continuous functional on $\mathscr{D}$, but that is very easy and is left to the reader. The same applies to other definitions in this section.

Definition 1.17 (Multiplication) If f is a distribution and h is a smooth $\mathbb{R} \to \mathbb{C}$ function (cf. Definition 1.2), we define the product of f and h to be the distribution $hf: \phi \mapsto \langle f,h\phi\rangle$ for all ϕ in $\mathscr{D}$. □

Note that if ϕ is a test function and h is smooth, then $h\phi$ is a test function, and therefore $\langle f,h\phi\rangle$ is well-defined. If h is not smooth, then neither is $h\phi$, hence $h\phi \notin \mathscr{D}$, and Definition 1.17 does not work. Thus we cannot define the product of a distribution with a function which is discontinuous or has a discontinuous derivative. Distributions cannot in general be multiplied. The difficulty can be seen in the following

Example 1.18 The **Heaviside function** H is defined by

$$H(x) = \begin{cases} 0 & \text{for} \quad x < 0, \\ \frac{1}{2} & \text{for} \quad x = 0, \\ 1 & \text{for} \quad x > 0, \end{cases}$$

and an almost identical function H_1 is defined by

$$H_1(x) = \begin{cases} 0 & \text{for} \quad x \leqslant 0, \\ 1 & \text{for} \quad x > 0. \end{cases}$$

H and H_1 are locally integrable, and they generate the same regular distribution $H: \phi \mapsto \int_0^\infty \phi(x)\,dx$. There is no distinction between H_1 and H in generalised function theory, and this reflects the fact that from the point of view of the physicist the distinction between them is artificial: one could never distinguish between H_1 and H experimentally. If Definition 1.17 were extended to discontinuous functions, we would have $\langle H_1\delta,\phi\rangle = 0$, $\langle H\delta,\phi\rangle = \frac{1}{2}\phi(0)$. Now, any definition of the product of two distributions would have to agree with Definition 1.17 in the case that one of the distributions was regular. Since H_1 and H generate the same distribution $\boldsymbol{H}$, we would have two different values for $\langle \boldsymbol{H}\delta,\phi\rangle$, and thus an inconsistent theory. This shows that it is impossible to define the product of $\boldsymbol{\delta}$ with a discontinuous function. However, one can extend Definition 1.17 so as to allow multiplication by functions which are not smooth; see Problem 1.4. □

Definition 1.17 has the property that if $\boldsymbol{f}$ is a regular distribution generated by a function f, and h is smooth, then the distribution $\boldsymbol{hf}$ is the same as the distribution generated by the locally integrable function hf – they are both given by $\phi \mapsto \int h(x)f(x)\phi(x)\,\mathrm{d}x$. In other words, the definition of the product of a generalised function and an infinitely differentiable function is consistent with the ordinary rule for multiplying functions.

We leave to the reader the (easy) task of verifying that the usual rules of algebra (distributive law, etc.) are satisfied by our definition of multiplication, and proceed to consider the idea of equality for generalised functions. Clearly two distributions are equal if they are the same functional on $\mathscr{D}$, that is, $\boldsymbol{f} = \boldsymbol{g}$ if and only if $\langle \boldsymbol{f},\phi\rangle = \langle \boldsymbol{g},\phi\rangle$ for all $\phi\in\mathscr{D}$. For ordinary functions we can talk about two functions being equal at a point. For generalised functions this does not make sense since there is no such thing as the 'value' of a generalised function at a point x. However, we can form an idea of the behaviour of a generalised function over an interval (a,b) by considering its action on test functions which vanish outside (a,b) and therefore give no information about the distribution outside (a,b). □

Definition 1.19 Two distributions $\boldsymbol{f}$ and $\boldsymbol{g}$ are said to be **equal on** (a,b) if $\langle \boldsymbol{f},\phi\rangle = \langle \boldsymbol{g},\phi\rangle$ for all $\phi\in\mathscr{D}$ such that $\text{supp}(\phi)\subset(a,b)$. □

Thus, for example, $\boldsymbol{\delta} = \boldsymbol{0}$ on $(0,a)$ for any $a > 0$, or indeed on $(0,\infty)$. Here $\boldsymbol{0}$ denotes the distribution $\phi \mapsto 0$ for all ϕ, which is generated by the function $z(x) \equiv 0$. In this case it is natural to say that the delta function has the value zero for all $x > 0$, and the next definition legalises that usage.

Definition 1.20 If $\boldsymbol{f}$ is a distribution and g is a locally integrable function, $\boldsymbol{f}$ is said to **take values** $g(x)$ **on an interval** (a,b) if $\boldsymbol{f} = \boldsymbol{g}$ on (a,b) (in the sense of Definition 1.19). In this context we sometimes use $\boldsymbol{f}(x)$ as an alternative notation for $\boldsymbol{f}$, and write $\boldsymbol{f}(x) = g(x)$ for $a < x < b$.

Example 1.21 $\boldsymbol{\delta}$ takes the value 0 for $x > 0$ and $x < 0$. In the notation explained above, this statement can be written $\delta(x) = 0$ for $x \neq 0$. But it should be remembered that $\delta(x)$ does not stand for the value of $\boldsymbol{\delta}$ at a point x; statements containing the notation $\boldsymbol{f}(x)$ should be interpreted strictly according to Definitions

1.19 and 1.20. (At the same time, there is little harm in thinking of $f(x)$ as if it were the value of a function at a point, provided that you realise that it is only a convention and liable to be misleading. For example, $\delta(0)$ really is meaningless, since there is no ordinary function g such that δ takes values $g(x)$ on an interval including 0.) □

We now define operations analogous to simple changes of variable for ordinary functions.

Definition 1.22 (Change of Variable) For any distribution f and any real a, a new distribution f_{+a} is defined by $\langle f_{+a},\phi\rangle = \langle f,\phi(x-a)\rangle$. We call f_{+a} the **translation** of f by a. When the notation $f(x)$ is used for f (as explained in Definition 1.20), $f(x+a)$ is a convenient notation for the translation of f. □

If f is a regular distribution generated by a function f, then it is easy to see that f_{+a} is generated by the function $f(x+a)$. Thus $f(x+a)$ is a consistent and sensible notation. Of course, when we write $f(x+a)$, we mean the function $x \longmapsto f(x+a)$; similarly, in Definition 1.22, strictly speaking we should not write $\langle f,\phi(x+a)\rangle$ since $\phi(x+a)$ is not a function but a number, but should write $\langle f,\psi\rangle$ where $\psi : x \longmapsto \phi(x+a)$. It is pedantic and clumsy to insist on keeping to these rules all the time, however, and we shall assume that the reader can understand from the context whether $f(x)$ stands for a number or a function.

Definition 1.23 (Change of Variable) For any distribution f and any real $a \neq 0$, a new distribution $f_{.a}$ is defined by $\langle f_{.a},\phi\rangle = \langle f,\phi(x/a)\rangle/|a|$. When the notation $f(x)$ is used for f, $f(ax)$ is a convenient notation for $f_{.a}$. □

If f is a regular distribution generated by a function f, then it is easy to see that $f_{.a}$ is generated by the function $f(ax)$. Thus $f(ax)$ is a sensible and consistent notation. If f is a distribution which takes values $g(x)$ on (a,b) (in the sense of Definition 1.20), then $f(x+c) = g(x+c)$ on $(a-c, b-c)$ and $f(cx) = g(cx)$ on $(a/c,b/c)$ if $c > 0$ or on $(b/c,a/c)$ if $c < 0$.

Example 1.24 $\langle\delta(x-a),\phi\rangle = \langle\delta(x),\phi(x+a)\rangle = \phi(a)$. Thus $\delta(x-a)$ is a distribution which picks out the value at a of a test function to which it is applied. This is sometimes called the 'sifting property' of the delta function. We also have $\delta(x-a) = 0$ for $x \neq a$. Similarly, $\langle\delta(ax),\phi\rangle = \langle\delta(x),\phi(x/a)\rangle/|a| = \phi(0)/|a|$. Hence

$$\delta(ax) = (1/|a|)\delta(x) \quad \text{for} \quad a \neq 0 \quad . \qquad \square \quad (1.3)$$

We now proceed to the differential calculus of distributions. Our general plan for defining operations on distributions is to begin with locally integrable functions, find an expression for the operation applied to the corresponding regular distribution, and then generalise to all distributions. If f is a differentiable function, f the corresponding regular distribution, and $\mathrm{d}f/\mathrm{d}x$ the distribution generated by f', we have

$$\langle \mathrm{d}f/\mathrm{d}x,\phi\rangle = \int_{-\infty}^{\infty} f'\phi\,\mathrm{d}x = -\int_{-\infty}^{\infty} f\phi'\,\mathrm{d}x$$

on integration by parts (there are no boundary terms because ϕ vanishes at infinity). For regular distributions corresponding to differentiable functions, we thus have

$$\langle \mathrm{d}f/\mathrm{d}x,\phi\rangle = -\langle f,\phi'\rangle \quad . \tag{1.4}$$

We shall use (1.4) as a definition of differentiation for any distribution. But first we must show that the right hand side of (1.4) is always a distribution.

Proposition 1.25 For any distribution f, the functional $\phi \longmapsto -\langle f,\phi'\rangle$ is a distribution.

Proof If $\phi\in\mathscr{D}$, then $\phi'\in\mathscr{D}$, so $\phi \longmapsto -\langle f,\phi'\rangle$ defines a functional on $\mathscr{D}$. It is clearly linear. If $\phi_n \to \Phi$ in $\mathscr{D}$, then $\phi_n' \to \Phi'$ in $\mathscr{D}$ (by the standard theorem on differentiating uniformly convergent sequences), hence $-\langle f,\phi_n'\rangle \to -\langle f,\Phi'\rangle$, which proves that the functional $\phi \longmapsto -\langle f,\phi'\rangle$ is continuous. □

We can now define differentiation for generalised functions.

Definition 1.26 The **derivative** of a generalised function f is the generalised function f' defined by $\langle f',\phi\rangle = -\langle f,\phi'\rangle$ for all $\phi\in\mathscr{D}$. □

Equation (1.4) shows that the distribution generated by the derivative f' of a function f is the same as the derivative of the distribution f; these two possible ways of interpreting the symbol f', for a differentiable function f, are identical. Our new definition of differentiation is consistent with ordinary calculus.

The advantage of our theory over ordinary calculus is that every generalised function is differentiable; this follows from Proposition 1.25. If f is a locally integrable function which is not differentiable, the distribution f' is called the **generalised derivative** of f. For example, a continuous but not differentiable function has a generalised derivative with a discontinuity; a function with a simple step discontinuity has a generalised derivative involving a delta function.

Example 1.27 $|x|$ is a locally integrable function, differentiable for all $x \neq 0$, but certainly not differentiable at 0. The generalised derivative is calculated as follows. For any test function ϕ,

$$\begin{aligned}\langle |x|',\phi\rangle &= -\langle |x|,\phi'\rangle \\ &= -\int_{-\infty}^{\infty} |x|\phi'(x)\,\mathrm{d}x \\ &= \int_{-\infty}^{0} x\phi'(x)\,\mathrm{d}x - \int_{0}^{\infty} x\phi'(x)\,\mathrm{d}x \\ &= -\int_{-\infty}^{0}\phi(x)\,\mathrm{d}x + \int_{0}^{\infty}\phi(x)\,\mathrm{d}x,\end{aligned}$$

integrating by parts and using the fact that ϕ vanishes at infinity. We define a function sgn(x) (read as 'sign of x') by

$$\operatorname{sgn}(x) = \begin{cases} -1 & \text{for} \quad x < 0 \\ 1 & \text{for} \quad x > 0 \end{cases} \quad . \tag{1.5}$$

It is unnecessary for our purposes to specify the value of sgn(0), since it generates the same distribution **sgn** for any choice of sgn(0). We now have, from the above,

$$\begin{aligned}\langle |x|',\phi\rangle &= \int_{-\infty}^{\infty} \operatorname{sgn}(x)\phi(x)\,\mathrm{d}x \\ &= \langle \mathbf{sgn},\phi\rangle\end{aligned}$$

for all $\phi\in\mathscr{D}$, so $|x|' = \mathbf{sgn}$. Note that since the generalised derivative is a distribution, we cannot speak of its value at a point, only over an interval. Thus the awkward question 'what is the value at $x = 0$ of the derivative of $|x|$?' has no meaning in our language. □

Example 1.28 Consider the Heaviside function defined in Example 1.4. We have $\langle H',\phi\rangle = -\int_0^{\infty}\phi'(x)\,\mathrm{d}x = \phi(0)$ since ϕ vanishes at infinity. Hence

$$H'(x) = \delta(x) \quad . \tag{1.6}$$

Example 1.29 The derivative of the delta function is defined by $\langle \delta',\phi\rangle = -\langle \delta,\phi'\rangle = -\phi'(0)$. Similarly, the n-th derivative of δ is given by $\langle \delta^{(n)},\phi\rangle = (-1)^n\phi^{(n)}(0)$. □

It is easy to show that the derivative of a sum of generalised functions equals the sum of the derivatives. The rule for differentiating a product needs a little more care. If h is a smooth function and f a distribution, their product is given by Definition 1.17, and we have, for any $\phi\in\mathscr{D}$,

$$\begin{aligned}\langle (hf)',\phi\rangle &= -\langle hf,\phi'\rangle \\ &= -\langle f,h\phi'\rangle \quad \text{by 1.17,} \\ &= -\langle f,(h\phi)' - h'\phi\rangle \\ &= \langle f',h\phi\rangle + \langle f,h'\phi\rangle \quad \text{by 1.26} \\ &= \langle hf',\phi\rangle + \langle h'f,\phi\rangle \quad \text{by 1.17,}\end{aligned}$$

showing that

$$(hf)' = hf' + h'f \quad .$$ □

We have now constructed the basic algebra and differential calculus of distributions. In the next section we consider distributional convergence.

1.4 Convergence of Distributions

In ordinary calculus, one first introduces the idea of the limit of a function, and then uses it to define differentiation. We have been able to develop the calculus of distributions without the idea of the limit of generalised functions; this is because we have defined convergence of test functions in $\mathscr{D}$, and based the differentiation of distributions on the properties of $\mathscr{D}$. However, it is useful also to define the idea of convergence for distributions. Just as for continuity, we cannot use an '$\epsilon - \delta$' definition; we therefore define convergence of distributions in terms of their action on $\mathscr{D}$.

Definition 1.30 A sequence (f_n) of distributions is said to be **convergent** if the sequence of numbers $(\langle f_n, \phi \rangle)$ is convergent for all $\phi \in \mathscr{D}$. □

Notice that this definition does not involve the existence of a limiting distribution towards which the sequence tends. It is framed strictly in terms of the sequence itself (unlike the usual definition of convergence in elementary analysis, which requires one to find the limit of a sequence before the definition can be used). However, it can be proved that if a sequence of distributions converges, then there is in fact a limit distribution.

Theorem 1.31 If (f_n) is a convergent sequence of distributions, then there is a distribution F such that $\langle f_n, \phi \rangle \to \langle F, \phi \rangle$ for all $\phi \in \mathscr{D}$. We write $f_n \to F$ as $n \to \infty$. □

Given a convergent sequence of distributions (f_n), we can always define a linear functional $\phi \longmapsto \lim(\langle f_n, \phi \rangle)$. This is the required distribution F, but to prove that it is a distribution, that is, to prove that it is a continuous functional, is not easy. We shall not need the theorem in this book, and for its proof we refer the reader to the standard text-books (see section 1.7).

The next theorem relates distributional convergence to ordinary convergence for regular distributions.

Theorem 1.32 (Distributional Convergence) If $F, f_1, f_2, \ldots$ are locally integrable functions such that $f_n \to F$ uniformly in each bounded interval, then $f_n \to F$ distributionally.

Proof Let ϕ be a test function with $\text{supp}(\phi) \subset (a,b)$, then

$$\langle f_n,\phi\rangle = \int_a^b f_n(x)\phi(x)\,\mathrm{d}x$$
$$\to \int_a^b F(x)\phi(x)\,\mathrm{d}x$$

by a standard theorem on the integration of uniformly convergent sequences of functions. Hence for any $\phi \in \mathscr{D}$, $\langle f_n,\phi\rangle \to \langle F,\phi\rangle$. □

It is not in general true that if $f_n \to F$ where f_n and F are locally integrable functions, then $f_n \to F$; the convergence must be uniform, as in Theorem 1.32 (or some other extra condition must be satisfied). This will be illustrated by Example 1.34 below. But first we must consider a simpler example.

Example 1.33 $d_n \to \delta$ as $n \to \infty$, where $d_n(x) = n/\pi(1+n^2x^2)$ is the sequence considered in section 1.1. This statement is usually written more concisely as

$$\frac{n}{\pi(1+n^2x^2)} \to \delta(x) \quad \text{as} \quad n \to \infty \quad . \tag{1.7}$$

To prove it, we must show that $\int d_n(x)\phi(x)\,\mathrm{d}x \to \phi(0)$ as $n \to \infty$, for $\phi \in \mathscr{D}$. We have, using the fact that $\int_{-\infty}^{\infty} d_n(x)\,\mathrm{d}x = 1$,

$$\left|\int_{-\infty}^{\infty} d_n(x)\phi(x)\,\mathrm{d}x - \phi(0)\right| = \left|\int_{-\infty}^{\infty} d_n(x)[\phi(x)-\phi(0)]\,\mathrm{d}x\right|$$
$$\leqslant \left|\phi(0)\int_{-\infty}^{a} d_n(x)\,\mathrm{d}x\right| + \left|\int_a^b d_n(x)[\phi(x)-\phi(0)]\,\mathrm{d}x\right| + \left|\phi(0)\int_b^{\infty} d_n(x)\,\mathrm{d}x\right|$$

where (a,b) is an interval containing $\text{supp}(\phi)$, and $a < 0 < b$. By means of the substitution $nx = y$, the first and third integrals here are easily shown to tend to zero as $n \to \infty$. In the second, use the mean value theorem to write $|\phi(x) - \phi(0)| \leqslant M|x|$, where $M = \max|\phi'(x)|$; the integral is then $\leqslant M\int_a^b |xd_n(x)|\,\mathrm{d}x$, which is easily evaluated exactly, and shown to tend to zero as $n \to \infty$. This completes the proof of (1.7).

Example 1.34

$$d_n'(x) = -\frac{2}{\pi}\frac{n^3x}{(1+n^2x^2)^2} \to \delta'(x) \quad \text{as} \quad n \to \infty \quad . \tag{1.8}$$

This is easily deduced from (1.7): $\langle d_n',\phi\rangle = \int_{-\infty}^{\infty} d_n'\phi\,\mathrm{d}x = -\int_{-\infty}^{\infty} d_n\phi'\mathrm{d}x$ (integrating by parts), $\to -\phi'(0) = \langle\delta',\phi\rangle$ by (1.7). Notice that $d_n'(x) \to 0$ pointwise as $n \to \infty$ for all x. Thus if a sequence of locally integrable functions converges to another locally integrable function, in the sense of ordinary pointwise convergence, it does not follow that it converges distributionally to the corresponding generalised function. This is not to be regarded as a weakness of distribution theory. On the contrary, if the reader sketches the graphs of the functions d_1', d_2', d_3', he will see that although $d_n' \to 0$ is a true statement, it does not represent the true state of affairs as clearly as (1.8) does. □

Equation (1.8) is a special case of the following theorem.

Theorem 1.35 (Termwise Differentiation) If $F, f_1, f_2, \ldots$ are distributions such that $f_n \to F$ as $n \to \infty$, then $f_n' \to F'$.

Proof

$$\begin{aligned}\langle f_n',\phi\rangle &= -\langle f_n,\phi'\rangle \\ &\to -\langle F,\phi'\rangle \\ &= \langle F',\phi\rangle \quad .\end{aligned}$$

□

Using Theorem 1.35, (1.8) follows immediately from (1.7). Notice how much simpler this theorem is than the corresponding result for ordinary functions: in that case, (f_n) must converge uniformly and the differentiated sequence also must be proved to converge uniformly before one can be sure that $f_n' \to F'$. In distribution theory, any convergent sequence can be differentiated without qualms. Similarly, a convergent infinite series (which we define in the usual way as the sequence of partial sums) can be differentiated term by term. This is one of the benefits that we reap from the rather stringent conditions defining the space $\mathscr{D}$ and convergence in $\mathscr{D}$.

Our outline of the basic framework of distribution theory is now complete. Essentially all the usual results of calculus hold in the generalised theory; for instance, one can prove that the limit of a product or a sum is the product or sum of the limits, and so on. Similarly one can define convergence of a distribution which depends on a continuously varying parameter t as well as for distributions depending on an integer parameter n. All this will be found in the systematic text-books listed in section 1.7.

1.5 Further Developments

In this section we shall introduce another piece of notation for simplifying work with generalised functions, and then consider some important examples of singular distributions.

Notation 1.36 For any distribution f and test function ϕ, the symbol $\int_{-\infty}^{\infty} f(x)\phi(x)\,\mathrm{d}x$ stands for the number $\langle f,\phi\rangle$. □

One can tell whether the integration sign stands for a genuine integral or the action of a functional on a test function by whether the integrand contains ordinary functions or generalised functions f. Many authors make no distinction between the notations for ordinary and generalised functions. This is a reasonable policy because whenever confusion is possible, it is harmless: that is, the two alternative interpretations of a formula lead to the same result in the end. Thus, if f is a regular distribution generated by the function f, then $\int f(x)\phi(x)\,dx = \int f(x)\phi(x)\,dx$ by definition of f. If you look back at Definitions 1.22, 1.23, and 1.26, you will see that they are constructed so that if the integral notation for distributions is used, they correspond to the usual rules for simple changes of variable and integration by parts (the boundary terms being zero because ϕ has bounded support). Therefore, using the integral notation one can manipulate the integrals just as if they stood for genuine integration, while feeling assured that the rules of distribution theory are being strictly obeyed. Thus, for instance, we can write the first result in Example 1.24 as

$$\int_{-\infty}^{\infty}\delta(x-a)\phi(x)\,dx = \phi(a) \quad , \tag{1.9}$$

and prove it by a change of variable in the integral, $x - a = y$, so that the left hand side of (1.9) becomes $\int\delta(y)\phi(y+a)\,dy = \phi(a)$. Equation (1.9) is the key property of the delta function in the informal treatment usual in physics or mathematical-methods books.

We shall now consider some examples of singular distributions. So far we have met the delta function and its derivatives, which are singular according to Definition 1.15. That definition does not quite correspond to the use of the word 'singular' in ordinary analysis. The function $\log|x|$ has what is generally called a singularity at $x = 0$. But it is locally integrable (its indefinite integral is $x\log|x| - x$, which is a continuous function), and generates a regular distribution. Its derivative $1/x$, however, is not locally integrable, and does not generate a regular distribution. We must therefore consider more carefully the derivative of the generalised function $\mathbf{log}|\boldsymbol{x}|$.

It must surely correspond in some way to the integral $\int_{-\infty}^{\infty}(\phi(x)/x)\,dx$. The difficulty is that the integrand has a singularity so strong that it must be excised from the domain and the integral defined by a limiting process. The result of such a process is called an **improper integral.** Set $I(\epsilon,\delta) = \int_{-\infty}^{-\epsilon}[\phi(x)/x]\,dx + \int_{\delta}^{\infty}[\phi(x)/x]\,dx$. If $I(\epsilon,\delta)$ tends to a definite limit as ϵ and δ independently tend to zero (through positive values), then the improper integral $\int\phi(x)\,dx/x$ is said to converge to that limit. In general this will not be the case, unless $\phi(0) = 0$. But if we take $\delta = \epsilon$, and then let $\epsilon \to 0$, we find that $I(\epsilon,\epsilon)$ tends to a limit. Taking $\delta = 2\epsilon$, say, and then letting $\epsilon \to 0$ gives a different limit, and thus many different values can be assigned to $\int\phi(x)\,dx/x$ by proceeding to the limit in different ways. There is one obvious way of choosing a particular value from all these possibilities, and that is to take the symmetrical case, $\epsilon = \delta$. The value obtained in this way (if it exists) is called the principal value of the integral. We must now prove that $I(\epsilon,\epsilon)$ does tend to a limit as $\epsilon \to 0$.

Notation 1.37 $\epsilon \downarrow 0$ means ϵ tends to zero taking positive values only.

Lemma 1.38 For any test function ϕ, $\int_{-\infty}^{-\epsilon}[\phi(x)/x]\,dx + \int_{\epsilon}^{\infty}[\phi(x)/x]\,dx \to -\int_{-\infty}^{\infty}\phi'(x)\log|x|\,dx$ as $\epsilon\downarrow 0$.

Proof Integrating by parts gives

$$\int_{-\infty}^{-\epsilon}[\phi(x)/x]\,dx + \int_{\epsilon}^{\infty}[\phi(x)/x]\,dx =$$
$$\phi(-\epsilon)\log|\epsilon| - \int_{-\infty}^{-\epsilon}\phi'(x)\log|x|\,dx - \phi(\epsilon)\log|\epsilon| - \int_{\epsilon}^{\infty}\phi'(x)\log|x|\,dx \quad .$$

The integrals containing $\log|x|$ are convergent as $\epsilon\downarrow 0$. And $[\phi(-\epsilon)-\phi(\epsilon)]\log|\epsilon| = -\phi'(\theta\epsilon)2\epsilon\log|\epsilon|$ for some θ between -1 and 1, and this $\to 0$ as $\epsilon\downarrow 0$ because ϕ' is bounded. Hence the above equation implies $(\int_{-\infty}^{-\epsilon} + \int_{\epsilon}^{\infty})[\phi(x)/x]\,dx \to -(\int_{-\infty}^{0} + \int_{0}^{\infty})\phi'(x)\log|x|\,dx$. □

Definition 1.39 If f is a function defined for $x \neq 0$, we define the **principal value** of $\int_{-\infty}^{\infty}f(x)\,dx$ by $P\int_{-\infty}^{\infty}f(x)\,dx = \lim_{\epsilon\downarrow 0}\{\int_{-\infty}^{-\epsilon}f(x)\,dx + \int_{\epsilon}^{\infty}f(x)\,dx\}$ whenever the limit exists. □

Lemma 1.38 shows that the limit exists when $f(x) = \phi(x)/x$, and that

$$P\int[\phi(x)/x]\,dx = -\int_{-\infty}^{\infty}\log|x|\,\phi'(x)\,dx \quad . \tag{1.10}$$

We can now define a distribution corresponding to $1/x$, in the same way as for locally integrable functions, but with $\int x^{-1}\phi(x)\,dx$ given a definite value by the principal-value rule.

Definition 1.40 P/x denotes the functional $\phi \mapsto P\int_{-\infty}^{\infty}x^{-1}\phi(x)\,dx$. □

Lemma 1.38 shows that this is defined for all $\phi\in\mathscr{D}$, and P/x is thus a linear functional on $\mathscr{D}$. One can prove directly that it is a continuous functional, and therefore a distribution. But it is easier to note from (1.10) that

$$\langle P/x,\phi\rangle = -\langle\log|x|,\phi'\rangle \quad , \tag{1.11}$$

and therefore

$$(\log|x|)' = P/x \quad ; \tag{1.12}$$

the continuity of the functional P/x now follows from Proposition 1.25. Equation (1.12) shows that P/x is the natural choice for a generalised function corresponding to $1/x$. Other choices are possible, corresponding to different ways of resolving the singularity in the integrand, as discussed above, but they are less useful, and can in any case be expressed in terms of P/x (see Problem 1.10).

We can define distributions corresponding to other negative powers; but the principal value cannot be used to assign a definite value to $\int_{-\infty}^{\infty}dx\,\phi(x)/x^n$, because it does not exist if $n > 1$. The integral is truly divergent for $n > 1$. We therefore define negative powers directly as derivatives of $\log|x|$.

Definition 1.41 For any integer $n > 1$, we define the distribution $\mathbf{x}^{-n}$ to be the n-th derivative of

$$\frac{(-1)^{n-1}}{(n-1)!} \log|\mathbf{x}| \quad . \qquad \square$$

Some authors use a special notation, such as Pfx^{-n}, as a reminder that $\int \mathbf{x}^{-n}\phi(x)\,\mathrm{d}x$ is not the same thing as $\int x^{-n}\phi(x)\,\mathrm{d}x$ (the latter integral being meaningless in general). In our notation the difference between $\mathbf{x}^{-n}$ and x^{-n} is enough to indicate whether the distribution or the ordinary function is being considered.

We can interpret equation (1.11) as follows: to evaluate the improper integral $\int\phi(x)x^{-1}\,\mathrm{d}x$, integrate it by parts as if it were a convergent integral. The result is the convergent integral $-\int\phi'(x)\log|x|\,\mathrm{d}x$. The integration by parts is not justified, of course, but this procedure gives the result (1.11) of our rigorous definitions, and can therefore be regarded as a formal procedure for obtaining the results of the correct theory. Similarly, Definition 1.41 for the case $n = 2$ is equivalent to $\int_{-\infty}^{\infty}\mathbf{x}^{-2}\phi(x)\,\mathrm{d}x = -\int_{-\infty}^{\infty}\log|x|\,\phi''(x)\,\mathrm{d}x$, and this can be interpreted as saying that to 'evaluate' the divergent integral $\int_{-\infty}^{\infty}x^{-2}\phi(x)\,\mathrm{d}x$ one should integrate it by parts formally twice until it becomes convergent, and regard the resulting convergent integral as the 'value' of the divergent integral. The same rule applies to other negative powers. This formal procedure was invented by the mathematician Hadamard (1932) long before the development of distribution theory, as a convenient device for dealing with divergent integrals appearing in the theory of wave propagation; he called it "extracting the finite part of the divergent integral", and developed a set of rules for manipulating the finite part. In the theory of distributions, Hadamard's finite part appears as simply a particular case of a singular distribution, and no special rules are needed.

Many other singular distributions can be defined, but we shall not pursue the topic. We close this section with an important property of $\mathbf{P/x}$. If it is to be an analogue of the function $1/x$, it ought to satisfy

$$\mathbf{x}.\mathbf{P/x} = \mathbf{1} \quad , \tag{1.13}$$

where $\mathbf{1}$ denotes the regular distribution generated by the constant function with value 1. (1.13) is easy to prove, using Definition 1.17: $\int \mathbf{x}\,\mathbf{P/x}\,\phi(x)\,\mathrm{d}x = P\int x\,\phi(x)/x\,\mathrm{d}x = \int\phi(x)\,\mathrm{d}x = \langle \mathbf{1},\phi\rangle$; the integrand is regular, so the principal value is just the ordinary integral. However, the converse of (1.13) is less straightforward.

In ordinary analysis, if a function f satisfies $xf(x) = 1$, it follows that $f(x) = 1/x$ for $x \neq 0$; no conclusion can be drawn about the value of $f(0)$. In this sense the solution of the equation $xf(x) = 1$ contains an undetermined constant. A somewhat similar result holds for generalised functions, but in this case we cannot speak of the value of f at the point $x = 0$. The nearest thing in distribution theory to a function localised at a point is the delta function, and the analogue of the statement at the beginning of this paragraph is the following.

Theorem 1.42 A distribution f satisfies $xf = 1$ if and only if

$$f(x) = P/x + C\delta(x) \tag{1.14}$$

for some constant C.

Proof $x\delta(x) = 0$ (since $\int x\delta(x)\phi(x)\,dx = 0$ for all $\phi \in \mathscr{D}$). So if f satisfies (1.14), then $xf = 1$ (using (1.13)). Conversely, suppose $xf = 1$. Choose a test function θ such that $\theta(0) = 1$. Then for any $\phi \in \mathscr{D}$, $\phi(x) - \phi(0)\theta(x) \to 0$ as $x \to 0$, and it is easy to show that the function $\widetilde{\phi}(x) = [\phi(x) - \phi(0)\theta(x)]/x$ is well behaved at $x = 0$, and is in fact a test function. Now,

$$\begin{aligned}\int f(x)\phi(x)\,dx &= \int f(x)[x\widetilde{\phi}(x) + \phi(0)\theta(x)]\,dx \\ &= \int \widetilde{\phi}(x)\,dx + \phi(0)\int f(x)\theta(x)\,dx \quad ,\end{aligned}$$

since $xf = 1$;

$$= P\int dx/x\,[\phi(x) - \phi(0)\theta(x)] + \phi(0)\int f(x)\theta(x)\,dx$$

since the principal value of a convergent integral equals its ordinary value;

$$\begin{aligned}&= P\int dx/x\,\phi(x) + \phi(0)\int [f(x) - P/x]\,\theta(x)\,dx \\ &= \int [P/x + C\delta(x)]\,\phi(x)\,dx \quad ,\end{aligned}$$

where the constant C depends only on f and θ, which was chosen arbitrarily at the beginning, and not on ϕ. This completes the proof. □

1.6 Fourier Series and the Poisson Sum Formula

In the last section we saw how distribution theory can be used to give a meaning to certain divergent integrals, the underlying idea being to integrate formally by parts often enough to turn the divergent integral into a convergent one. Similar methods can be used to give a value to certain divergent series. Divergent Fourier series in particular can often be made convergent by integration.

Example 1.43 The series

$$\sum_1^\infty \cos(2n\pi x) \tag{1.15}$$

diverges when x is an integer, oscillates when x is half an odd integer, and fails to converge whenever x is rational. However, it is the series obtained by differentiating

$$\sum_1^\infty [\sin(2n\pi x)]/2n\pi = f(x) \tag{1.16}$$

term by term (an operation not permissible in classical analysis since the series converges nonuniformly). This is the Fourier series of the function f which is periodic with period 1 and defined for $0 < x < 1$ by $f(x) = (1 - 2x)/4$; see Fig. 2. The series (1.16) therefore converges pointwise but nonuniformly. To obtain

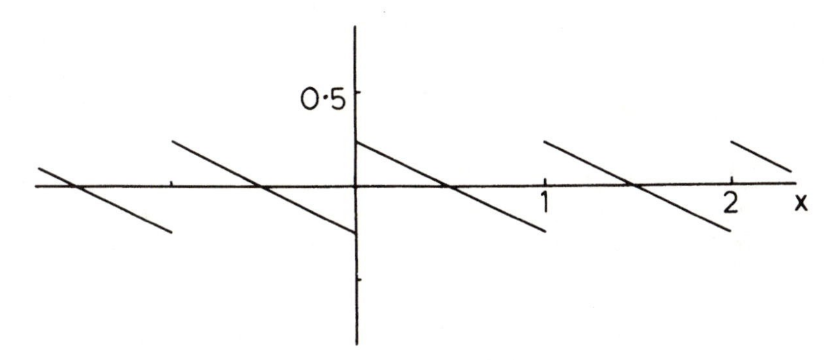

Fig. 2. The sum of the Fourier series (1.16).

a uniformly convergent series, we integrate again, that is, we note that (1.16) is the series obtained by differentiating

$$-\sum_1^\infty [\cos(2n\pi x)]/(2n\pi)^2 = F(x) \tag{1.17}$$

termwise. This series converges uniformly for all x (by the M-test), and the partial sums are continuous and therefore locally integrable; hence it follows from Theorem 1.32 that it converges distributionally. In other words, the series of distributions $-\Sigma_1^\infty(2n\pi)^{-2}\cos(2n\pi x)$ converges to the distribution F generated by the classical sum $F(x)$ of the series (1.17). Now, a Fourier series can be integrated term by term, even if it does not converge uniformly (see Titchmarsh (1939) for a proof). Applying this theorem to (1.16) shows that $F' = f$. It now follows from Theorem 1.35 that

$$\sum_1^\infty [\sin(2n\pi x)]/2n\pi = f(x) \quad ,$$

and $$\sum_1^\infty \cos(2n\pi x) = f'(x) \quad .$$

It is clear from Fig. 2 that $f'(x) = -\frac{1}{2}$ for all x except the integers, and f has a jump discontinuity of $\frac{1}{2}$ at each integer. Each discontinuity in f contributes a delta function to f', so finally we have

$$\sum_1^\infty \cos(2n\pi x) = -\tfrac{1}{2} + \tfrac{1}{2}\sum_{-\infty}^{\infty} \delta(x-k) \quad . \tag{1.18}$$

Thus we have given a meaning to the divergent series $\Sigma_1^\infty \cos(2n\pi x)$. We note in passing that (1.18) gives the following useful representation of the delta function:

$$\delta(x) = 1 + 2\sum_1^\infty \cos(2n\pi x) \quad \text{for} \quad -1 < x < 1 \quad , \tag{1.19}$$

all but one of the delta functions in (1.18) being zero in (−1,1). □

A similar procedure will give a value to any Fourier series whose coefficients grow like a power of n. Using the exponential form for neatness, we consider the series $\Sigma_{-\infty}^\infty b_n e^{inx}$. If p is an integer such that $n^{2-p} b_n \to 0$ as $n \to \pm\infty$, it can easily be shown that $\Sigma_{-\infty}^\infty b_n e^{inx}$ converges distributionally to $b_0 + g^{(p)}(x)$, where the superscript denotes the p-th derivative, and $g(x) = \Sigma' b_n (in)^{-p} e^{inx}$, Σ' denoting the sum from $-\infty$ to ∞ excluding $n = 0$.

We now give an application of (1.18). Using the exponential form of the series, we may rewrite it as

$$\sum_{-\infty}^{\infty} \delta(x-k) = \sum_{-\infty}^{\infty} e^{2n\pi ix}$$

or

$$\sum_{-\infty}^{\infty} \delta\left(\frac{y}{\lambda} - k\right) = \sum_{-\infty}^{\infty} e^{2n\pi iy/\lambda}$$

for any real constant $\lambda \neq 0$. Applying this to any test function ϕ gives

$$\sum_{k=-\infty}^{\infty} \int \phi(y) \delta\left(\frac{y}{\lambda} - k\right) dy = \sum_{n=-\infty}^{\infty} \int \phi(y) e^{2n\pi iy/\lambda} dy \quad .$$

Using formula (1.9) on the left hand side, and recognising the right hand side as the Fourier transform of ϕ (the reader unfamiliar with Fourier transforms should refer to section 3.1), we have

$$|\lambda| \sum_{-\infty}^{\infty} \phi(k\lambda) = \sum_{-\infty}^{\infty} \Phi(2n\pi/\lambda) \tag{1.20}$$

where Φ is the Fourier transform of ϕ. This is called the **Poisson sum formula**; it is a useful device for transforming a series from one form into another form which may be more useful. It takes a particularly elegant form when $\lambda = 1$: the sum of the values of any function at the integers equals the sum of the values of its Fourier transform at integral multiples of 2π.

Example 1.44 Applying (1.20) to the function e^{-y^2} gives $|\lambda| \Sigma_{-\infty}^{\infty} e^{-k^2\lambda^2} = \sqrt{\pi}\, \Sigma_{-\infty}^{\infty} e^{-n^2\pi^2/\lambda^2}$. Neither of these sums can be evaluated exactly, and numerical methods must be used. For large values of $|\lambda|$ the series on the left converges rapidly, but for small $|\lambda|$ it converges very slowly; the series on the right converges rapidly for small $|\lambda|$, and can be used to evaluate the sum when the series on the left is useless. It is quite often possible to use the Poisson formula in this way to improve the convergence of a series.

Example 1.45 Applying the Poisson formula to the series $\sum_{-\infty}^{\infty} 1/(1 + a^2 n^2)$ turns it into a series which is easily summed exactly, giving the value $(\pi/a)\coth(\pi/a)$ for the series. The details are left to the reader. This example is representative of a whole class of series which can be evaluated by using (1.20); see Problem 1.12 for more examples. □

The Poisson sum formula appears in the solution of certain boundary-value problems for partial differential equations. The reader familiar with electrostatics may remember that problems involving point charges in boxes can be solved by two methods: separation of variables leads to a solution in the form of an infinite series, and the method of images leads to a solution in the form of a different infinite series, corresponding to an infinite set of image charges. These two series turn out to be related to each other by the Poisson transformation, and as in Example 1.44, one series converges well in circumstances where the other converges badly. These matters are discussed in Morse & Feshbach, and Wallace.

Finally, we should remark that although we have derived (1.20) by means of distribution theory, it was discovered long before the delta function was invented. For a rigorous discussion along classical lines, see Titchmarsh (1937), where some remarkable generalisations of the formula will be found (see also Problems 1.14–1.16).

1.7 Summary and References

In section 1.1 the δ-function was introduced by physical reasoning, as a function whose graph is an infinitely high and narrow spike. Dirac is the classical reference for this; Van der Pol & Bremmer describe the prehistory of the delta function, its origins in nineteenth century mathematics.

In sections 1.2, 1.3, and 1.4 we introduced the space of test functions, and defined distributions as continuous linear functionals on that space. We showed how to construct regular distributions corresponding to locally integrable functions, and defined operations such as change of variable, differentiation, convergence etc. for distributions in general, by analogy with the regular case. We showed that every distribution is differentiable, and every convergent sequence can be differentiated term by term. This theory is due to Schwartz (1950); for other accounts of it, see Stakgold (1967 and 1979), Zemanian (1965), Shilov, Gel'fand & Shilov, Korevaar, and Vladimirov. For different approaches to distribution theory, see Erdélyi, Hoskins, Korevaar, Lighthill, and Jones.

In section 1.3 we defined the product of a distribution and a smooth function, and noted the serious difficulties which prevent one from multiplying distributions by discontinuous functions or other distributions. Progress has recently been made in overcoming these difficulties; see Colombeau.

In section 1.5 we introduced notation which allows us to manipulate distributions much as if they were ordinary functions, and then considered the principal value of an integral, and the distributions corresponding to negative powers; see the text-books mentioned above. In section 1.6 we showed how to sum divergent Fourier series, and derived the Poisson sum formula. For classical methods of summing divergent series, see Hardy; for the classical theory of the Poisson sum formula see Titchmarsh (1937).

PROBLEMS 1

Sections 1.2 and 1.3

1.1 a, b are constants; show that

(a) $x^n\delta^{(m)}(x) = \begin{cases} 0 & \text{if} \quad m<n \\ (-1)^n n!\delta(x) & \text{if} \quad m=n \\ \dfrac{(-1)^n m!}{(m-n)!}\delta^{(m-n)}(x) & \text{if} \quad m>n; \end{cases}$

(b) $\sin(at)\delta'(t) = -a\delta(t)$;

(c) $e^{at}\delta^{(n)}(t-b) = e^{ab}\sum_{r=0}^{n}\binom{n}{r}(-a)^{n-r}\delta^{(r)}(t-b)$, where $\binom{n}{r}$ denotes the binomial coefficient $n!/r!(n-r)!$.

1.2 Formulate definitions of odd and even distributions, and prove the usual properties (e.g., derivative of an even function is an odd function and vice versa, product of two odd functions is even, etc.).

1.3 A distribution is said to have rank r if it is the r-th derivative of a regular distribution, but not the s-th for any $s<r$. It is said to have infinite rank if it is not of rank r for any r. Show that $\delta^{(n)}$ has rank $n+1$. Show that $\phi \mapsto \phi(0)+\phi'(1)+\ldots+\phi^{(n)}(n)+\ldots$ defines a distribution, and that its rank is infinite. (It can be proved that every distribution which is zero outside a finite interval (in the sense of Definition 1.20) has finite rank; see Korevaar, or Zemanian (1965), who uses the term 'order' instead of 'rank'.)

1.4 Extension of Multiplication. Although, in general, distributions cannot be multiplied by non-smooth functions, distributions of finite rank (see Problem 1.3) can be dealt with as follows.

Let K^r be the class of ordinary functions k continuously differentiable $r-1$ times, and such that $k^{(r-1)}$ is differentiable except at isolated points, and $k^{(r)}$ is locally integrable. For example, the function $|x|$ belongs to K^1. We define the product of any distribution of rank r with a member of K^r as follows. If f is of rank r, with $f=g^{(r)}$ where g is a locally integrable function, we define the distribution hf by $\langle hf,\phi\rangle = (-1)^r\int_{-\infty}^{\infty}g(x)[h\phi]^{(r)}\,dx$ for all $\phi\in\mathscr{D}$.

(a) Verify that when h is smooth, this agrees with Definition 1.17.

(b) If $h\in K^{n+1}$, show that $h\delta^{(n)} = h(0)\delta^{(n)} - nh'(0)\delta^{(n-1)} + \ldots + (-1)^n h^{(n)}(0)\delta$.

(c) Simplify the expression $xe^{|x|}\delta'(x)$.

1.5 Show that no test function (except the identically zero function) can be an analytic function in the sense of complex variable theory.

Sections 1.4 and 1.5

1.6 Show that $\sin(kx)/(\pi x) \to \delta(x)$ as $k \to \infty$. Sketch the graphs of the first few terms of the sequence, and compare with Fig. 1.

1.7 Formulate a definition of convergence, analogous to Def. 1.30, for distributions which depend on a real parameter; i.e., if f_t is a distribution for each real t, give a definition of '$f_t \to F$ as $t \to a$'. Prove from your definition that if $f_t \to F$ and $g_t \to G$, then $bf_t + cg_t \to bF + cG$ for any constants b, c.

1.8 Using Problem 1.7 we can define differentiation differently, in a way similar to elementary calculus. For any distribution f, prove that $[f(x+h) - f(x)]/h$ converges as $h \to 0$ (where $f(x+h)$ is given by Definition 1.22). The limit is called the derivative of f; prove that it always agrees with Definition 1.26.

1.9 Using Problem 1.7,

(a) show that $x/(x^2 + a^2) \to P/x$ as $a \to 0$;

(b) show that $(1 - \cos Rx)/x \to P/x$ as $R \to \infty$ (you must first show that $(1 - \cos Rx)/x$ is locally integrable and hence generates a distribution).

1.10 Using Problem 1.9(a), show that $1/(x \pm i\alpha) \to P/x \mp i\pi\delta(x)$ as $\alpha \downarrow 0$. This is called the **Plemelj formula.** (It can be obtained very easily by complex contour integration ($i\pi\delta(x)$ corresponds to half a residue at $x = 0$), but that method does not prove convergence in the sense of distribution theory because of Problem 1.5.)

Section 1.6

1.11 Prove the following generalisation of the Poisson sum formula:
$|b|\sum_{-\infty}^{\infty}\phi(a+nb) = \sum_{-\infty}^{\infty} e^{-2n\pi ia/b}\,\Phi(2n\pi/b)$ where Φ is the Fourier transform of ϕ.

1.12 Using the Poisson sum formula:

(a) Show that $\sum_{-\infty}^{\infty} 1/(1 + a^2n^2) = (\pi/a)\coth(\pi/a)$,

(b) Show that $\sum_{1}^{\infty} \cos(na)/(1 + n^2) = (\pi/2)\cosh(\pi - a)/\sinh\pi - \frac{1}{2}$ for $0 < a < 2\pi$,

(c) Evaluate $\sum_{1}^{\infty} 1/(n^4 + a^4)$. By taking the limit $a \to 0$, show that $\Sigma\, n^{-4} = \pi^4/90$.

1.13 Obtain analogues of the Poisson sum formula involving the Fourier cosine or (much harder) sine transform.

1.14 Prove the following beautiful extension of the Poisson sum formula to finite sums: if f is a well behaved function with $f(\alpha) = f(\beta) = 0$, then $\sum_{\alpha \leqslant n \leqslant \beta} f(n) = \sum_{-\infty}^{\infty} \int_{\alpha}^{\beta} f(y) \cos 2\pi k y \, dy$, where the sum on the left is over all integers in the interval $[\alpha, \beta]$; α, β need not themselves be integers.

1.15 Starting from the cosine form of the Poisson sum formula, and integrating by parts, show that $\sum_{0}^{\infty} f(n) = \frac{1}{2} f(0) + \int_0^{\infty} f(x)\, dx + \sum_1^{\infty} \int_0^{\infty} f'(x) \sin 2n\pi x \, dx / n\pi$. Use further integrations by parts to obtain $\Sigma_0^{\infty} f(n) = \int_0^{\infty} f(x)\, dx + \frac{1}{2} f(0) - \frac{1}{12} f'(0) + \frac{1}{720} f'''(0) - \frac{1}{30240} f'''''(0) + \ldots$, which is called the **Euler-MacLaurin sum formula**; see Lanczos (1966) or Whittaker & Watson for further information. If $\int_0^{\infty} f(x)\, dx$ can be evaluated exactly, the formula can be used to evaluate $\Sigma f(n)$ approximately. Apply this technique to $\Sigma(a+n)^{-2}$ where $a > 0$,

1.16 Contemplate the amazing formula below. It is quoted, without proof, by Titchmarsh (1937), who gives the reference to the original paper by Ramanujan.

$$\sqrt{\alpha}[f(\alpha) - f(5\alpha) - f(7\alpha) + f(11\alpha) + f(13\alpha) - f(17\alpha) - \ldots]$$
$$= \sqrt{\beta}[F(\beta) - F(5\beta) - F(7\beta) + \ldots]$$

for any numbers α, β such that $\alpha\beta = \pi/6$; here $F(x) = \sqrt{2/\pi} \int_0^{\infty} f(t) \cos(xt)\, dt$. The numbers appearing in the series are those integers which are relatively prime to 6.

More problems will be found on page 375.

Chapter 2

Differential Equations and Green's Functions

In this chapter we study ordinary differential equations for generalised functions. The simplest differential equation of all is $u' = f$ where f is a given function. Its solution is given at once by integration. We have not yet defined integration for generalised functions; we do so in section 2.1. Section 2.2 discusses the theory of linear differential equations for generalised functions, and the rest of the chapter deals with Green's functions, which provide a systematic method for solving inhomogeneous equations and for converting differential equations to integral equations – a procedure which is an essential step in the application of the theory of later chapters to problems of applied mathematics.

We shall not use bold type to distinguish generalised functions in this chapter. We hope that the reader no longer needs that notational crutch to understand whether ordinary or generalised functions are under discussion.

2.1 The Integral of a Distribution

We define the integral of a generalised function as an anti-derivative, that is, a generalised function whose derivative is the original function.

Definition 2.1 A distribution F is an **integral** of a distribution f if $F' = f$. □

This is the analogue of the indefinite integral of elementary calculus. We know that not all ordinary functions have integrals: if $f: \mathbb{R} \to \mathbb{R}$ is severely discontinuous (consult analysis text-books for more precise statements) then there is no differentiable function F such that $F' = f$. For generalised functions, however, the situation is, as usual, more satisfactory. We shall see that every generalised function has an integral.

This is not immediately obvious, because Definition 2.1 is nonconstructive: it tells us how to recognise an indefinite integral of f when we see one, but not how to construct one. However, it is easy to construct $\langle F,\phi\rangle$ if ϕ is itself the derivative of some test function $\Phi \in \mathscr{D}$. For then

$$\langle F,\phi\rangle = \langle F,\Phi'\rangle = -\langle F',\Phi\rangle = -\langle f,\Phi\rangle \quad . \tag{2.1}$$

For any given f, the right hand side of this equation can be found, and hence $\langle F,\phi\rangle$

can be calculated, provided that $\phi = \Phi'$ for some $\Phi \in \mathscr{D}$. How can we find such a Φ? If $\text{supp}(\phi) \subset [a,b]$, we can set $\Phi(x) = \int_a^x \phi(t)\,\mathrm{d}t$, and then Φ is smooth, $\Phi' = \phi$, and $\Phi(x) = 0$ for $x < a$. But in general $\Phi(x) \neq 0$ for $x > b$; Φ does not have bounded support, and so $\Phi \notin \mathscr{D}$. We have $\Phi(x) = 0$ for $x > b$ if and only if $\int_a^b \phi(x)\,\mathrm{d}x = 0$. Thus (2.1) can be used to calculate F only on the subset of $\mathscr{D}$ consisting of functions with zero integral. To complete the construction of F, and thus prove that every distribution is integrable, we must somehow extend (2.1) to the whole of $\mathscr{D}$. We shall do this by showing how to decompose any $\phi \in \mathscr{D}$ into a function with zero integral, to which (2.1) can be applied, plus a remainder term, which, as we shall see, leads to something like the arbitrary constant of integration of elementary calculus.

Theorem 2.2 (Existence of Integrals) For any distribution f there are infinitely many distributions F such that $F' = f$, and the difference between any two of them is a constant.

Proof Let Z be the set of test functions ϕ such that $\int_{-\infty}^{\infty} \phi(x)\,\mathrm{d}x = 0$. We shall show how to split any $\phi \in \mathscr{D}$ into a member of Z plus a remainder. Pick a $\phi_0 \in \mathscr{D}$ such that $\int_{-\infty}^{\infty} \phi_0(x)\,\mathrm{d}x = 1$. For any $\phi \in \mathscr{D}$ define a function ψ by

$$\psi(x) = \phi(x) - k\phi_0(x) \quad , \tag{2.2a}$$

where $$k = \int_{-\infty}^{\infty} \phi(x)\,\mathrm{d}x \quad . \tag{2.2b}$$

Then $\psi \in \mathscr{D}$ by 1.5, and integrating (2.2a) shows that $\psi \in Z$. Write $\Psi(x) = \int_{-\infty}^{x} \psi(y)\,\mathrm{d}y$; then $\Psi(x) = 0$ for $x < a$ and for $x > b$, where $[a,b]$ is an interval containing $\text{supp}(\psi)$. Hence $\Psi \in \mathscr{D}$, and we may therefore define a functional F on $\mathscr{D}$ by

$$\langle F,\phi\rangle = -\langle f,\Psi\rangle \quad . \tag{2.3}$$

In other words, $\langle F,\phi\rangle$ is to be calculated by splitting ϕ into a part ψ lying in Z and a remainder, and then using (2.1) applied to ψ, ignoring the remainder, $k\phi_0$. That is, we take $\langle F,\phi_0\rangle$ to be zero.

We must verify that the F defined by (2.3) is an integral of f. It is easy to verify that F is a distribution, that is, a continuous linear functional. To show that $F' = f$, note that if θ' is the derivative of a test function, then $\int_{-\infty}^{\infty} \theta'(x)\,\mathrm{d}x = 0$, so that the k in (2.2) vanishes. Hence we have, by 1.26,

$$\begin{aligned} \langle F',\theta\rangle &= -\langle F,\theta'\rangle \quad , \\ &= \langle f,\theta\rangle \quad , \text{ for all } \theta \in \mathscr{D}, \end{aligned}$$

using (2.3) and (2.2). This shows that F is an integral of f.

The above procedure gives many integrals of the distribution f, depending on the choice of ϕ_0. We now show that all possible integrals are obtained by adding constants to one of them. Let F and G be two integrals of f. Then for any $\phi \in \mathscr{D}$ we have

$$\langle F - G, \phi \rangle = \langle F - G, k\phi_0 + \psi \rangle$$

where ψ and k are defined by (2.2),

$$= k\langle F - G, \phi_0 \rangle + \langle F - G, \Psi' \rangle$$

$$= kC - \langle F' - G', \Psi \rangle \quad ,$$

where $C = \langle F - G, \phi_0 \rangle$ is a constant, independent of ϕ. Hence

$$\langle F - G, \phi \rangle = C \int \phi(x)\,\mathrm{d}x - \langle f - f, \Psi \rangle$$

$$= \langle C, \phi \rangle \quad .$$

Thus $F - G$ is a constant. Conversely, it is obvious that if F is an integral of f, then so is $F + C$ for any constant C. □

We now have an existence theorem for the differential equation $u' = f$. In the next section we shall consider the general linear equation.

2.2 Linear Differential Equations

We shall consider n-th order linear equations of the form $a_0(x)u^{(n)} + a_1(x)u^{(n-1)} + \ldots + a_n(x)u = f$, where $u^{(r)}$ denotes the r-th derivative of u.

Notation 2.3 D will stand for the differentiation operator, and $L = a_0 \mathrm{D}^n + a_1 \mathrm{D}^{n-1} + \ldots + a_n$, where a_i are smooth functions (Definition 1.2). □

Our equation can now be written

$$Lu = f \quad . \tag{2.4}$$

We must insist that the coefficients a_i be smooth in order that the product of the distribution $\mathrm{D}^i u$ and the function a_i be well-defined according to Definition 1.17. If one uses the extended multiplication rule of Problem 1.4, then one can deal with equations with non-smooth coefficients; see Problem 2.11.

There are various kinds of solution of (2.4), regarded as an equation for a generalised function u. We classify them as follows.

Definition 2.4

(i) Any distribution satisfying (2.4) is called a **generalised solution.**

(ii) A **classical solution** of (2.4) is an ordinary function which is differentiable n times and satisfies it (and therefore generates a regular distribution which satisfies (2.4) in the generalised sense).

(iii) **A weak solution** is an ordinary function which may not be n times differentiable, and therefore not a classical solution, but which generates a regular distribution which is a generalised solution.

(iv) **A distributional solution** is a singular distribution satisfying (2.4). □

A moment's thought will show that every solution is either a classical, a weak, or a distributional solution, and that if f in (2.4) is a singular distribution, then all solutions must be distributional or weak. The remarkable thing is that admitting generalised functions can produce new solutions of even entirely classical equations, that is, equations containing no singular distributions.

Example 2.5 The equation $xu' = 0$ has a classical solution u = constant. But $u(x) = c_1 H(x) + c_2$ is a weak solution for any constants c_1, c_2. Proof: $u'(x) = c_1 \delta(x)$, hence $\langle xu', \phi\rangle = \langle c_1 \delta, x\phi\rangle$ by Definition 1.17, and this is zero. We thus have nonclassical solutions which violate the 'rule' that a first-order equation has a general solution with just one arbitrary constant. This phenomenon is well known in the classical theory of differential equations (see, for example, Schwarzenberger), and this weak solution is not really anything new. However, we are entitled to consider it a proper generalised solution of our equation, whereas classically it is not differentiable, therefore not a (classical) solution, and the classical mathematician must go through some contortions in order to discuss it. □

Example 2.6 The equation $x^2 u' = 0$ has the same classical and weak solutions as $xu' = 0$, but also has distributional solutions $u = c_1 \delta(x) + c_2 H(x) + c_3$ for any constants c_1, c_2 and c_3. These solutions have no analogues in the classical theory. □

It is rather worrying that perfectly conventional differential equations can acquire new and singular solutions when generalised functions are allowed. It is natural to ask whether such new distributional solutions exist for all differential equations (with the uneasy feeling that if the answer is yes, then we will probably have to rework the theory of linear differential equations completely). Fortunately, the answer is no. Distributional solutions of classical differential equations only arise when the leading coefficient vanishes, and they have singularities at the zeros of $a_0(x)$. The elementary theory of linear equations, and very many applications, are concerned with equations with nonzero leading coefficient, for which the following theorem assures us that the only solutions are the classical ones. Distribution theory adds nothing new in this case.

Theorem 2.7 If $a_0(x) \neq 0$ for all x, and f is a continuous function, then every generalised solution of (2.4) is a classical solution. □

The proof of this theorem is quite long; we refer the reader to Zemanian (1965), p. 162. We shall not discuss the case of vanishing a_0, but turn our attention to equations with a singular right hand side.

2.3 Fundamental Solutions of Differential Equations

The rest of this chapter describes a general method for solving inhomogeneous equations. The basic idea is quite simple. Consider the equation

$$Lu = f \tag{2.4}$$

where L is the differential operator $a_0 D^n + \ldots + a_n$. If f is a sufficiently well-behaved function, we have

$$f(x) = \int_{-\infty}^{\infty} \delta(x-y) f(y)\, dy \quad . \tag{2.5}$$

Strictly speaking, this equation should be interpreted according to 1.36: the integral sign is a notation for the action of the functional δ on a test function f. From an intuitive point of view, however, we can think of (2.5) as saying that the function $f(x)$ is the superposition (that is, sum or integral) of an infinite number of delta functions of x, centred on all possible points y, the 'amplitude' or coefficient of the delta function centred on y being the number $f(y)$. For a more detailed analysis of this idea, see Problem 2.4.

Now, (2.4) is a linear equation, which means that if $f = a_1 f_1 + a_2 f_2$ then the solution of (2.4) is $a_1 u_1 + a_2 u_2$ where u_i is the solution of the equation $Lu_i = f_i$. Extending this idea to the continuous superposition (2.5), we can expect the solution of (2.4) to be a superposition (in this case an integral) of solutions of the equation

$$Lu(x) = \delta(x-y) \quad . \tag{2.6}$$

In (2.6) L is a differential operator with respect to x, and y is a parameter in the equation. We think of the solution as a function of x, but it will also depend on the parameter y; we therefore rename it $w(x;y)$. Our argument then suggests that the solution u of (2.4) is obtained from w by the same superposition integral as that in (2.5) giving $f(x)$ in terms of $\delta(x-y)$:

$$u(x) = \int_{-\infty}^{\infty} w(x;y) f(y)\, dy \quad , \tag{2.7}$$

where w satisfies

$$Lw(x;y) = \delta(x-y) \quad . \tag{2.8}$$

The point of all this is that if we can solve (2.8) and find w, then we can immediately write down the solution of any inhomogeneous equation with the same left-hand side, just by evaluating the integral (2.7). We have obtained a general formalism for solving inhomogeneous equations.

Doubts immediately arise about the procedure outlined above. The argument was entirely nonrigorous and unconvincing – it was intended to suggest, not to convince. Furthermore, to speak of writing down 'the' solution of a differential equation is unsatisfactory; there are many solutions, satisfying different boundary conditions. We must reconsider the subject more carefully.

We shall now construct a rigorous version of the above argument. We shall prove that the function u given by the formula (2.7) is indeed a solution of (2.4).

Definition 2.8 Any solution of (2.8), where L involves differentiation with respect to x, is called a **fundamental solution** for the operator L. □

w is called a fundamental solution because it is, according to (2.7), a fundamental building-block for constructing solutions of inhomogeneous equations. It is not obvious whether it is a weak or a distributional solution, or indeed, whether (2.8) has any solutions at all. We shall show that there is always a fundamental solution, by constructing one. First we consider an example.

Example 2.9 Let $Lu = u'' + 2ku' + \omega^2 u$. The equation $Lu = f$ is the equation of motion of an oscillator of natural frequency depending on ω and damping proportional to k, with an applied force proportional to $f(t)$; we are thinking of u as a function of time, so $u' = \mathrm{d}u/\mathrm{d}t$. A fundamental solution $w(t;s)$ satisfies

$$w'' + 2kw' + \omega^2 w = \delta(t-s) \quad ; \tag{2.9}$$

it describes the motion of the system when the external force is an impulse concentrated at time $t = s$. Now, when solving differential equations, or indeed, solving any problem, it is permissible to use any methods at all, no matter how dubious, provided that once the solution has been found it can be proved to satisfy all the conditions of the problem. We shall therefore disregard theoretical subtleties when calculating w, but verify at the end that the function we have constructed really is a fundamental solution.

Accordingly, we interpret (2.9) as implying that

$$w'' + 2kw' + \omega^2 w = 0 \quad \text{for} \quad t \neq s \quad , \tag{2.10}$$

where $'$ denotes a t-derivative. If we take

$$w(t;s) = 0 \quad \text{for} \quad t < s \quad , \tag{2.11}$$

then the equation is satisfied for $t < s$. For $t > s$, w is a nonzero solution of (2.10) determined by conditions at $t = s$ as follows.

The function w must be continuous at $t = s$, for otherwise w' would be the derivative of a discontinuous function and hence contain a delta function, and w'' would involve δ' and (2.9) would therefore not be satisfied. Hence w is continuous at $t = s$, and we have

$$w(s+;s) = w(s-;s) = 0 \tag{2.12}$$

where $w(s \pm;s)$ denotes $\lim_{\epsilon \downarrow 0} w(s \pm \epsilon;s)$. Now, w' is discontinuous at $t = s$, giving a delta function in w'' so that (2.9) is satisfied. To determine the magnitude of the discontinuity in w', we integrate (2.9) with respect to t from $s - \epsilon$ to $s + \epsilon$, which gives $[w'(s+\epsilon;s) - w'(s-\epsilon;s)] + 2k[w(s+\epsilon;s) - w(s-\epsilon;s)] + \omega^2 \int_{s-\epsilon}^{s+\epsilon} w(t;s)\,dt = 1$. We now take the limit $\epsilon \downarrow 0$. Because w is continuous, the second and third terms tend to zero as $\epsilon \downarrow 0$. Hence $w'(s+;s) - w'(s-;s) = 1$, which, using (2.11) gives

$$w'(s+;s) = 1 \quad . \tag{2.13}$$

We now use (2.12) and (2.13) as initial conditions to determine w for $t > s$. The solution of (2.10) for $t > s$ which satisfies (2.12) and (2.13) is

$$w(t;s) = \Omega^{-1} e^{-k(t-s)} \sin[\Omega(t-s)] \quad \text{for} \quad t > s \tag{2.14}$$

where $\quad \Omega = (\omega^2 - k^2)^{\frac{1}{2}} \quad .$

We have assumed that $\omega > k > 0$, but similar formulas apply when $\omega < k$ or $\omega = k$. Equations (2.11) and (2.14) give a fundamental solution for L. It is not unique, of course; any solution of the homogeneous equation $Lw = 0$ can be added, and the resulting function will still satisfy (2.9).

Finally, we note again that this solution has been found by nonrigorous methods. In the theorem below we shall prove that the result is rigorously correct. □

We now consider the general n-th order equation.

Theorem 2.10 (Existence of Fundamental Solutions) If $a_0(x) \neq 0$ for all x, and $a_0, \ldots, a_n$ are smooth functions, then there is a fundamental solution for $L = \Sigma_0^n a_i(x) D^{n-i}$. All fundamental solutions for L are obtained by adding solutions of the homogeneous equation $Lu = 0$ to any one of them, and they are all weak solutions of (2.8).

Proof We first give a procedure for constructing a function $w(x;y)$, and then prove that the function so constructed is a solution of (2.8). The procedure is essentially that used in the above example. As in (2.11), we take

$$w(x;y) = 0 \quad \text{for} \quad x < y \quad . \tag{2.15}$$

Now, $w^{(n)}$ must have a singularity no worse than a delta function. Hence $w^{(n-1)}$ must at worst be discontinuous, and all the lower order derivatives must be continuous. In view of (2.15), this implies

$$w(y+;y) = w'(y+;y) = \ldots = w^{(n-2)}(y+;y) = 0 \tag{2.16}$$

We need one more condition to determine w for $x > y$. It is found by integrating $Lw = \delta(x - y)$ from $x = y - \epsilon$ to $x = y + \epsilon$ and then letting $\epsilon \downarrow 0$. All the terms on the left hand side except the first give zero in the limit $\epsilon \downarrow 0$, because all the derivatives up to $w^{(n-2)}$ are continuous. Hence we have

$$a_0(y)[w^{(n-1)}(y+;y) - w^{(n-1)}(y-;y)] = 1 \quad , \tag{2.17}$$

which, in view of (2.15), gives

$$a_0(y)w^{(n-1)}(y+;y) = 1 \quad . \tag{2.18}$$

We now define w for $x > y$ to be the solution of $Lw = 0$ which satisfies the initial conditions (2.16) and (2.18); the classical existence theorem for ordinary differential equations guarantees that there is always such a solution.

We now prove that the function w constructed above is a weak solution of (2.8). We must show that for any $\phi \in \mathscr{D}$, $\langle Lw,\phi\rangle = \langle\delta(x-y),\phi\rangle = \phi(y)$. We have $\langle a_0(x)w^{(n)} + \ldots + a_n(x)w,\phi\rangle = \langle w,(-1)^n(a_0\phi)^{(n)} + \ldots + a_n\phi\rangle = \int_{-\infty}^{\infty} w(x;y)[(-1)^n(a_0\phi)^{(n)} + \ldots + a_n\phi]\,dx$ where the integral here is a genuine integral (not to be interpreted as in 1.36) because the distribution w is regular. Hence by (2.15) $\langle Lw,\phi\rangle = \int_y^\infty w(x;y)[(-1)^n(a_0\phi)^{(n)} + \ldots + a_n\phi]\,dx$.

We can now integrate each term by parts often enough to transfer all the differentiations from ϕ to w. All the boundary terms from the upper limit of integration vanish because ϕ is a test function, and those from the lower limit vanish by (2.16) except the first, so we have $\langle Lw,\phi\rangle = a_0(y)w^{(n-1)}(y;y)\phi(y) + \int_y^\infty (Lw)\phi\,dx = \phi(y)$ using (2.18) and the fact that $Lw = 0$ for $x > y$ by definition. Hence w is a generalised solution of (2.8), that is, a fundamental solution for L. It is a continuous function, but its n-th derivative does not exist at $x = y$, so it is not a classical but a weak solution of (2.8).

Clearly, if w is a fundamental solution, then so is the sum of w and any solution of the homogeneous equation $Lu = 0$. Conversely, if v is any fundamental solution other than the one constructed above, then $L(w - v) = 0$, so the function $h = w - v$ satisfies the homogeneous equation; thus every fundamental solution can be obtained by adding a solution of $Lh = 0$ to any one of them. Theorem 2.7 assures us that h is a classical solution, hence a regular distribution, so every fundamental solution is a regular distribution, and hence a weak solution of (2.8). □

The last part of this Theorem shows that fundamental solutions for ordinary differential equations of the type considered here are always ordinary functions (or, what amounts to the same thing, regular distributions). It is therefore possible to discuss fundamental solutions entirely in terms of classical mathematics (see Remark 2.14 below). That approach, however, breaks down when applied to partial differential equations. We shall see in the next chapter that fundamental solutions for the wave equation involve delta functions, and therefore must be interpreted distributionally.

2.4 Green's Functions

Linear differential equations are usually associated with boundary conditions which, except in the case of eigenvalue problems, determine a single solution. According to section 2.3, a fundamental solution can be used to solve an inhomogeneous equation by means of equation (2.7), but the solution will not in general satisfy the given boundary conditions. However, if we choose the fundamental solution which itself satisfies the boundary conditions, and if the conditions are of the homogeneous type described in Definition 2.11, then the particular solution of (2.4) given by the formula (2.7) will also satisfy the conditions. That particular fundamental solution is called Green's function.

Definition 2.11 Consider an n-th order differential operator $L = \Sigma_{i=0}^{n} a_i(x)\mathrm{D}^{n-i}$ with a_0 non-vanishing and a_i smooth for each i; suppose a set of boundary conditions is given in the form of n linear homogeneous equations involving the function and its first $n - 1$ derivatives. **Green's function** for L with those boundary conditions is a fundamental solution which satisfies those conditions.

Example 2.12 Consider the operator $L = \mathrm{D}^2 + k^2$, with boundary conditions $u(0) = u(1) = 0$. This is a particular case of the operator of Example 2.9, and we could obtain Green's function by applying our boundary conditions to the fundamental solutions obtained there. But we prefer to work from first principles, as follows.

Green's function $g(x;y)$ satisfies

$$g_{xx} + k^2 g = \delta(x - y) \quad , \tag{2.19}$$

$$g(0;y) = 0 \quad , \tag{2.20}$$

$$g(1;y) = 0 \quad , \tag{2.21}$$

using subscripts to stand for partial differentiation. For $x < y$, (2.19) gives $g_{xx} + k^2 g = 0$, and the solution of this which satisfies (2.20) is

$$g(x;y) = A \sin kx \quad \text{for} \quad x < y \quad , \tag{2.22}$$

where A is constant as far as x is concerned, but may depend on y. Similarly, the condition (2.21) applied to (2.19) for $x > y$ gives

$$g(x;y) = B \sin k(x - 1) \quad \text{for} \quad x > y \quad . \tag{2.23}$$

g is continuous at $x = y$, therefore

$$A \sin ky = B \sin k(y - 1) \quad . \tag{2.24}$$

To determine A and B we need another equation; it comes from the relation between g_x on the two sides of the point $x = y$. As in Example 2.9, we integrate (2.19) from $x = y - \epsilon$ to $x = y + \epsilon$ and let $\epsilon \downarrow 0$, giving

$$g_x(y+;y) - g_x(y-;y) = 1 \quad . \tag{2.25}$$

Using (2.22) and (2.23), we have $g_x(y-;y) = kA\cos ky$ and $g_x(y+;y) = kB\cos k(y-1)$, so (2.25) gives

$$kB\cos k(y-1) - kA\cos ky = 1 \quad . \tag{2.26}$$

Solving (2.24) and (2.26) for A and B finally determines g:

$$g(x;y) = \begin{cases} \sin kx \sin k(y-1)/k\sin k & \text{for } x<y \\ \sin ky \sin k(x-1)/k\sin k & \text{for } y<x \end{cases} \tag{2.27}$$

if $\sin k \neq 0$.

If $\sin k = 0$, then (2.24) and (2.26) are inconsistent, and there is no solution of (2.19) satisfying (2.20) and (2.21). Green's function does not exist when the parameter k^2 is an eigenvalue of the operator $-D^2$. We shall see that this is an example of a general rule about the existence of Green's function.

The symmetry of (2.27) is remarkable. It is sometimes used to compress (2.27) into the form $g(x;y) = \sin kX_< \sin k(X_> - 1)/k\sin k$, where $X_<$ denotes the smaller of x and y, and $X_>$ the greater. Green's function for very many physical problems (roughly speaking, those with no frictional energy dissipation) is symmetric, as we shall see. Physically this means that the response at a point x to a unit source or disturbance at the point y is the same as the response at y to a unit source at x, a statement familiar to engineers as the 'Reciprocity Principle'. □

We shall now generalise these results to the Sturm-Liouville problem (see Appendix G).

Theorem 2.13 (Existence of Green's Function) For a regular Sturm-Liouville operator $L = D[p(x)D] + q(x)$ with separated end-point conditions, Green's function exists if and only if zero is not an eigenvalue. When it exists it is a symmetric function.

Proof We shall show how to construct Green's function $g(x;y)$. Let $[a,b]$ be the interval on which the Sturm-Liouville equation applies. The general theory of differential equations tells us that there is a solution u_1 (not identically zero) of the equation $Lu = 0$ which satisfies the boundary condition at $x = a$, and that $c_1 u_1(x)$ is a solution for any c_1 independent of x. Similarly we have a family of solutions $c_2(y)u_2(x)$ satisfying the boundary condition at $x = b$. We construct a function

$$g(x;y) = \begin{cases} c_1(y)u_1(x) & \text{for } x<y \\ c_2(y)u_2(x) & \text{for } x>y \end{cases} \quad , \tag{2.28}$$

where c_1 and c_2 are determined by the conditions that g be continuous and g_x have a discontinuity at $x = y$ given by (2.17):

$$\left.\begin{aligned} c_1(y)u_1(y) - c_2(y)u_2(y) &= 0 \quad , \\ c_1(y)u_1'(y) - c_2(y)u_2'(y) &= -1/p(y) \quad . \end{aligned}\right\} \tag{2.29}$$

These equations can be solved for c_1 and c_2 if and only if the determinant of their coefficients is nonzero, that is,

$$u_1(y)u_2'(y) - u_2(y)u_1'(y) \neq 0 \quad . \tag{2.30}$$

Now, an easy calculation shows that $[p(u_1u_2' - u_2u_1')]' = 0$ if u_1 and u_2 satisfy $Lu = 0$; hence

$$p(u_1u_2' - u_2u_1') = k \tag{2.31}$$

for some constant k. It follows that $u_1u_2' - u_2u_1'$ either vanishes nowhere or vanishes everywhere. But $u_1u_2' - u_2u_1'$ is proportional to $(u_2/u_1)'$, and hence is identically zero if and only if u_1 and u_2 are proportional (or one of them is zero, which is not the case here). But if they are proportional, then each of them satisfies both boundary conditions, because each condition is satisfied by one of the functions. They are therefore eigenfunctions of L with eigenvalue zero. Conversely, if L has zero as an eigenvalue, then the eigenfunction must appear in the family c_1u_1 of solutions of $Lu = 0$ satisfying the condition at $x = a$, hence u_1 must be such an eigenfunction and must satisfy the condition at $x = b$ too. Hence the two families of solutions c_1u_1 and c_2u_2 coincide, so u_1 and u_2 are proportional and the determinant in (2.30) vanishes. We conclude that (2.29) can be solved for c_1 and c_2 if and only if zero is not an eigenvalue of L.

Assuming that zero is not an eigenvalue, we can solve (2.29) for c_1 and c_2 and substitute in (2.28); we get

$$g(x;y) = \begin{cases} Ku_1(x)u_2(y) & \text{if} \quad x < y \\ Ku_1(y)u_2(x) & \text{if} \quad y < x \quad , \end{cases} \tag{2.32}$$

where we have used (2.31), writing $k^{-1} = K$. This function has been constructed by the method which we have proved (Theorem 2.10) to lead to a fundamental solution, and it satisfies the boundary conditions, hence it is Green's function. It is clearly symmetric, that is, $g(x;y) = g(y;x)$. □

Remark 2.14 Throughout this chapter we have assumed that our equations have infinitely differentiable coefficients. This limitation is forced on us by the structure of the distribution theory that we are using (but see Problem 2.11). Green's function for ordinary differential equations, however, is an ordinary continuous function, and can be discussed completely within the framework of classical analysis; the equations are not then required to have smooth coefficients. Thus, Green's function for the Sturm-Liouville operator $L = \mathrm{D}p\mathrm{D} + q$, with only the usual conditions of continuity of p' and q, is defined to be the function g constructed in the proof of Theorem 2.13; one then proves by classical analysis, differentiating under the integral sign, that $\int g(x;y)f(y)\,dy$ satisfies $Lu = f$. For an account of this approach to Green's function, see, for example, Courant-Hilbert vol. 1.

2.5 Applications of Green's Function

In the last section we defined Green's function, and showed how to calculate it. We shall now prove that it can be used to solve inhomogeneous equations in the way outlined at the beginning of section 2.3.

Proposition 2.15 (Solution of Inhomogeneous Equations) Let $g(x;y)$ be Green's function for an operator L with boundary conditions as described in Definition 2.11, and let f be a locally integrable function. Then the function

$$u(x) = \int_{-\infty}^{\infty} g(x;y)f(y)\,dy \tag{2.33}$$

is a weak solution of the equation $Lu = f$ (2.34)

and satisfies the given boundary conditions, provided that f vanishes at infinity fast enough for $\int(D^r g)f\,dy$ to converge uniformly for any $r \leqslant n - 1$.

Remark The case of equations on finite intervals (such as Sturm-Liouville systems) can be included in this theorem by defining f to vanish outside that interval. The integral in (2.33) becomes a finite integral, and uniform convergence follows easily.

Proof u clearly satisfies the boundary conditions, as is seen by direct substitution, using the uniform convergence to calculate derivatives. It need not be a classical solution of (2.34), because if f is a discontinuous function then u, like g, may not be n times differentiable. To show that it is a weak solution we take an arbitrary $\phi \in \mathscr{D}$ and prove that $\langle Lu,\phi\rangle = \langle f,\phi\rangle$. We have

$$\langle Lu,\phi\rangle = \langle u, \Sigma_{i=0}^{n}(-D)^{n-i}a_i\phi\rangle$$

by 1.26. Define an operator L^* by $L^*u = \Sigma_{i=0}^{n}(-D)^{n-i}(a_iu)$,

then $\langle Lu,\phi\rangle = \langle u,L^*\phi\rangle$

$$= \int_{-\infty}^{\infty}\{L^*\phi(x)\int_{-\infty}^{\infty} g(x;y)f(y)\,dy\}\,dx \quad ,$$

where both integrals can be interpreted as genuine integrals (i.e., not according to 1.36) since u is a locally integrable function. The x-integration is over a finite interval because ϕ has bounded support, and the y-integral converges uniformly by hypothesis. Hence we can invert the order of integration:

$$\langle Lu,\phi\rangle = \int\{f(y)\int g(x;y)L^*\phi(x)\,dx\}\,dy$$

$$= \int f(y)\langle g,L^*\phi\rangle\,dy$$

where g is regarded as a generalised function of x, with y as a parameter. Hence

$$\begin{aligned}\langle Lu,\phi\rangle &= \int f(y)\langle Lg,\phi\rangle\,\mathrm{d}y\\ &= \int f(y)\langle\delta(x-y),\phi\rangle\,\mathrm{d}y\\ &= \int f(y)\phi(y)\,\mathrm{d}y\\ &= \langle f,\phi\rangle \quad .\end{aligned}$$

Thus u is a weak solution of $Lu = f$. □

Example 2.16 Consider the equation

$$u'' + 2ku' + \omega^2 u = f(t) \tag{2.35}$$

for $t > 0$, with $u(0) = u'(0) = 0$, and suppose that $\omega > k > 0$. We found a fundamental solution for this equation in Example 2.9; it is given by (2.11) and (2.14), and it satisfies our boundary conditions; therefore Green's function for $\mathrm{D}^2 + 2k\mathrm{D} + \omega^2$ with the conditions $u(0) = u'(0) = 0$ is

$$g(t;s) = \begin{cases} 0 \quad \text{for} \quad t < s \quad , \\ \Omega^{-1}\mathrm{e}^{-k(t-s)}\sin\Omega(t-s) \quad \text{for} \quad t > s \quad , \end{cases} \tag{2.36}$$

where $\Omega = (\omega^2 - k^2)^{\frac{1}{2}}$. This is not a symmetric function of s and t, reflecting the fact that our equation and boundary conditions do not form a Sturm-Liouville system. The formula (2.33) for the solution of (2.35) gives $u(t) = \Omega^{-1}\int_0^t \mathrm{e}^{k(s-t)}$ $\sin\Omega(t-s)f(s)\,\mathrm{d}s$, or, after changing the variable of integration from s to $t-s$,

$$u(t) = \Omega^{-1}\int_0^t \mathrm{e}^{-ks}\sin(\Omega s)f(t-s)\,\mathrm{d}s \quad . \tag{2.37}$$

It is easy to verify that the integral here converges uniformly if f is bounded, say, so the application of Proposition 2.15 is justified.

This problem can be interpreted physically as the problem of finding the displacement $u(t)$ of an oscillator under the influence of a force $f(t)$ for $t > 0$, given that it was at rest and at equilibrium at $t = 0$. The formula (2.37) expresses clearly the idea that the displacement u at time t depends on the value of the force f at all previous times, the contribution of the force at time $t - s$ having a weighting factor $\mathrm{e}^{-ks}\sin(\Omega s)$ which falls off exponentially as s increases – in other words, the most recent values of the force are most important in determining the solution at time t. With these zero initial conditions, the force f is the sole cause of the motion, and the Green's function (2.36) is the tool for calculating the displacement u in terms of the force f which causes it. For this reason, the Green's function for boundary conditions of the type $u(0) = u'(0) = 0$ is sometimes called the 'causal' Green's function. □

Formula (2.33) sheds some light on the fact that Green's function does not exist if zero is an eigenvalue of L. We can think of L as a transformation of one function into another, and the equation $Lu = f$ as saying that L maps u into f. Now, the solution (2.33) of the equation $Lu = f$ gives a procedure for transforming f into u, by multiplying by g and integrating. The integral operation on the right hand side of (2.33) thus gives the inverse of the differential operator L. But not every transformation can be inverted. If L transforms two different functions u_1, u_2 into the same function f, then it is clearly impossible to invert L: given f, we cannot pick a unique u such that $Lu = f$, since u_1 and u_2 have equal claims. Green's function is the essential ingredient of the inverse transformation, so we expect that it will fail to exist whenever L transforms two different functions into the same function. But if zero is an eigenvalue of L, then there is a non-zero function v such that $L(cv) = 0$ for any constant c. Thus there are many different functions all transformed into the zero function by L, so L is not invertible, and we do not expect there to be a Green's function. This explains the zero-eigenvalue condition of Theorem 2.13. The idea of integral transformations being inverses of differential operators will be explored more carefully in Chapter 9.

Apart from giving explicit solutions of inhomogeneous equations, Green's function has another application which we shall be using repeatedly in later chapters. It can be used to transform an equation from one form into another, which may be easier to solve. The idea is illustrated by the following example.

Example 2.17 Consider the Sturm-Liouville eigenvalue problem $Lu = \lambda u$ for $a < x < b$, with boundary conditions at a and b. To solve this equation and find the eigenvalues is in general quite difficult. But we can transform the problem by using the formula (2.33), treating λu as if it were an inhomogeneous term; if g is Green's function for L (assuming that zero is not an eigenvalue), then $Lu = \lambda u$ is equivalent to

$$u(x) = \lambda \int_a^b g(x;y) u(y)\, dy \quad . \tag{2.38}$$

This is not an explicit solution, of course; it is an integral equation, that is, an equation in which the unknown function appears under an integral sign. □

This example is typical of many ordinary and partial differential equations which can be transformed into integral equations by means of a Green's function. At first sight the integral equation seems no easier to solve then the original differential equation, but it has many advantages. For one thing, every solution of (2.38) satisfies the boundary conditions; they are built into the integral equation, whereas they must be added as extra conditions to the differential equation $Lu = \lambda u$. A deeper reason for preferring the integral equation is that in many respects integral operators have nicer properties than differential operators. The main theme of Chapters 5–10 is the solution of equations involving various types of operations, and it usually turns out that the results apply to integral equations but not directly to differential equations. Green's function is thus the keystone of this book; formulas like (2.38) provide the link between the differential equations arising in applied mathematics and the theorems of functional analysis which apply to integral equations.

2.6 Summary and References

We began by constructing the distributional version of the indefinite integral in section 2.1, and then went on to the general theory of linear differential equations in section 2.2. We distinguished between classical solutions, weak solutions (ordinary functions which satisfy the equation in a generalised or integral sense, though they are not classical solutions), and distributional solutions (which involve singular distributions), and noted that for well-behaved equations, the only solutions are classical solutions. For all this, see Stakgold; for the proof that well-behaved equations have only classical solutions, see Zemanian (1965), p. 162.

In sections 2.3 and 2.4 we defined a fundamental solution for a differential operator L to be a solution of the equation $Lu = \delta(x - y)$, and Green's function to be a fundamental solution satisfying a particular set of homogeneous boundary conditions. We showed that for a Sturm-Liouville system, Green's function exists if and only if zero is not an eigenvalue, and if it exists it is continuous and symmetric; it is a weak solution of the equation $Lu = \delta$. In section 2.5 Green's function was applied to the solution of inhomogeneous equations and to the transformation of differential equations into integral equations. Problem 2.7 shows how Green's function can also be used to solve the homogeneous equation with inhomogeneous boundary conditions. Our discussion follows Stakgold (1967); accounts of Green's function from other points of view can be found in Kreider et al, Courant-Hilbert I, Coddington & Levinson, Zemanian (1965), Lanczos (1961), and many books with titles like "Mathematical Methods of Physics".

PROBLEMS 2

Sections 2.1 and 2.2

2.1 Prove that if a distribution f satisfies $xf = 0$, then $f = C\delta$ for some constant C.

2.2 Find the general solution of the equation $x^k f'(x) = 0$, where k is a positive integer.

2.3 Show that the general solution of $xf' + f = 0$ is $f(x) = c_1\delta(x) + c_2 P/x$.

Sections 2.3, 2.4 and 2.5

2.4 Given real numbers $h > 0$ and a, define a function $\Delta(x;a,h)$ by $\Delta(x;a,h) = 0$ if $|x - a| > h/2$ and $\Delta(x;a,h) = 1/h$ if $|x - a| < h/2$.

(a) Show that $\Delta(x;a,h) \to \delta(x - a)$ as $h \to 0$.

(b) Show that for any continuous function f, $\sum_{n=-\infty}^{\infty} f(nh)\Delta(x;nh,h)h \to f(x)$ as $h \to 0$.
Illustrate this result with a diagram.

(c) Discuss the relation of this problem to the first paragraph of section 2.3.

2.5 Find Green's function for $D^2 + k^2$ with boundary conditions $u'(0) = u'(1) = 0$.

2.6 (For readers familiar with Bessel functions.) Use Green's function to write down the solution of $(xy')' + xy = f(x)$ for $a < x < b$, with $y(a) = y(b) = 0$; assume that $0 < a < b$.

2.7 This problem shows how Green's function can be used to deal with inhomogeneous boundary conditions. Consider the equation $u'' + u = f(x)$ with boundary conditions $u(0) = a$, $u(1) = b$. We transform the problem into one with homogeneous boundary conditions by taking any function $h(x)$ which satisfies the given conditions (for example, $h(x) = a + (b - a)x$) and writing $u = h + v$. Then v satisfies the equation $v'' + v = F$, with boundary conditions $v(0) = v(1) = 0$, where $F = f - h'' - h$ is a known function. Solve this equation by means of Green's function, and thus obtain the solution u of the original problem.

This technique is clearly applicable to other boundary-value problems.

2.8 Find Green's function for the operator D^n for $x > 0$, with boundary conditions $y(0) = y'(0) = y''(0) = \ldots = y^{(n-1)}(0) = 0$. Hence express the n-fold repeated integral of a function $f(x)$ as a single integral. (The formula obtained can be extended to non-integral n, giving a definition of integrating a "fractional number of times". See Ross, or Oldham & Spanier.)

2.9 What can you do about solving $Lu = f$ when zero is an eigenvalue? (See, e.g., Stakgold under the heading "modified Green's function".)

2.10 The vertical deflection $y(x)$ of a horizontal girder of length a, clamped at both ends and carrying a load-distribution $f(x)$, satisfies $D^4 y = cf$, $y(0) = y'(0) = y(a) = y'(a) = 0$; c is a constant, related to the strength of the girder. Use Green's function to find y in terms of f. Is the resulting formula better (that is, more useful) than that obtained by direct integration of the equation $D^4 y = cf$?

2.11 The theory of Green's function given in this chapter can be extended to equations with non-smooth coefficients by means of the multiplication rule given in Problem 1.4. What conditions must the functions p and q satisfy in order that $[DpD + q]g$ should make sense according to that multiplication rule?

More problems will be found on page 375.

Chapter 3

Fourier Transforms and Partial Differential Equations

The main aim of this chapter is to construct Green's functions for partial differential equations. We also take the opportunity of discussing the distributional Fourier transform, which is a useful technique for solving partial differential equations, and in particular for calculating Green's functions. Section 3.1 is a very brief sketch of Fourier transform theory; the reader who is quite unfamiliar with Fourier transforms should also consult books such as Dettman or Sneddon which describe applications of the transform. Sections 3.2 and 3.3 discuss the distributional theory of the Fourier transform. In section 3.4 we prepare for the subject of partial differential equations by considering generalised functions of several variables, and sections 3.5 and 3.6 discuss Green's functions for the Laplace and wave equation respectively.

3.1 The Classical Fourier Transform

Definition 3.1 A function $f: \mathbb{R} \to \mathbb{C}$ is called **absolutely integrable** if $\int_{-\infty}^{\infty} |f(x)|\, dx$ exists.

Examples 3.2 Every test function is absolutely integrable. Every continuous function which tends to zero at infinity faster than $|x|^{-(1+a)}$ for $a > 0$ is absolutely integrable. No polynomial is absolutely integrable (except the trivial identically zero function).

Definition 3.3 For any absolutely integrable function f, we define its **Fourier transform** $\tilde{f}$ by

$$\tilde{f}(k) = \int_{-\infty}^{\infty} f(x) e^{ikx}\, dx \quad . \qquad \square \quad (3.1)$$

The convergence of the integral in (3.1) follows at once from the fact that f is absolutely integrable. In fact it converges uniformly with respect to k.

Definition 3.4. A function $f: \mathbb{R} \to \mathbb{C}$ is called **piecewise smooth** if all its derivatives exist and are continuous except (possibly) at a set of points $x_1, x_2, \ldots$ such that any finite interval contains only a finite number of the x_i, and if the function and all its derivatives have, at worst, finite jump discontinuities there. A function is called **n times piecewise continuously differentiable** if the above condition is satisfied with 'all its' replaced by 'its first n'.

Theorem 3.5 (Inversion Theorem) If f is absolutely integrable, continuous, and piecewise smooth, then

$$f(x) = (1/2\pi)\int_{-\infty}^{\infty}\tilde{f}(k)e^{-ikx}\,dk \quad . \qquad \square \quad (3.2)$$

This is an example of a whole class of theorems, which all have (3.2) or some closely related formula as their conclusion, but impose different conditions on f. The proofs are not easy, and the theory of Fourier transforms is an important branch of mathematical analysis; see, e.g., Titchmarsh (1937). The important thing is the reciprocal nature of the transformation: (3.2) shows that f is related to $\tilde{f}$ in the same way that $\tilde{f}$ is related to f, apart from the minus sign and the factor of 2π.

The following results are easily proved.

Proposition 3.6 Let f,ϕ be absolutely integrable and piecewise smooth;

(a) $\tilde{\tilde{f}}(x) = 2\pi f(-x)$ if f is continuous ;

(b) $\tilde{f'}(k) = -ik\tilde{f}(k)$ if f is differentiable ;

(c) $(\tilde{f})' = i(\widetilde{xf})$ if xf is absolutely integrable ;

(d) set $f_a(x) = f(x-a)$, then $\tilde{f_a}(k) = e^{ika}\tilde{f}(k)$;

(e) $\int_{-\infty}^{\infty}f(x)\tilde{\phi}(x)\,dx = \int_{-\infty}^{\infty}\tilde{f}(x)\phi(x)\,dx$;

(f) if f is an even function, so is $\tilde{f}$; if f is an odd function so is $\tilde{f}$.

Remark One must be careful with the notation. $\tilde{f'}(k)$ means the function obtained by transforming f', and is not to be confused with $(\tilde{f})'$. Similarly $(\widetilde{xf})$ is the transform of the function xf, more precisely, of the function whose value at x is $xf(x)$.

Proof (a) is just a restatement of (3.2). Formula (b) follows from (3.1), with f replaced by f', by integrating by parts. (c) is obtained by differentiating (3.1); the integral converges uniformly, so that differentiation under the integral sign is allowed. (d) follows from (3.1) by an obvious change of variable. (e) is easily proved by expressing each side as a double integral and changing the order of integration; this is allowed because the integrals converge uniformly, f and ϕ being absolutely integrable. (f) is easily proved by means of the change of variable $x \to -x$ in (3.1). □

Fourier transforms are useful for solving differential equations, because according to 3.6(b) differentiation is transformed into the algebraic operation of multiplication by k; a differential equation is thus transformed into an algebraic equation. One can solve a differential equation for an unknown function f by transforming it into an algebraic equation for $\widetilde{f}$, solving to find $\widetilde{f}$, and then recovering f by using (3.2) or some other method to invert the Fourier transform.

The transform has a clear physical interpretation, particularly when f is a function of time. The usual notation is then

$$\hat{f}(\omega) = \int_{-\infty}^{\infty} f(t)e^{-i\omega t}\,dt \quad , \tag{3.3}$$

leading to

$$f(t) = (1/2\pi)\int_{-\infty}^{\infty} \hat{f}(\omega)e^{i\omega t}d\omega \quad ; \tag{3.4}$$

we write $\hat{f}$ instead of $\widetilde{f}$ because of the difference in sign: $\hat{f}(\omega) = \widetilde{f}(-\omega)$. Equation (3.4) can be interpreted as saying that any (reasonable) function f can be regarded as a superposition (that is, a sum or an integral) of an infinite number of sinusoidal oscillations, with different frequencies ω, the amount of the component with frequency ω being $\hat{f}(\omega)/2\pi$. Equation (3.4) is called the spectral resolution of the function f, and $\hat{f}$ is called the spectral density, because splitting a beam of light into a spectrum by means of a prism is the classic example. A light wave can be represented mathematically by a function $f(t)$ giving the magnitude of the electric field; sinusoidal functions correspond to pure colours (monochromatic light). Equation (3.4) then says that any light beam is a superposition of monochromatic beams, and a prism (or a water droplet in the case of rainbows) is a device for separating these different monochromatic components.

The above is an extremely incomplete sketch of some aspects of the Fourier transform. For our purposes, the main feature of the theory is its inadequacy. The theory is restricted to absolutely integrable functions. The most familiar functions are polynomials and algebraic, trigonometric, and exponential functions. Unfortunately, polynomials, trigonometric functions, and exponential functions are not absolutely integrable, and therefore do not have Fourier transforms in the classical sense. The integral in (3.1) fails to converge even when f is such a basic function as a constant. This is a very unsatisfactory state of affairs.

We have seen in Chapter 1 how to interpret some divergent integrals using distribution theory. We might therefore expect to get a better theory of the Fourier transform in terms of generalised functions. We shall present such a theory in section 3.3, but it is not a straightforward matter.

The natural way to define the Fourier transform of a generalised function f, following the method used to define operations on generalised functions in Chapter 1, would be to start from Proposition 3.6(e) and define $\widetilde{f}$ by $\langle\widetilde{f},\phi\rangle = \langle f,\widetilde{\phi}\rangle$ for all $\phi\in\mathscr{D}$. Unfortunately, the right hand side of this equation does not make sense, because $\widetilde{\phi}$ is not in general a test function; the Fourier transform of a function of bounded support is not usually itself a function of bounded support (see Problem 3.1). In order to make the definition work, we must use a new space of test functions, say $\mathscr{S}$, with the property that Fourier transforms of functions in $\mathscr{S}$ are themselves in $\mathscr{S}$. We will then be able to define transforms of functionals on

$\mathscr{S}$ by the above method. The precise definition of $\mathscr{S}$ is a delicate matter, because if we impose too stringent requirements on the behaviour of the functions at infinity, then their transforms will not satisfy those conditions, and therefore will not belong to the space; this was our difficulty with the space $\mathscr{D}$. On the other hand, if we impose too weak conditions on the behaviour at infinity of functions in $\mathscr{S}$, then the Fourier transforms will not be smooth, and therefore again will not belong to the space $\mathscr{S}$. There is essentially only one way to balance these conflicting tendencies, and that is to frame the definitions as in the following section.

Readers who are interested in useful techniques but not in mathematical details may omit section 3.2 and the beginning of section 3.3, and read from Definition 3.20 onwards, without worrying about the precise meaning of "distributions of slow growth".

3.2 Distributions of Slow Growth

In this section we shall construct another version of distribution theory, using exactly the same ideas as in Chapter 1, but with technical modifications to fit Fourier transforms into the theory. The reader may well ask why we did not introduce this theory in the first place, instead of introducing the space $\mathscr{D}$ in Chapter 1 and abandoning it in Chapter 3. The reason is that although the theory of this section is superior in that it permits Fourier transformation, it is less powerful in that it does not include such a wide range of generalised functions as the theory of Chapter 1.

Generalised functions in our new sense will be defined as continuous linear functionals on a space of functions defined so as to satisfy the requirements discussed at the end of the last section.

Definition 3.7 A **function of rapid decay** is a smooth function $\phi: \mathbb{R} \to \mathbb{C}$ such that

$$x^n\phi^{(r)}(x) \to 0 \quad \text{as} \quad x \to \pm\infty \quad \text{for all} \quad n, r \geqslant 0 \quad . \tag{3.5}$$

The set of all functions of rapid decay is called $\mathscr{S}$.

Example 3.8 $P(x)e^{-ax^2}$ is a function of rapid decay for any number $a > 0$ and any polynomial P.

Proposition 3.9 (a) Every test function is a function of rapid decay. That is, $\mathscr{D} \subset \mathscr{S}$.

(b) If $\phi, \psi \in \mathscr{S}$, then $a\phi + b\psi \in \mathscr{S}$ for any constants a, b.

(c) If $\phi \in \mathscr{S}$, then $x^n\phi^{(r)}(x) \in \mathscr{S}$ for all n, $r \geqslant 0$.

(d) If $|x^n\phi^{(r)}(x)|$ is a bounded function for each n, $r \geqslant 0$, then ϕ is a function of rapid decay.

(e) Every function of rapid decay is absolutely integrable.

Proof (a) and (b) are obvious; (c) is obvious after a moment's thought (any combination of differentiation and multiplication by powers of x applied to $x^n\phi^{(r)}$ can be expressed as a combination of differentiation and powers of x applied to ϕ). (d) is very easy to prove: we must show that for all n and r, $x^n\phi^{(r)}(x)\to 0$, which follows immediately by writing it as $x^{-1}.x^{n+1}\phi^{(r)}(x) = x^{-1}$. (bounded function) $\to 0$ as $x\to\pm\infty$. Finally, (e) follows immediately from consideration of Examples 3.2; functions of rapid decay vanish at infinity faster than any power of x, hence in particular faster than $|x|^{-(1+a)}$ for $a > 0$, hence are absolutely integrable. □

We now prove that the space $\mathscr{S}$ contains the Fourier transforms of all its members.

Proposition 3.10 If $\phi\in\mathscr{S}$ then $\widetilde{\phi}\in\mathscr{S}$.

Proof 3.9(e) shows that ϕ has a Fourier transform; 3.9(c) shows that $x^n\phi(x)$ is absolutely integrable; applying 3.6(c) n times shows that $\widetilde{\phi}$ is differentiable n times for any n. Now we must consider the behaviour of $k^n\widetilde{\phi}^{(r)}(k)$ for large k. Applying 3.6(c) r times, we have

$$\begin{aligned} |k^n\widetilde{\phi}^{(r)}(k)| &= |k^n\textstyle\int_{-\infty}^{\infty}(ix)^r\phi(x)e^{ikx}\,dx| \\ &= |\textstyle\int_{-\infty}^{\infty}x^r\phi(x)(d/dx)^n e^{ikx}\,dx| \\ &= |\textstyle\int_{-\infty}^{\infty}e^{ikx}(d/dx)^n[x^r\phi(x)]\,dx| \quad , \end{aligned}$$

integrating by parts n times

$$\leqslant \int |(d/dx)^n[x^r\phi(x)]|\,dx \quad .$$

Now 3.9(c) and 3.9(e) guarantee the convergence of this last integral; it is independent of k, and thus gives a bound for $k^n\widetilde{\phi}^{(r)}$. It follows from 3.9(d) that $\widetilde{\phi}$ is a function of rapid decay. □

Definition 3.11 (Convergence) If $\Phi, \phi_1, \phi_2, \phi_3, \ldots$ are functions of rapid decay, we say $\phi_m\to\Phi$ in $\mathscr{S}$ if, for all integers r and n, $x^n\phi_m^{(r)}(x)\to x^n\Phi^{(r)}(x)$ as $m\to\infty$, uniformly in x. □

Examples 3.12 The sequence $\phi_m(x) = e^{-m(x^2+1)}$ converges to zero in $\mathscr{S}$. Proof: for each n and r, $x^n\phi_m^{(r)}(x) = e^{-m}p(x,m)e^{-mx^2}$ where p is a polynomial in x and m; $p(x,m)e^{-mx^2}$ is a bounded function of x and m, hence $x^n\phi_m^{(r)}(x)\to 0$ uniformly. On the other hand, the sequence $\psi_m(x) = m^{-1}e^{-m^3x^2}$ does not converge in $\mathscr{S}$, even though it is a uniformly convergent sequence of functions; it fails the test of Definition 3.11 with $n = 0, r = 1$, since $\psi_m'(x) = -2m^2xe^{-m^3x^2}$, which is not uniformly convergent.

Proposition 3.13 (a) If $\phi_m \to \Phi$ in $\mathscr{S}$, then $\phi'_m \to \Phi'$ and $P(x)\phi_m \to P(x)\Phi$ in $\mathscr{S}$ for any polynomial P.

(b) If $\phi_m, \Phi \in \mathscr{D}$ for all m, and $\phi_m \to \Phi$ in $\mathscr{D}$, then $\phi_m \to \Phi$ in $\mathscr{S}$.

Proof (a) follows straight from the definition. To prove (b), remember from Definition 1.8 that there is an interval $[a,b]$ outside which ϕ_m and Φ vanish. Let C be the larger of $|a|$ and $|b|$, then

$$|x^n[\phi_m^{(r)}(x) - \Phi^{(r)}(x)]| \leqslant C^n|\phi_m^{(r)}(x) - \Phi^{(r)}(x)| \quad \text{for all} \quad x \in [a,b] \quad ,$$

which tends to zero as $m \to \infty$ uniformly in x with n and C fixed. Hence Definition 3.11 is satisfied. □

We can now define a distribution in our new sense in the same way as in Definition 1.9.

Definition 3.14 **A distribution of slow growth** is a continuous linear functional on the space $\mathscr{S}$, that is, a linear functional which maps every convergent sequence in $\mathscr{S}$ into a convergent sequence in $\mathbb{C}$. □

Every functional on $\mathscr{S}$ is also a functional on $\mathscr{D}$ since all test functions in $\mathscr{D}$ also belong to $\mathscr{S}$. Proposition 3.13(b) shows that convergent sequences in $\mathscr{D}$ are also convergent in the sense of convergence in $\mathscr{S}$, and are therefore mapped into convergent sequences in $\mathbb{C}$ by any continuous functional on $\mathscr{S}$. Hence every continuous functional on $\mathscr{S}$ is also continuous on $\mathscr{D}$, therefore every distribution of slow growth is a distribution in the sense of Chapter 1. The converse is not true. Distributions of slow growth form a proper subset of the set of all distributions. This justifies calling them 'distributions' with a qualifying phrase. The term 'slow growth' will be explained below.

Examples 3.15 (a) Any polynomial $p(x)$ generates a distribution of slow growth $p: \phi \mapsto \int_{-\infty}^{\infty} p(x)\phi(x)\,dx$ for all $\phi \in \mathscr{S}$. The rapid decay of ϕ ensures that the integral converges, and we shall prove below that the functional is continuous (Theorem 3.18). This is an example of a regular distribution generated by a locally integrable function in the manner of Theorem 1.14.
(b) The regular distribution e^{-x} is not a distribution of slow growth. We cannot define a functional on $\mathscr{S}$ by the rule $\phi \mapsto \int_{-\infty}^{\infty} e^{-x}\phi(x)\,dx$ because this integral is not in general convergent: $\phi(x)$ might behave like $e^{-|x|/2}$ for large $|x|$, for example. □

Example 3.15(b) justifies the remark at the beginning of this section that the generalised functions considered here are more restricted than those of Chapter 1. Ordinary functions which grow too rapidly at infinity have no counterparts in the set of distributions of slow growth, though they have corresponding distributions in the ordinary sense. By 'grow too rapidly' we mean grow faster than any power of x. We express this more precisely by saying $f(x) = 0(x^n)$ (read "$f(x)$ is big 'oh' of x^n") if $f(x)$ for large x equals a bounded function times x^n. The 0 stands for order of magnitude; $f(x) = 0(x^n)$ means f is of the same order of magnitude as x^n, or smaller. We now express this more formally.

Definition 3.16 We write $f(x) = 0(x^n)$ as $|x| \to \infty$ if there exist numbers C and R such that $|f(x)| \leqslant C|x|^n$ whenever $|x| > R$. A **function of slow growth** is a locally integrable $\mathbb{R} \to \mathbb{C}$ function f such that $f(x) = 0(x^n)$ for some n as $|x| \to \infty$.

Examples 3.17 (a) Every n-th degree polynomial p is $0(x^n)$. Proof: if a is the coefficient of x^n in the polynomial p, then $p(x)/x^n \to a$ as $|x| \to \infty$; therefore there is an R such that $|p(x)/x^n| < 2a$ for $|x| > R$, so Definition 3.16 is satisfied. Hence every polynomial is a function of slow growth.
(b) e^{-x} is not a function of slow growth, because $x^{-n}e^{-x} \to \infty$ as $x \to -\infty$ for any $n \geqslant 0$. But e^{iax} is a function of slow growth if a is real (and x is a real variable). □

The similarity between Examples 3.17 and 3.15 is no accident, of course. It is generally true that functions of slow growth generate distributions of slow growth (which is why distributions of slow growth are so called), and functions which do not grow slowly, in our technical sense of the word, generate distributions which are not of slow growth.

Theorem 3.18 (Regular Distributions) To each locally integrable function of slow growth f there corresponds a distribution of slow growth f defined by

$$\langle f,\phi\rangle = \int_{-\infty}^{\infty} f(x)\phi(x)\,dx \quad . \tag{3.6}$$

Proof If f is locally integrable, (3.6) defines a functional on $\mathscr{S}$, provided that the integral converges at the upper and lower limits. If f is of slow growth, then $f(x) = 0(|x|^N)$ for large x, for some N, and (3.5), with $n = N + 2$ and $r = 0$, shows that the integrand in (3.6) tends to zero as $|x| \to \infty$ so fast that the integral converges. Equation (3.6) thus defines a functional on $\mathscr{S}$, and it is obviously linear. We must prove that it is continuous, that is, that $\langle f,\phi_n\rangle \to \langle f,\Phi\rangle$ whenever $\phi_n \to \Phi$ in $\mathscr{S}$.

If f is locally integrable and of slow growth, then there is an integer p such that $(1 + x^2)^{-p}f(x)$ is absolutely integrable: $p = n + 2$ would do, where n is as in Definition 3.16. Then

$$\langle f,\phi_m - \Phi\rangle = \int_{-\infty}^{\infty} f(x)(1 + x^2)^{-p}(1 + x^2)^p[\phi_m(x) - \Phi(x)]\,dx \quad ,$$

$$\therefore |\langle f,\phi_m - \Phi\rangle| \leqslant \max_x\{|(1 + x^2)^p[\phi_m(x) - \Phi(x)]|\}|\int_{-\infty}^{\infty} f(x)\,dx/(1 + x^2)^p| \quad ; \tag{3.7}$$

the max exists by (3.5). If $\phi_m \to \Phi$ in $\mathscr{S}$, then the maximum value of any power of x times $(\phi_m - \Phi)$ tends to zero as $m \to \infty$, hence so does any polynomial times $(\phi_m - \Phi)$, hence so does the right hand side of (3.7), the integral being independent of m. This proves the continuity of the functional f. □

This theorem provides us with a set of regular distributions in the same way as Theorem 1.14. The delta function and its derivatives are examples of singular distributions of slow growth. This is intuitively obvious: $\delta^{(r)}(x)$ vanishes for $x \neq 0$, so it grows as slowly as a function can possibly grow, namely, not at all; it is easy to prove that $\phi \mapsto \phi^{(r)}(0)$ is a continuous functional on $\mathscr{S}$, justifying the statement that $\delta^{(r)}$ is a distribution of slow growth.

We can define operations of addition, translation, multiplication by smooth functions, etc. just as in Chapter 1. Multiplication requires a little care: we can define the product of a distribution of slow growth with a smooth function as in Definition 1.17, but the result is a distribution of slow growth only if the smooth function does not increase too rapidly at infinity. The product of a distribution of slow growth and a smooth function of slow growth is always another distribution of slow growth.

Differentiation of distributions of slow growth is defined by $\langle f',\phi\rangle = -\langle f,\phi'\rangle$ as in Chapter 1. Every distribution of slow growth is differentiable by Proposition 1.25, and its derivative is also a distribution of slow growth. This again is intuitively reasonable: if a function grows slowly, its derivative cannot grow too rapidly, otherwise the function would have to increase rapidly too. It is not too difficult to justify this crude argument using Proposition 3.9(c); see Problem 3.3.

In this way, the theory of generalised functions of slow growth can be built up along very similar lines to the theory of ordinary generalised functions. We leave the details to the reader to work out or look up, and proceed to the theory of Fourier transforms, which gave the reason for introducing the space $\mathscr{S}$ in the first place.

3.3 Generalised Fourier Transforms

The aim of the previous section was to define a space $\mathscr{S}$ on which Fourier transforms of distributions can be defined as discussed at the end of section 3.1. We shall define the transform $\widetilde{f}$ of a generalised function f by $\langle \widetilde{f},\phi\rangle = \langle f,\widetilde{\phi}\rangle$ for $\phi \in \mathscr{S}$. But we must first prove that the functional $\widetilde{f}$ defined by this formula is a distribution.

Proposition 3.19 If f is a distribution of slow growth, then the functional $\widetilde{f}$: $\phi \mapsto \langle f,\widetilde{\phi}\rangle$ is a distribution of slow growth.

Proof Proposition 3.10 ensures that $\langle \widetilde{f},\phi\rangle$ is well-defined for all $\phi \in \mathscr{S}$, so f is a mapping of $\mathscr{S}$ to $\mathbb{C}$. It is clearly linear; we must show that it is continuous.

Let (ϕ_m) be a convergent sequence in $\mathscr{S}$ with $\phi_m \to \Phi$; we shall show that $(\widetilde{\phi}_m)$ is also convergent in $\mathscr{S}$. Proposition 3.6(b), (c) shows that for any function ϕ,

$$k^n(\widetilde{\phi})^{(r)} = i^{n+r}\widetilde{(x^r\phi)^{(n)}} \text{ for all } n,r.$$

Hence
$$\begin{aligned} |k^n(\widetilde{\phi}_m - \widetilde{\Phi})^{(r)}| &\leqslant \int |[x^r(\phi_m - \Phi)]^{(n)}|\, dx \\ &= \int (1+x^2)^{-1} |(1+x^2)[x^r(\phi_m - \Phi)]^{(n)}|\, dx \\ &\leqslant \max_x \{|(1+x^2)[x^r(\phi_m - \Phi)]^{(n)}|\} . \int (1+x^2)^{-1}\, dx \quad . \end{aligned}$$

But $\phi_m - \Phi \to 0$ in $\mathscr{S}$, therefore any combination of differentiations and powers of x multiplied by $\phi_m - \Phi$ tends to zero uniformly in x as $m \to \infty$. Hence the above expression tends to 0 as $m \to \infty$, so $k^n(\widetilde{\phi}_m - \widetilde{\Phi})^{(r)} \to 0$ uniformly, proving that $\widetilde{\phi}_m \to \widetilde{\Phi}$ in $\mathscr{S}$.

Now, f is a continuous functional, so $\langle f,\widetilde{\phi}_m\rangle \to \langle f,\widetilde{\Phi}\rangle$ as $m \to \infty$, hence $\langle \widetilde{f},\phi_m\rangle \to \langle \widetilde{f},\Phi\rangle$, showing that $\widetilde{f}$ is a continuous functional. □

We can now define the Fourier transform of a distribution of slow growth, also known as the generalised Fourier transform.

Definition 3.20 If f is a distribution of slow growth, its **Fourier transform** is the distribution of slow growth $\widetilde{f}$ defined by

$$\langle \widetilde{f},\phi\rangle = \langle f,\widetilde{\phi}\rangle \tag{3.8}$$

for all ϕ in $\mathscr{S}$. If f is a locally integrable function of slow growth, the distribution $\widetilde{f}$ is called the **generalised Fourier transform** of f. □

To justify calling this a generalisation of the ordinary Fourier transform, we must show that it coincides with the ordinary transform when applied to ordinary Fourier-transformable functions. More precisely, we must show that if f is a regular distribution of slow growth generated by a Fourier-transformable function f, then the generalised Fourier transform of f equals the distribution generated by $\widetilde{f}$.

Theorem 3.21 If f is an absolutely integrable function, then its Fourier transform $\widetilde{f}$ is a function of slow growth, and the distribution generated by $\widetilde{f}$ is the generalised Fourier transform of f.

Proof If f is absolutely integrable, then $\widetilde{f}$ as defined by (3.1) is a bounded function, since for all k, $|\widetilde{f}(k)| \leqslant \int_{-\infty}^{\infty} |f(x)|\, dx$. Therefore $\widetilde{f}$ is a function of slow growth. Write $\widetilde{f}$ for the distribution generated by $\widetilde{f}$ and (temporarily) write $(\widetilde{f})$ for the Fourier transform of the distribution f. We wish to show that $\widetilde{f} = (\widetilde{f})$. We have $\langle(\widetilde{f}),\phi\rangle = \langle f,\widetilde{\phi}\rangle = \int_{-\infty}^{\infty} f(x)\widetilde{\phi}(x)\, dx = \int \widetilde{f}(x)\phi(x)\, dx$ by Proposition 3.6(e), $= \langle \widetilde{f},\phi\rangle$ for all $\phi \in \mathscr{S}$. This completes the proof. □

Thus when f is a locally integrable function the two possible interpretations of the symbol $\widetilde{f}$ coincide, and we need not distinguish carefully between ordinary and generalised Fourier transforms.

Proposition 3.6 applies also to generalised transforms, and the proofs are straightforward. We shall prove 3.6(a) as an example. It is to be interpreted in the manner discussed after Definition 1.22 – $f(x)$ is just another notation for f, and $f(-x)$ denotes the distribution called $\underset{\sim}{f}_{.-1}$ in Definition 1.23. Then we have, for any $\phi \in \mathscr{S}$, $\langle \widetilde{\widetilde{f}}, \phi\rangle = \langle \widetilde{f}, \widetilde{\phi}\rangle$ by (3.8), $= \langle f, \widetilde{\widetilde{\phi}}\rangle = \langle f, 2\pi\phi(-x)\rangle$ by 3.6(a), $= \langle 2\pi f(-x), \phi\rangle$ using Definition 1.23, as required. The rest of Proposition 3.6 is extended to generalised functions in the same way.

We have now constructed a theory in which polynomials and sine and cosine functions have (generalised) Fourier transforms. However, e^x does not have a generalised Fourier transform, because it is not a function of slow growth. It is possible to construct a more powerful theory in which exponential functions do have transforms; it involves another new space of functions, and corresponding new classes of distributions, which are sometimes called 'ultradistributions'; we shall not consider them further, but refer the reader to Zemanian (1965) or Gel'fand & Shilov.

We now consider some examples.

Example 3.22 The simplest function is a constant. Write **1** for the distribution generated by the constant function whose value everywhere is 1. Then

$$\begin{aligned}\langle \widetilde{\mathbf{1}}, \phi\rangle &= \langle \mathbf{1}, \widetilde{\phi}\rangle \\ &= \int \widetilde{\phi}(x)\,\mathrm{d}x \\ &= \widetilde{\widetilde{\phi}}(0) \\ &= 2\pi\phi(0)\end{aligned}$$

using 3.6(a). Hence $\widetilde{\mathbf{1}} = 2\pi\delta$. (3.9)

Example 3.23 We write $\boldsymbol{x}$ for the distribution generated by the function $f(x) = x$. Then

$$\begin{aligned}\langle \widetilde{\boldsymbol{x}}, \phi\rangle &= \langle \boldsymbol{x}, \widetilde{\phi}\rangle \\ &= \int x\widetilde{\phi}(x)\,\mathrm{d}x \\ &= i\int \widetilde{(\phi')}\,\mathrm{d}x \quad \text{using 3.6(b),} \\ &= i\widetilde{\widetilde{(\phi')}}(0) \\ &= 2\pi i\phi'(0) \quad \text{using 3.6(a).}\end{aligned}$$

Hence $\widetilde{\boldsymbol{x}} = -2\pi i\delta'$. (3.10)

This result can also be deduced from (3.9) by means of 3.6(c), taking $f = 1$. Either method is easily extended to give

$$\widetilde{(x^n)} = 2\pi(-i)^n \delta^{(n)} \quad . \tag{3.11}$$

Example 3.24 The transform of δ is most efficiently deduced from (3.9) by means of 3.6(a): $2\pi\widetilde{\delta} = \widetilde{\widetilde{1}} = 2\pi 1$, hence

$$\widetilde{\delta} = 1 \quad . \tag{3.12}$$

This result is also easily derived directly, as follows:

$$\begin{aligned} \langle\widetilde{\delta},\phi\rangle &= \langle\delta,\widetilde{\phi}\rangle \\ &= \widetilde{\phi}(0) \\ &= \int \phi(x)\,\mathrm{d}x = \langle 1,\phi\rangle \end{aligned}$$

from which (3.12) follows.

Applying the inversion formula 3.6(a) to (3.11) gives

$$\widetilde{\delta^{(n)}} = (-i)^n k^n \quad . \tag{3.13}$$

Example 3.25 The transform of the shifted delta function $\delta(x - a)$ is of some interest. We have

$$\begin{aligned} \langle\widetilde{\delta(x-a)},\phi\rangle &= \langle\delta(x-a),\widetilde{\phi}\rangle \\ &= \widetilde{\phi}(a) \\ &= \int \mathrm{e}^{iau}\,\phi(u)\,\mathrm{d}u \quad , \end{aligned}$$

$$\therefore \widetilde{\delta(x-a)} = \mathrm{e}^{iak} \quad . \tag{3.14}$$

Applying the inversion formula 3.6(a) to (3.11), and using k as the Fourier transform variable, gives

$$\widetilde{\mathrm{e}^{iax}} = 2\pi\delta(k + a) \quad . \qquad \square \tag{3.15}$$

The formula (3.15), of which (3.9) is a special case, is very easily understood if the Fourier transform is interpreted as a spectral density as discussed in section 3.1. If the function e^{iax} is represented in the manner of (3.2) as a superposition of functions of the form e^{-ikx}, then clearly the contribution from all values of k except $k = -a$ is zero. The spectral density is concentrated on the point $k = -a$, and is thus a delta function of the form of (3.15). The inverse transform (3.12) can similarly be interpreted as saying that the function consisting of an infinitely thin and high spike at $x = 0$ can be represented as a superposition of sinusoidal or complex exponential functions, in which all frequencies occur with equal weight. These ideas have immediate application in diffraction theory and other parts of physics and engineering; see Arsac or Bracewell, for example.

The transforms of certain discontinuous functions will be needed later.

Example 3.26 The function sgn (defined in Example 1.27) is clearly a function of slow growth, and therefore generates a distribution of slow growth. Obviously

$$\mathrm{sgn}(x) = 2[H(x) - \tfrac{1}{2}] \tag{3.16}$$

where H is Heaviside's function defined in 1.18. We shall calculate the transform of **sgn** by a rather indirect method. We have $\mathbf{sgn}' = 2\delta$. Transforming this equation, and using 3.6(b) and (3.12), gives $-ik\widetilde{\mathbf{sgn}}(k) = 2$. In classical mathematics we would deduce that $\widetilde{\mathrm{sgn}} = 2i/k$. In distribution theory we must use Theorem 1.42 to deduce that $\widetilde{\mathbf{sgn}}(k) = 2i(P/k) + C\delta(k)$ for some constant C. To find C we use Proposition 3.6(f), where even and odd generalised functions are defined by $f(-x) = \pm f(x)$ respectively. Since $\widetilde{\mathbf{sgn}}$ and $\boldsymbol{P/k}$ are both odd and δ is even, we have the even distribution $C\delta$ equal to the odd distribution $\widetilde{\mathbf{sgn}} - 2i\boldsymbol{P/k}$. This can only be the case if they are both zero, so $C = 0$ and

$$\widetilde{\mathbf{sgn}}(k) = 2i\boldsymbol{P/k} \quad . \tag{3.17}$$

Example 3.27 The Fourier transform of $\boldsymbol{H}(x)$ can be immediately deduced from (3.17) by means of (3.16). Since the transform of 1 is $2\pi\delta$, according to (3.9), we have

$$\widetilde{\boldsymbol{H}}(k) = i(\boldsymbol{P/k}) + \pi\delta(k) \quad . \tag{3.18}$$

3.4 Generalised Functions of Several Variables

Since the aim of this chapter is the solution of partial differential equations, we must extend the theory above to functions of more than one variable. This is very easily done; essentially all distribution theory applies to functions of several variables if scalars are replaced by vectors at appropriate places in the argument. We shall sketch the theory very briefly.

To construct a generalised function of n variables we begin with test functions of n variables. We thus consider functions $\phi(x_1, x_2, \ldots, x_n)$, which we shall write as $\phi(\boldsymbol{x})$, assembling the n variables x_i into a vector variable $\boldsymbol{x}$. There is a conflict of notation here with our earlier use of bold type to denote distributions; we shall now drop that usage, as in Chapter 2, and reserve the use of bold type for vectors. We write $|\boldsymbol{x}| = (x_1^2 + \ldots + x_n^2)^{\frac{1}{2}}$ for the length of the vector. The analogue of a closed interval $[a,b]$ in one dimension is a sphere $\{\boldsymbol{x} : |\boldsymbol{x} - \boldsymbol{a}| \leqslant r\}$ for some $\boldsymbol{a}$, r.

The support of a function ϕ is the set of points $\boldsymbol{x}$ such that $\phi(\boldsymbol{x}) \neq 0$. A test function is then defined to be a smooth function whose support is contained in some finite sphere. Convergence in the space of test functions is defined as in 1.8 with the interval $[a,b]$ replaced by a sphere and ordinary derivatives replaced by partial derivatives. An n-dimensional distribution is then defined as a continuous linear functional on the space of test functions of n variables. We use the notation

$$\langle f,\phi\rangle = \int f(\boldsymbol{x})\phi(\boldsymbol{x})\, \mathrm{d}^n\boldsymbol{x} \tag{3.19}$$

for the action of the distribution f on the test function ϕ; $\int F\, \mathrm{d}^n\boldsymbol{x}$ stands for the multiple integral $\int \ldots \int F\, \mathrm{d}x_1 \ldots \mathrm{d}x_n$. A locally integrable function is one whose integral over every bounded region (that is, over every region which is inside some sphere) exists. Every locally integrable function f generates a distribution defined by (3.19), now interpreted as defining the left hand side by means of the integral on the right. Distributions obtained this way are called 'regular', and others are called 'singular'.

Examples 3.28 (a) The (ordinary) function $|\boldsymbol{x}|^{-1} = (x_1^2 + \ldots + x_n^2)^{-\frac{1}{2}}$ has a singularity at $\boldsymbol{x} = \boldsymbol{0}$. However, if $n \geqslant 3$ then the integral of $|\boldsymbol{x}|^{-1}$ over any bounded region is finite (you should verify this for $n = 3$ by integrating over a sphere centred on $\boldsymbol{0}$), so it is a locally integrable function. It thus generates a regular generalised function $|\boldsymbol{x}|^{-1}$ if $n \geqslant 3$. In two-dimensional space and (as we have seen in Chapter 1) in one-dimensional space the generalised functions corresponding to $|\boldsymbol{x}|^{-1}$ need a rather more careful discussion, and we shall not consider them further.

(b) The ordinary function $\log|\boldsymbol{x}|$ is locally integrable for all n, though it has a singularity at $\boldsymbol{x} = \boldsymbol{0}$ (you should verify this as in (a)). It therefore generates a regular distribution $\log|\boldsymbol{x}|$.

Example 3.29 The analogue of the delta function in n dimensions is the distribution $\delta(\boldsymbol{x})$ mapping any test function ϕ into $\phi(\boldsymbol{0})$. We have

$$\int \delta(\boldsymbol{x} - \boldsymbol{a})\phi(\boldsymbol{x})\, \mathrm{d}^n\boldsymbol{x} = \phi(\boldsymbol{a}) \quad . \tag{3.20}$$

It is sometimes convenient to write $\delta(\boldsymbol{x})$ in the form of a product $\delta(x_1)\delta(x_2)\ldots \delta(x_n)$. The multiple integrals in (3.20) can then be evaluated one at a time, the integration with respect to x_r having the effect of replacing x_r in ϕ by a_r. We regard this as a notational device for obtaining the correct value (3.20) for integrals containing the n-dimensional delta function, though full accounts of this subject develop a mathematical machinery in which the product representation of the delta function has a rigorous foundation. □

Remark 3.30 In ordinary multidimensional calculus there are well-known rules for changing variables in multiple integrals, corresponding to changes, for example, from Cartesian to polar coordinates. Similar rules can be developed in distribution theory, but we shall content ourselves with the warning that one must be careful when using polar coordinates. It is wrong, for instance, to represent the three-dimensional delta function in polar coordinates by a simple-minded analogue of the product representation mentioned in 3.29: replacing $\delta(\boldsymbol{x})$ by $\delta(r)\delta(\theta)\delta(\phi)$ would give the wrong value for integrals. See, for example, Stakgold (1967) section 5.5 for discussion of the right way to represent the delta function in curvilinear coordinates. □

We must now discuss Fourier transforms in n dimensions. If f is an absolutely integrable function of n variables, we can temporarily regard it as a function of x_1 with $x_2, \ldots, x_n$ held fixed, and then take its Fourier transform, which is a function of a new variable, k_1 say, together with $x_2, \ldots$; now we regard this new function as a function of x_2 with all the other variables held fixed, and take its transform, obtaining a function of $k_1, k_2, x_3, \ldots, x_n$; this procedure is repeated n times. This gives a function which we call $\tilde{f}(k_1, k_2, \ldots, k_n)$, and

$$\tilde{f}(k_1, \ldots k_n) = \int \mathrm{d}x_n \mathrm{e}^{ik_n x_n} \int \mathrm{d}x_{n-1} \ldots \int \mathrm{d}x_1 \mathrm{e}^{ik_1 x_1} . f(x_1, \ldots x_n)$$
$$= \int \mathrm{d}^n x \mathrm{e}^{i(k_1 x_1 + \ldots + k_n x_n)} f(x) \quad .$$

It is now natural to assemble the n variables $k_1, \ldots, k_n$ into a vector k, so that $k_1 x_1 + \ldots + k_n x_n$ can be written $k \cdot x$, and

$$\tilde{f}(k) = \int \mathrm{e}^{ik\cdot x} f(x) \, \mathrm{d}^n x \quad . \tag{3.21}$$

The inversion formula follows from (3.2) applied n times:

$$f(x) = (1/2\pi)^n \int \mathrm{e}^{-ik\cdot x} \tilde{f}(k) \, \mathrm{d}^n k \quad . \tag{3.22}$$

If one of the variables represents time, the conventional notation is as follows. We write the function as $f(x,t)$, and define its transform with respect to all its variables by

$$\hat{\tilde{f}}(k,\omega) = \int \mathrm{d}^n x \int \mathrm{d}t \, \mathrm{e}^{i(k\cdot x - \omega t)} f(x,t) \quad . \tag{3.23}$$

The inversion formula is then

$$f(x,t) = (1/2\pi)^{n+1} \int \mathrm{d}^n k \int \mathrm{d}\omega \, \mathrm{e}^{-i(k\cdot x - \omega t)} \hat{\tilde{f}}(k,\omega) \quad . \tag{3.24}$$

It is not necessary to transform all the variables, of course; it is often convenient to transform with respect to time only, say. Examples will be found in standard accounts of integral transforms for applied mathematicians.

The formulae of Proposition 3.6 are easily carried over to the n-dimensional case. We note here for future reference that the analogue of 3.6(b) is

$$\widetilde{\partial f/\partial x_r} = -ik_r \tilde{f} \quad , \tag{3.25}$$

and consequently $\widetilde{\nabla^2 f} = -|k|^2 \tilde{f}$. (3.26)

If f is a function of time, as in (3.23), we have

$$\widehat{\widetilde{\partial f/\partial t}} = i\omega \hat{\tilde{f}} \quad . \tag{3.27}$$

Finally we must discuss the Fourier transform of generalised functions of several variables. The theory of section 3.3 is extended to n dimensions in the same way as the theory of Chapter 1, as outlined at the beginning of this section. The (uninteresting) details can be found in the standard texts (see section 3.7). The result is that generalised functions which grow at infinity no faster than a polynomial have Fourier transforms, and the properties of the ordinary transform, such as equations (3.24)–(3.27), hold for the generalised transform when interpreted appropriately. The n-dimensional delta function has the transform

$$\widetilde{\delta(x)} = 1 \quad , \tag{3.28}$$

with inverse $\widetilde{1} = \int e^{-ik\cdot x}\, d^n x = (2\pi)^n \delta \quad .$ (3.29)

Equation (3.28) is extended to the shifted delta function in the following way, analogous to (3.14):

$$\int \delta(x-a) e^{ik\cdot x}\, d^n x = \widetilde{\delta(x-a)} = e^{ia\cdot k} \quad . \tag{3.30}$$

We have now assembled the tools we need for partial differential equations.

3.5 Green's Function for the Laplacian

The most important linear partial differential equations are Laplace's equation $\nabla^2 u = 0$, with its inhomogeneous version, Poisson's equation $\nabla^2 u = -f$; the wave equation $u_{tt} - \nabla^2 u = f$, where f is again an inhomogeneous term and subscripts denote partial derivatives; and the diffusion equation $u_t - \nabla^2 u = f$. We saw in Chapter 2 that the key to solving inhomogeneous boundary-value problems is Green's function. For a partial differential equation, as for an ordinary differential equation, Green's function is defined as the solution of the equation with homogeneous boundary conditions and the right hand side replaced by a delta function. In the next two sections we shall construct Green's functions for the Laplace and wave equations; the diffusion equation is left as an exercise.

Consider then the three-dimensional Poisson equation

$$\nabla^2 u = \partial^2 u/\partial x_1^2 + \partial^2 u/\partial x_2^2 + \partial^2 u/\partial x_3^2 = -f(x) \quad . \tag{3.31}$$

The solution u of (3.31) can be interpreted as the steady-state temperature distribution in a uniform medium, produced by heat sources distributed through space with strength $f(x)$ at the point x (negative values of f corresponding to points where heat is being extracted). There are many other physical applications of (3.31). We shall construct a fundamental solution for the operator $-\nabla^2$, that is, a solution of the equation

$$-\nabla_x^2\, w(x;y) = \delta(x-y) \quad , \tag{3.32}$$

where the subscript indicates which of the two vector variables of w is being differentiated. Having obtained the fundamental solution, we shall then consider the question of boundary conditions.

In constructing the fundamental solution we shall again use the principle that it is permissible to use any methods to find solutions of an equation, provided that the result is then justified by rigorous arguments. Accordingly we shall solve (3.32) by Fourier transforms, and manipulate the Fourier integrals with complete disregard for the niceties of distribution theory; but the solution will then be honestly proved to satisfy (3.32).

Write $v(k;y) = \int w(x;y)e^{ik\cdot x}\,d^3x$. Then the Fourier transform of equation (3.32), using the results (3.26) and (3.30), gives $|k|^2 v = e^{ik\cdot y}$. The inversion formula (3.22) now gives

$$\begin{aligned} w(x;y) &= (2\pi)^{-3}\int e^{-ik\cdot x} e^{ik\cdot y}\,|k|^{-2}\,d^3k \\ &= (2\pi)^{-3}\int e^{ik\cdot\xi}/(k_1^2+k_2^2+k_3^2)\,d^3k \end{aligned}$$

where $\xi = y - x$. We evaluate this integral using polar coordinates in k-space with axis along ξ. In other words, we change variables in the integral from k_1, k_2, k_3 to k, θ, ϕ, defined by

$$\begin{aligned} k &= \sqrt{k_1^2+k_2^2+k_3^2} = |k| \quad , \\ \theta &= \cos^{-1}[k\cdot\xi/|k|\,|\xi|] \quad , \\ \phi &= \tan^{-1}(k_2'/k_1') \quad , \end{aligned}$$

where k_1', k_2', k_3' are the Cartesian coordinates of the vector k with respect to new axes of which the third is parallel to ξ.

Then $k\cdot\xi = k\xi\cos\theta$, where $\xi = |\xi|$, and

$$\begin{aligned} w(x;y) &= (2\pi)^{-3}\int_0^\infty k^2\,dk\int_0^\pi \sin\theta\,d\theta\int_0^{2\pi} d\phi\, e^{ik\xi\cos\theta}/k^2 \\ &= (2\pi)^{-2}\int_0^\infty (2\sin k\xi)/(k\xi)\,dk \quad . \end{aligned}$$

Consulting tables of definite integrals, or evaluating the integral by complex contour integration, gives

$$\int_0^\infty (\sin k\xi)/k\,dk = \pi/2 \quad \text{if} \quad \xi > 0 \quad .$$

Hence
$$w(x;y) = \frac{1}{4\pi|x-y|} \tag{3.33}$$

for $x \neq y$.

This result should look familiar to anyone who has studied electrostatics: it is essentially the potential at the point x produced by a unit point charge at the point y. This is what might have been expected physically, since $\delta(x-y)$ is just the charge-density corresponding to a unit point charge at y – it is zero for $x \neq y$ and its integral over any region including y is 1. The formula for the solution of (3.31) in terms of the fundamental solution is

$$u(x) = (1/4\pi)\int f(y)/|x-y|\,d^3y \quad ,$$

the physical interpretation of which is that the potential due to a charge-distribution $f(x)$ is obtained by superposition (meaning summation or integration) of potentials due to the elementary point charges of which the distribution may be considered to be composed.

Now we must prove that (3.33) really gives a fundamental solution, despite its nonrigorous derivation. Elementary calculation shows that $\nabla^2 w = 0$ for $x \neq y$, as expected from (3.32), and w is clearly singular at $x = y$. We must show that the singularity is just sufficient to make w a generalised solution of (3.32). That is, we must prove that

$$\langle -\nabla_{\boldsymbol{x}}^2(1/4\pi|\boldsymbol{x}-\boldsymbol{y}|),\phi\rangle = \phi(\boldsymbol{y}) \tag{3.34}$$

for any test function ϕ. We note that $|\boldsymbol{x}-\boldsymbol{y}|^{-1}$ is not defined for $\boldsymbol{x} = \boldsymbol{y}$, and certainly not differentiable there; but although when considered as an ordinary function it is singular, yet it is locally integrable and generates a regular distribution, in fact the distribution obtained by applying a translation through $\boldsymbol{y}$ to Example 3.28(a). The operator $\nabla_{\boldsymbol{x}}^2$ in (3.34) is to be understood in terms of generalised derivatives $\Sigma_1^3 \partial^2/\partial x_r^2$, where the generalised derivative is defined, as in Definition 1.26, by

$$\langle \partial f/\partial x_r,\phi\rangle = -\langle f,\partial\phi/\partial x_r\rangle \quad .$$

With this understanding, we can prove the following.

Theorem 3.31 The distribution $1/4\pi|\boldsymbol{x}-\boldsymbol{y}|$ is a fundamental solution for the operator $-\nabla^2$ in three dimensions.

Proof We must prove (3.34). Using the above definition of partial differentiation applied to $|\boldsymbol{x}-\boldsymbol{y}|^{-1}$ we have

$$\langle -\nabla^2(1/4\pi|\boldsymbol{x}-\boldsymbol{y}|),\phi\rangle = -(1/4\pi)\langle |\boldsymbol{x}-\boldsymbol{y}|^{-1},\nabla^2\phi\rangle \quad .$$

Since $|\boldsymbol{x}-\boldsymbol{y}|^{-1}$ is a regular distribution, its action on a test function is given by the integral of the product. So

$$\langle -\nabla^2(1/4\pi|\boldsymbol{x}-\boldsymbol{y}|),\phi\rangle = -(1/4\pi)\int_G \nabla^2\phi|\boldsymbol{x}-\boldsymbol{y}|^{-1}\,\mathrm{d}^3\boldsymbol{x}$$

where G is a sphere containing the support of ϕ; we choose G large enough to contain $\boldsymbol{y}$. Now, the integrand is ill-behaved at $\boldsymbol{x} = \boldsymbol{y}$. It is desirable to exclude this point from the region of integration, so we define G_ϵ to be the region G with a sphere of radius ϵ centred on $\boldsymbol{y}$ removed, where ϵ is any number small enough for that sphere to lie entirely inside G (see Fig. 3). Then G_ϵ approaches G as $\epsilon \downarrow 0$, and

$$\langle -\nabla^2(1/4\pi|\boldsymbol{x}-\boldsymbol{y}|),\phi\rangle = -1/4\pi \lim_{\epsilon\downarrow 0}\int_{G_\epsilon}(\nabla^2\phi)/|\boldsymbol{x}-\boldsymbol{y}|\,\mathrm{d}^3\mathrm{x} \quad . \tag{3.35}$$

We now apply **Green's theorem**, which says that for any region V bounded by a surface S with unit outward normal vector $\boldsymbol{n}$,

$$\int_V (g\nabla^2 h - h\nabla^2 g)\,\mathrm{d}V = \int_S (g\nabla h - h\nabla g)\cdot\boldsymbol{n}\,\mathrm{d}S \tag{3.36}$$

for any functions g, h which are twice continuously differentiable in V. This theorem is a direct consequence of the divergence theorem; see any book on vector analysis. The functions $|x - y|^{-1}$ and ϕ satisfy the conditions for Green's theorem in G_ϵ, though not in G. The boundary of G_ϵ consists of two parts, the outer

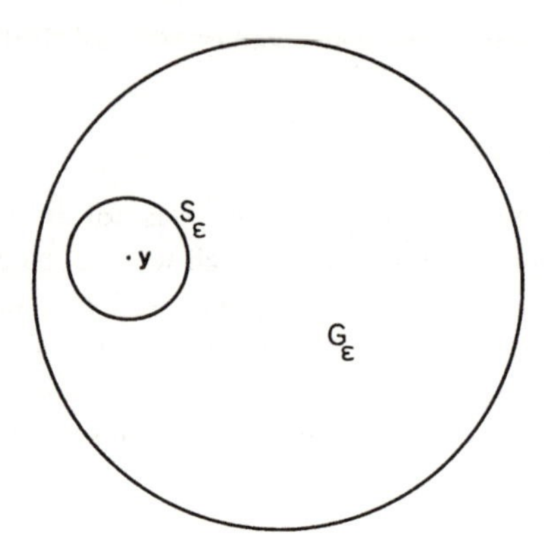

Fig. 3. The region G_ϵ.

boundary of G, on which ϕ and its derivatives vanish, and a spherical surface S_ϵ of radius ϵ centred on y. Applying Green's theorem to the integral in (3.35), noting that $\nabla^2 |x - y|^{-1} = 0$ in G_ϵ and that ϕ and $\nabla\phi$ vanish on the outer boundary of G, we have

$$\langle -\nabla^2(1/4\pi|x-y|), \phi\rangle = \operatorname*{Lim}_{\epsilon \downarrow 0}(-1/4\pi)\int_{S_\epsilon}\{|x-y|^{-1}\nabla\phi - \phi\nabla|x-y|^{-1}\}\cdot n \,\mathrm{d}S$$

$$= \operatorname*{Lim}_{\epsilon \downarrow 0}(-1/4\pi)\int_{S_\epsilon}\{|x-y|^{-1}\nabla\phi + \phi|x-y|^{-3}(x-y)\}\cdot n \,\mathrm{d}S$$

where n is a unit normal vector to S_ϵ pointing outwards from G_ϵ, that is, towards y.

Now, for x on S_ϵ, $x = y + \epsilon\nu$, where ν is a unit vector pointing outwards from y (see Figure 4). So $\nu = -n$, and $|x - y| = \epsilon$. Let θ, ψ be spherical polar angles for ν. Then $\mathrm{d}S = \epsilon^2 \sin\theta \,\mathrm{d}\theta \,\mathrm{d}\psi$, and

$$\langle -\nabla^2(1/4\pi|x-y|), \phi\rangle = \operatorname*{Lim}_{\epsilon \downarrow 0}(1/4\pi)\iint \sin\theta \,\mathrm{d}\theta \,\mathrm{d}\psi \{\epsilon\nu\cdot\nabla\phi + \phi\}$$

where ϕ in the integral is $\phi(x) = \phi(y + \epsilon\nu)$. Intuitively it is clear that the limit of the first term is zero, and the second gives $\phi(y)$ as required. The proof is easy. Expand $\phi(y + \epsilon\nu)$ in Taylor series, or use the mean value theorem, remember that the integration is with respect to $\nu = (\sin\theta\cos\psi, \sin\theta\sin\psi, \cos\theta)$ so that functions of y only can be taken outside the integral, and get

$$\langle -\nabla^2(1/4\pi|x-y|), \phi\rangle = \operatorname*{Lim}_{\epsilon \downarrow 0}(1/4\pi)\{\phi(y)\iint \sin\theta \,\mathrm{d}\theta \,\mathrm{d}\psi + O(\epsilon)\}$$

$$= \phi(y) \quad .$$

This completes the proof that (3.33) is a fundamental solution. □

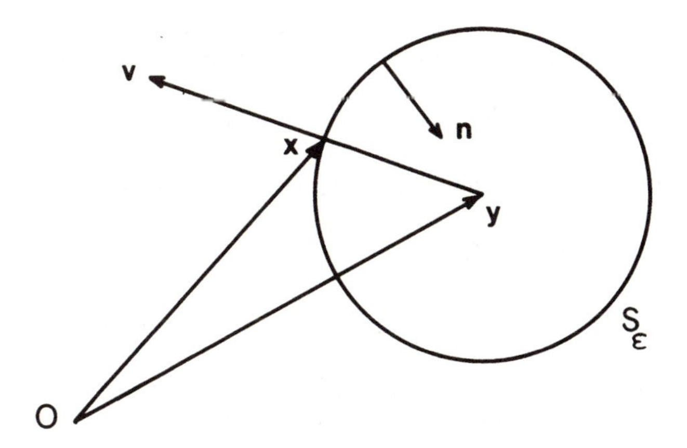

Fig. 4. Definition sketch for the proof of Theorem 3.31.

The fundamental solution (3.33) satisfies the boundary condition that it vanish at large distances; it is therefore the Green's function corresponding to this boundary condition. Many other kinds of boundary condition arise naturally in practice. A particularly important kind arises in Dirichlet's problem, that is, the problem of finding a solution of the Laplace or Poisson equation in a finite region V bounded by a surface S, given the values of the solution on S. Other conditions are possible; see Problem 3.13. Green's function for such problems is constructed by adding a solution of the homogeneous Laplace equation to the fundamental solution (3.33), so it is important to know under what conditions solutions of Laplace's equation can be found.

Theorem 3.32 (Existence theorem for Laplace's Equation) For any reasonably well-behaved closed surface S enclosing a volume V, there exists one and only one solution of Laplace's equation taking any given reasonably well-behaved set of values on S. If the given values vary continuously on S, the solution is continuous on the region consisting of V together with S. □

This theorem is stated rather vaguely because there are many possible interpretations of "reasonably well-behaved", and correspondingly many theorems. Proving such existence theorems is a major task, practically a branch of mathematics in its own right. For a brief sketch of one approach to the theory, see Chapter 12; for an account of the classical theory, see Courant-Hilbert Vol. II, Ch. IV, § 4. For practical purposes it can be assumed that Dirichlet's problem has a classical solution unless the surface S has a sharp spike or other pathological behaviour.

Given a Dirichlet problem on a region V with boundary S, we define the **Dirichlet-problem Green's function** for the region V to be the fundamental solution for $-\nabla^2$ which vanishes on S. Just as for ordinary differential equations, Green's function satisfies homogeneous boundary conditions, but it can be used to solve the more general Dirichlet problem in which nonzero values on S are given; see problem 3.14.

We now show that this Green's function always exists.

Theorem 3.33 (Dirichlet-Problem Green's Function) For any reasonable surface S enclosing a volume V, there exists a Green's function for $-\nabla^2$ in V satisfying the condition that it vanish on S. It is a symmetric function.

Proof We consider the fundamental solution $w(\boldsymbol{x};\boldsymbol{y}) + v(\boldsymbol{x};\boldsymbol{y})$, where w is given by (3.33) and v is a solution of Laplace's equation $\nabla^2 v = 0$. Theorem 3.32 shows that there exists a solution v whose values on S are minus the values of w on S, provided that $\boldsymbol{y}$ is inside V and not on S so that w is a continuous function of $\boldsymbol{x}$ for $\boldsymbol{x}$ on S. Then $w + v$ vanishes on S, and is therefore the required Green's function.

It remains to show that the Green's function $g = w + v$ is symmetric. Consider the following quick and dirty argument. Apply Green's theorem, equation (3.36), to the functions $g(\boldsymbol{x};\boldsymbol{y})$ and $g(\boldsymbol{x};\boldsymbol{z})$. Since g vanishes for $\boldsymbol{x}$ on S, we have

$$\begin{aligned} 0 &= \int_V \{g(\boldsymbol{x};\boldsymbol{y})\nabla_x^2 g(\boldsymbol{x};\boldsymbol{z}) - g(\boldsymbol{x};\boldsymbol{z})\nabla_x^2 g(\boldsymbol{x};\boldsymbol{y})\}\, \mathrm{d}V \\ &= -\int_V \{g(\boldsymbol{x};\boldsymbol{y})\delta(\boldsymbol{x}-\boldsymbol{z}) - g(\boldsymbol{x};\boldsymbol{z})\delta(\boldsymbol{x}-\boldsymbol{y})\}\, \mathrm{d}V \\ &= -g(\boldsymbol{z};\boldsymbol{y}) + g(\boldsymbol{y};\boldsymbol{z}) \quad , \end{aligned}$$

which shows that $g(\boldsymbol{y};\boldsymbol{z}) = g(\boldsymbol{z};\boldsymbol{y})$ as required. It is easy to pick holes in this argument: for example, Green's theorem applies to twice-differentiable functions, while g is not even continuous. But it can be made respectable by the method used in the proof of Theorem 3.31: apply Green's theorem in the region V_ϵ obtained by removing from V two spheres of radius ϵ centred on $\boldsymbol{y}$ and $\boldsymbol{z}$. Then Green's function is well-behaved and satisfies Laplace's equation in V_ϵ, so the only non-zero terms in the statement of Green's theorem are integrals over the surface of the two spheres of radius ϵ. Remembering that $g(\boldsymbol{x};\boldsymbol{y}) = 1/4\pi|\boldsymbol{x}-\boldsymbol{y}| +$ a bounded function, and taking the limit as $\epsilon\downarrow 0$ as in the proof of Theorem 3.31 (the details are left to the reader) we recover the same result as by the quick method above. □

Proposition 3.34 If f is bounded and integrable on a region V enclosed by a reasonable surface, then the function

$$u(\boldsymbol{x}) = \int_V g(\boldsymbol{x};\boldsymbol{y})f(\boldsymbol{y})\, \mathrm{d}^3\boldsymbol{y} \quad ,$$

where g is the Dirichlet-problem Green's function, is a weak solution of the equation $\nabla^2 u = -f$ in V.

Proof Let ϕ be a test function with $\mathrm{supp}(\phi) \subset V$. We must show that $\langle\nabla^2 u,\phi\rangle = -\langle f,\phi\rangle$. We have

$$\begin{aligned}\langle \nabla^2 u,\phi\rangle &= \langle u,\nabla^2\phi\rangle \\ &= \int_V \{\int_V d^3y\, g(x;y) f(y)\nabla^2\phi(x)\}\, d^3x \\ &= \int f(y)\langle \nabla_x^2 g(x;y),\phi\rangle\, d^3y \\ &= -\int f(y)\phi(y)\, d^3y = -\langle f,\phi\rangle\end{aligned}$$

because g is a fundamental solution. Changing the order of the x and y integration is justified because the single integrals converge uniformly. □

If f is reasonably well-behaved (e.g., piecewise continuously differentiable), then Green's function gives a classical solution, not just a weak solution, of $\nabla^2 u = -f$; see section 9.3. The Dirichlet Green's function can also be used to solve the Dirichlet boundary-value problem; see Problem 3.14.

We conclude this section with a general property of Green's function for the Dirichlet problem. It is a mathematical property which will be needed later, when we apply functional-analytic methods to partial differential equations, but it has a clear and elegant physical interpretation.

The function $1/4\pi|x-y|$ is, as we have seen, a fundamental solution for the negative Laplacian; it satisfies the boundary condition that it vanish at infinity. It may be interpreted as the equilibrium temperature distribution produced in an infinite uniform medium by a point source of heat, with the condition that the temperature is zero at large distances; call this problem A. The Dirichlet-problem Green's function in a region V, with surface S, can be interpreted as the equilibrium temperature produced by a point source when the temperature on S is forced to be zero; call this problem B. In both cases, the temperature is positive. We can continuously transform problem B into problem A by moving the boundary S, on which the temperature is forced to vanish, out to infinity. It is very plausible that the presence of the boundary at freezing-point has a depressing effect on the temperature, and that when that boundary is removed to infinity, the effect is to raise the temperature. This suggests that the Green's function for problem B is everywhere less than the infinite-space Green's function $1/4\pi|x-y|$. We now prove this.

Proposition 3.35 (Bounds for Green's Function) If $g(x;y)$ is the Dirichlet-problem Green's function for $-\nabla^2$ in the region $V \subset \mathbb{R}^3$ enclosed by a reasonable surface S, then $0 < g(x;y) < 1/4\pi|x-y|$ for $x,y \in V, x \neq y$.

Proof We use the 'maximum principle' which says that if $\nabla^2\phi = 0$ in V, then the maximum and the minimum values of ϕ both occur on the boundary S. A few moments' thought about the heat flow interpretation will show that this is plausible, and proofs will be found in books on potential theory or partial differential equations (e.g., Courant-Hilbert Vol. II, or Stakgold). Now

$$g(x;y) = 1/(4\pi|x-y|) + v(x;y) \quad ,$$

where ν satisfies Laplace's equation in V and $\nu(x;y) = -1/4\pi|x-y|$ for x on S. Thus $\nu < 0$ on S, and the maximum principle shows that $\nu < 0$ everywhere, hence $g < 1/4\pi|x-y|$.

We now use the maximum principle to show that $g > 0$. Let X, Y be any given vectors in V with $X \neq Y$. Since ν is continuous and bounded while $|x-Y|^{-1} \to \infty$ as $x \to Y$, there is a $\delta > 0$ such that $g(x;Y) > 0$ whenever $|x-Y| < \delta$. Now let V_ϵ be the region obtained from V by removing a sphere of radius ϵ centred on Y, where ϵ is a number small enough for that sphere to lie inside V. If we take $\epsilon < \delta$, then $g(x;Y) > 0$ when x is on the inner boundary of V_ϵ, and $g(x;Y) = 0$ on the outer boundary. It follows from the maximum principle that $g(x;Y) > 0$ for all $x \in V_\epsilon$. By taking $\epsilon < |X-Y|$ we can ensure that $X \in V_\epsilon$, hence $g(X;Y) > 0$ for any $X, Y \in V$. □

3.6 Green's Function for the Three-dimensional Wave Equation

The equation $u_{tt} - \nabla^2 u = f$ (3.37)

describes the propagation of pressure disturbances in air, that is, sound waves, generated by sources of sound with density function $f(x)$. It also describes the propagation of electromagnetic waves generated by electric current with density $f(x)$. We shall find a fundamental solution satisfying

$$(\partial^2/\partial t^2 - \nabla_x^2)w(x,t;y,s) = \delta(x-y)\delta(t-s) \quad . \tag{3.38}$$

We solve this by using the Fourier transform, as in section 3.5. Set

$$\nu(k,\omega;y,s) = \int d^3x \int dt\, e^{i(k\cdot x - \omega t)} w(x,t;y,s) \quad .$$

Then Fourier-transforming (3.38) gives

$$-(\omega^2 - |k|^2)\nu = e^{i(k\cdot y - \omega s)} \quad ,$$

so $$w(x,t;y,s) = (1/2\pi)^4 \int d^3k \int d\omega\, e^{i(k\cdot\xi - \omega\tau)}/(|k|^2 - \omega^2)$$

where $\xi = y - x$ and $\tau = s - t$. Again we evaluate this using spherical polar coordinates in k-space; the angular integrals are easily evaluated as in section 3.5, giving

$$w(x,t;y,s) = \frac{1}{4\pi^3\xi}\int_0^\infty k\,dk \int_{-\infty}^\infty d\omega\, \frac{\sin k\xi\, e^{-i\omega\tau}}{k^2 - \omega^2} \quad . \tag{3.39}$$

Consider the ω-integral

$$I = \int_{-\infty}^\infty e^{-i\omega\tau} d\omega/(k^2 - \omega^2) \quad . \tag{3.40}$$

We shall evaluate it by contour integration; the reader unfamiliar with complex contour integrals should skip to equation (3.41). The integrand has poles on the real axis. Any contour of integration which goes from $-\infty$ to ∞ avoiding these poles will give a solution of the equation (3.38). There are many choices: the contour can pass above the poles, below them, above one and below the other, or wind round them several times. Different choices will give different values for the integral, and therefore different fundamental solutions, satisfying different boundary conditions.

We take one of the simplest choices for the contour of integration: a contour going under both poles (see Figure 5). Then if $\tau > 0$ we can complete the contour

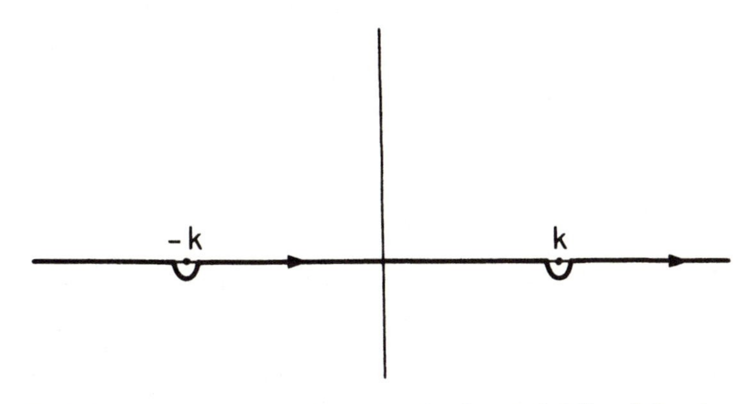

Fig. 5. Integration contour in the ω-plane for the retarded Green's function.

with a large semicircle in the lower half-plane, the integral along which vanishes as the radius of the semicircle tends to infinity, and then Cauchy's theorem gives $I = 0$ for $\tau > 0$. If $\tau < 0$ we can complete the contour with a large semicircle in the upper half-plane, and then the residue theorem gives $I = -(2\pi/k)\sin(k\tau)$ for $\tau < 0$. Putting these two results together, we have

$$I = -(2\pi/k)\sin(k\tau)H(-\tau) \tag{3.41}$$

if the integral is along a contour going under the poles; H here is Heaviside's function (Example 1.18). Hence (3.39) gives the following expression for the fundamental solution, which we shall now call g_{ret}, for reasons that will become clear shortly:

$$\begin{aligned} g_{\text{ret}}(x,t;y,s) &= -[H(-\tau)/2\pi^2\xi]\int_0^\infty \sin k\xi \sin k\tau \, dk \\ &= -[H(-\tau)/8\pi^2\xi]\int_{-\infty}^\infty [\cos k(\xi-\tau) - \cos k(\xi+\tau)]\, dk \\ &= -[H(-\tau)/4\pi\xi]\,[\delta(\xi-\tau) - \delta(\xi+\tau)] \end{aligned}$$

using the real part of (3.9). Finally we note that $H(-\tau)\delta(\xi-\tau) = 0$ for $\xi > 0$ since for any τ one of the factors is bound to be zero, and that $H(-\tau)\delta(\xi+\tau) = \delta(\xi+\tau)$ since the Heaviside function is 1 whenever the delta function is nonzero, so that

$$g_{\text{ret}}(x,t;y,s) = \frac{\delta(|x-y|+s-t)}{4\pi|x-y|} \quad . \tag{3.42}$$

Again, we have manipulated generalised functions in a cavalier fashion; the result can be rigorously justified, though we shall not pause to do so here.

We conclude this chapter with a discussion of the properties of this fundamental solution. Its most obvious feature is that it is even more singular than (3.33). Ordinary differential equations have fundamental solutions which are continuous functions; Laplace's equation is an example of the class of elliptic partial differential equations whose fundamental solutions are locally integrable functions with mild singularities; the wave equation is an example of the class of hyperbolic partial differential equations whose fundamental solutions can be singular distributions (though Problem 3.17 shows that they need not be).

A fundamental solution can be interpreted as the response to a localised point source. Thus $w(\boldsymbol{x},t;\boldsymbol{y},s)$ is the response at the point $\boldsymbol{x}$ at time t to a disturbance which is localised at the point $\boldsymbol{y}$ at time s, that is, a point source at $\boldsymbol{y}$ whose strength is zero for all times except s, and which flashes on and off instantaneously at time s. In the electromagnetic interpretation one might think of a small spark occurring at time s at the point $\boldsymbol{y}$. The particular fundamental solution (3.42) is zero for all times except $t = s + |\boldsymbol{x} - \boldsymbol{y}|$, at which time it has a delta-function pulse. At points $\boldsymbol{x}$ close to $\boldsymbol{y}$, the pulse occurs only slightly later than the spark; the further $\boldsymbol{x}$ is from $\boldsymbol{y}$, the later the pulse occurs, and the delay time just equals the distance from the point of observation $\boldsymbol{x}$ to the source point $\boldsymbol{y}$. In this way the fundamental solution beautifully expresses the fact that disturbances from a localised source spread outwards with constant velocity (equal to one with our choice of units) in all directions: (3.42) represents a spherical wave. The denominator $|\boldsymbol{x} - \boldsymbol{y}|$ in (3.42) means that the strength of the wave decreases as it spreads out. This is in accordance with the principle of conservation of energy: as a wave expands and covers a greater area, its strength must decrease if its total energy is to remain constant.

In deriving (3.42) we made an arbitrary choice of integration contour. If we had taken the contour lying above both poles instead of the one lying below, we would have obtained a different fundamental solution, namely

$$g_{\text{adv}}(\boldsymbol{x},t;\boldsymbol{y},s) = \delta(t - s + |\boldsymbol{x} - \boldsymbol{y}|)/(4\pi|\boldsymbol{x} - \boldsymbol{y}|) \quad . \tag{3.43}$$

Mathematically this fundamental solution is just as satisfactory as (3.42), but its physical properties are a little different. It says that the response at time t to a source flashing on and off instantaneously at time s is zero except at time $t = s - |\boldsymbol{x} - \boldsymbol{y}|$. In other words, the disturbance occurs *before* the spark that produces it. (3.43) is the fundamental solution which satisfies the boundary condition that it vanish for all $t > s$, which forces the disturbance to occur before (in advance of) the source producing it; it is therefore called the **advanced Green's function** for the wave equation. The solution (3.42) is called the **retarded Green's function**, because the effect occurs later than the cause producing it; it is the fundamental solution which satisfies the boundary condition that it vanish for all $t < s$. Common experience tells us that effects follow their causes, so we use the retarded Green's function, not the advanced one, to compute solutions of the inhomogeneous wave equation. But philosophically this is perhaps a little unsatisfying. Why does Nature arbitrarily choose g_{ret} and reject g_{adv}? A more symmetric and therefore psychologically more appealing choice would be $\frac{1}{2}(g_{\text{ret}} + g_{\text{adv}})$, which can easily be shown to be another fundamental solution. It corresponds to an integration contour in the ω-plane which is exactly along the real axis, and deals with the singularities by taking the principal value. There is in fact a satisfactory theory of wave propagation and radiation using this symmetric Green's function instead of the retarded one (Wheeler and Feynman (1945)); and yet no theory can be completely symmetric, for after all, the flow of time (whatever that means) proceeds definitely in one direction only. There are difficult philosophical issues here; discussions will be found in Whitrow, and Zeman.

There is one more property of the retarded Green's function (3.42) which is worth noting. A delta-function source produces a delta-function electric field travelling outwards as a spherical wave. Two delta-functions sparks in succession at the same point will produce an electric field consisting of two sharply-defined waves travelling without interfering with each other, so that an observer at any point will see two sharply defined pulses. Similarly, any sequence of sparks, and indeed any time-behaviour whatsoever of a point source, will produce a field at a distance whose time-variation exactly reproduces the time-variation of the source. In other words, when we look at an object from a distance, we can see exactly how it is behaving (if our eyes are good enough), just as clearly as if we were very close to it. This fact is so familiar that it seems obvious and not worth mentioning. Yet it is not universally true. If the wave equation (3.37) is modified in almost any way, its Green's function is no longer a delta function. In particular, the retarded Green's function for the wave equation in two dimensions is zero for $t < s$ but is nonzero for almost all $t > s$ – see Problem 3.17. This means that in a two-dimensional world, if you look at a spark from a distance you will see it suddenly appear, but instead of instantly disappearing again it gradually dies away. If you look at a moving object in a two-dimensional world, it will appear to smear itself out as it moves, rather than moving cleanly from one point to another, and superimposed on your vision of the object will be ghostly traces of all other objects which have ever occupied that position, the waves from which continue propagating outwards indefinitely. Hearing would also be difficult in a two-dimensional world, as sounds would reverberate indefinitely – it would be like living in an echo chamber. It can be shown that the same is true for any even-dimensional space. One is tempted to surmise that intelligent animals like us would be unlikely to develop in a two or four-dimensional world where perception and communication are so much more difficult, and that intelligence in the form we know it owes its existence to the delta function in the Green's function for the three-dimensional wave equation.

3.7 Summary and References

In section 3.1 we outlined the theory of the classical Fourier transform; more information will be found in Sneddon, and Lanczos (1966); for applications see also Arsac, and Bracewell; for the mathematical theory see Titchmarsh (1937). We also explained the difficulties in applying the transform directly to generalised functions, which led us to construct a new space of test functions in section 3.2, with a corresponding new class of distributions. In section 3.3 we defined the Fourier transform for these distributions of slow growth, and worked out several examples. The theory is outlined in Arsac, and Stakgold; more complete and deeper accounts will be found in Gel'fand & Shilov, and Zemanian (1968). Lighthill and Jones reach equivalent results from a different point of view. These books also cover the material in section 3.4 on distributions and transforms in several variables.

In section 3.5 we used the Fourier transform to obtain Green's function for Laplace's equation. We proved that Green's function always exists, is symmetric, and can be used to give a weak solution of Poisson's equation. In section 3.6 we discussed Green's functions for the wave equation. Rigorous (yet readable) treatments of these matters from the classical and distributional viewpoints respectively are given by Sobolev and Vladimirov. Other useful material will be found in Stakgold, Courant-Hilbert, Morse & Feshbach, Dennery & Krzywicki, Dettman, and Wallace.

PROBLEMS 3

Sections 3.1 & 3.2

3.1 Calculate the Fourier transform of the function f defined by $f(x) = 1$ for $|x| < 1, f(x) = 0$ for $|x| > 1$. Sketch the graph of $|\widetilde{f}|$. Is $\widetilde{f}$ absolutely integrable? Does Theorem 3.5 apply? (Note that f is not a test function, but can be turned into a test function by rounding off the corners slightly. It is plausible (and true) that this makes only a small change in $\widetilde{f}$, and we thus have a more or less concrete example of a test function whose transform is not a test function.)

3.2 Show that the Fourier transform $\widetilde{\widetilde{f}}$ of the function $\widetilde{f}$ of Problem 3.1 exists for $|x| \neq 1$, and 3.6(a) holds except at the points $x = \pm 1$ where f is discontinuous. What happens for $x = \pm 1$?

3.3 If f is a distribution of slow growth, show that $\phi \mapsto -\langle f,\phi'\rangle$ defines a distribution of slow growth (a result very similar to 1.25). This justifies the statement near the end of section 3.2 that every distribution of slow growth has a derivative which is a distribution of slow growth.

Section 3.3

3.4 Calculate the generalised Fourier transforms of $\sin(ax)$, $\cos(ax)$, $x^n \sin(ax)$, $x^n \cos(ax)$, x^{-n}, and any other functions you can think of. Here a is real and n is a positive integer.

3.5(a) Prove that $H(x)e^{-\epsilon x} \to H(x)$ distributionally as $\epsilon \downarrow 0$. Evaluate the Fourier transform of $H(x)e^{-\epsilon x}$, and use the result of Problem 3.8 to deduce the transform of $H(x)$. (The Plemelj formula, Problem 1.10, may be useful.)

(b) Use the formula of Proposition 3.6(c) to deduce the Fourier transform of $xH(x)$ from that of H. Extend the result to $x^n H(x)$ by repeated application of 3.6(c).

3.6 Resolve the following paradox. A man, call him X, sits in a dark room wearing coloured spectacles. The only source of light is a flashbulb, which is timed to fire at 6 pm, producing an instantaneous pulse of light of intensity $\delta(t-6)$ at time t; thus the intensity is zero for $t < 6$. Now, by 3.25, $\delta(t-6) = (1/2\pi)\int_{-\infty}^{\infty} e^{i\omega(t-6)}\,d\omega = (1/\pi)\int_0^{\infty} \cos\omega(t-6)\,d\omega$, which means that the light from the flashbulb is a superposition of monochromatic wavetrains of all frequencies ω. X's coloured spectacles only transmit light of frequencies between ω_1 and ω_2. Hence the light reaching X's eyes is $(1/\pi)\int_{\omega_1}^{\omega_2} \cos\omega(t-6)\,d\omega$, which is easily calculated and is nonzero for $t < 6$. But this is nonsense. It means that by virtue of his coloured spectacles, X is able to see the light before it is produced. It is easy to devise arrangements by which X is able to see into the future (for example, he could have a row of light bulbs, one for each horse running in the Derby, arrange for his friend Y to switch on the bulb corresponding to the winner, observe this bulb using his coloured spectacles before the race is run, and make a small fortune). What is wrong with the above argument?

The answer has nothing to do with the peculiarities of the delta function; exactly the same argument works if $\delta(t-6)$ is replaced by an ordinary function such as $\Delta(t;6,1)$ (see Problem 2.4). The point is quite a subtle one.

3.7 f is an absolutely integrable $\mathbb{R} \to \mathbb{R}$ function. F is an approximation to it whose graph is a straight line segment in each interval (x_{i-1}, x_i) for $i = 1, \ldots n$; F is continuous, F' has a jump of A_i at each x_i, and $F(x) = 0$ for $x < x_0$ or $x > x_n$. Show that $\Sigma A_i = \Sigma x_i A_i = 0$, and that $\tilde{F}(k) = (-1/k^2)\Sigma(A_i e^{ikx_i})$. Show that $|\tilde{f}(k) - \tilde{F}(k)| \leqslant \int_{-\infty}^{\infty} |f(x) - F(x)|\,dx$. How would you choose the polygon F so as to get the best approximation to $\tilde{f}$?

3.8 If f_n and f are distributions of slow growth, and $f_n \to f$ distributionally as $n \to \infty$, prove that $\tilde{f}_n \to \tilde{f}$. This justifies transforming a series term by term.

Section 3.5

3.9(a) Obtain a two-dimensional form of Green's theorem by applying equation (3.36) to functions g and h which depend only on x and y, in a region V which is a cylinder whose generators are in the z-direction and whose cross-section is an arbitrary curve C in the xy plane.

(b) Prove, in the manner of Theorem 3.31, that $(1/2\pi)\log|\boldsymbol{x}-\boldsymbol{y}|$ is a fundamental solution for the operator ∇^2 in two dimensions.

3.10 Another method of finding fundamental solutions for Laplace's equation. Look for a solution of the form $w(\boldsymbol{x};\boldsymbol{y}) = W(\boldsymbol{x}-\boldsymbol{y})$, where W satisfies $\nabla^2 W(\boldsymbol{z}) = 0$ for $\boldsymbol{z} \neq 0$. Assume (symmetry) that W depends only on $|\boldsymbol{z}| = r$, not on the angular co-ordinates. Then $\nabla^2 W = 0$ is a second-order ordinary differential equation in r. Solve this equation. Then simultaneously justify the solution and determine the unknown constants it contains by the method used in the proof of Theorem 3.31. Carry out this programme in two and in three dimensions (you will find in each case that one of the two arbitrary constants is determined, the other can be given any value – fundamental solutions are nonunique).

3.11 The 'Helmholtz equation' $\nabla^2 u + k^2 u = 0$, where k is a positive constant, appears when one looks for solutions of the wave equation (3.37) of the special form $u(\boldsymbol{r})e^{ikt}$. Use the method of Problem 3.10 to obtain a fundamental solution for the operator $\nabla^2 + k^2$ in three dimensions and (more difficult) in two dimensions. (Remember that a second-order ordinary differential equation $y'' + a(x)y' + b(x)y = 0$ can be simplified by writing $y(x) = c(x)v(x)$ and choosing $c(x)$ so that the resulting equation for v has no term in v'.)

3.12 The 'biharmonic equation' $\nabla^4 u = f$ arises in elasticity theory; $\nabla^4 u$ means $\nabla^2(\nabla^2 u)$.

(a) Prove the following analogue of Green's Theorem (Equation (3.36)):

$$\int_V (\phi\nabla^4\psi - \psi\nabla^4\phi)\,\mathrm{d}V = \int_S \mathrm{d}\boldsymbol{S}\cdot[\phi\nabla(\nabla^2\psi) - \psi\nabla(\nabla^2\phi) + \nabla^2\phi\nabla\psi - \nabla^2\psi\nabla\phi].$$

(b) Use the method of Problem 3.10 to obtain a fundamental solution for the biharmonic equation in three dimensions, using the result of (a) above in place of Green's Theorem.

3.13 The 'Neumann Problem' is a variant of the Dirichlet problem: it is the problem of finding a solution of $\nabla^2\phi = 0$ in a region V when the values not of ϕ but of $\boldsymbol{n}\cdot\nabla\phi$ are given on the surface S bounding V, where $\boldsymbol{n}$ is the outwards unit normal vector to S. There is a theorem like Theorem 3.32 saying that a solution of the Neumann problem exists, provided that the given values g of $\boldsymbol{n}\cdot\nabla\phi$ on S satisfy $\int g\,\mathrm{d}S = 0$. Use the divergence theorem to show that this is a necessary condition for the existence of a solution. Hence show that no fundamental solution for $-\nabla^2$ in a region V satisfies the condition that the normal component of its gradient vanish on the boundary. In other words, there is no Green's function for the Neumann problem.

By the reasoning in section 2.5, we might expect the homogeneous Neumann problem to have non-zero solutions. Verify this. (The difficulty of the non-existence of the Neumann Green's function can be circumvented; see Problem 2.9.)

3.14 The Dirichlet Green's function $g(\boldsymbol{x};\boldsymbol{y})$ for a region V bounded by a surface S can be used to solve the Dirichlet boundary-value problem $-\nabla^2 u = q(\boldsymbol{x})$ in V, $u(\boldsymbol{x}) = b(\boldsymbol{x})$ for $\boldsymbol{x}$ on S, where q is a given function on V, and b a given function on S. By applying Green's theorem to the functions g and u, in the region obtained from V by removing a small sphere centred on $\boldsymbol{x}$, show that

$$u(\boldsymbol{x}) = \int_V g(\boldsymbol{x};\boldsymbol{y})q(\boldsymbol{y})\,\mathrm{d}^3y - \int_S [\boldsymbol{n}\cdot\nabla_y g(\boldsymbol{x};\boldsymbol{y})]\, b(\boldsymbol{y})\,\mathrm{d}S$$

where $\boldsymbol{n}$ is a unit normal vector to S and ∇_y denotes the gradient with respect to the variable $\boldsymbol{y}$.

Section 3.6

3.15 Derive the expression (3.43) for the advanced Green's function.

3.16. Investigate the fundamental solutions for the wave equation obtained by taking a contour of integration in (3.40) passing over one pole and under the other.

3.17 Calculate the retarded Green's function for the wave equation in one and (more difficult) two dimensions.

3.18 Using the retarded Green's function to solve the inhomogeneous wave equation (3.37), and evaluating one of the integrals, obtain the solution $u(\boldsymbol{x},t) = \int \mathrm{d}^3y f(\boldsymbol{y},t - |\boldsymbol{x}-\boldsymbol{y}|)/4\pi|\boldsymbol{x}-\boldsymbol{y}|$. Verify that this satisfies (3.37), by evaluating its derivatives.

3.19 Calculate Green's function for the one-dimensional diffusion equation $u_t - u_{xx} = f$ with the boundary condition $u \to 0$ as $x \to \pm\infty$. Write it in the form $g(x,t;y,s) = G(x-y,T)$ where $T = t - s$. Show that $G(x-y,T) - G(x+y,T)$ is Green's function for the diffusion equation in the region $x > 0$ with the conditions $u = 0$ at $x = 0$ and $u \to 0$ as $x \to \infty$. (This can be interpreted physically in terms of images by reflection, a technique familiar to students of electrostatics.) For what problem is $G(x-y,T) + G(x+y,T)$ the Green's function?

Show that Green's function for the region $0 < x < a$, with conditions $u(0,t) = u(a,t) = 0$ for all t, is $\Sigma_{-\infty}^{\infty} G(x-y-2na,T) - \Sigma_{-\infty}^{\infty} G(x+y-2na,T)$. Interpret this in terms of images. Use Poisson's sum formula (section 1.6) to obtain a different representation of this Green's function, and investigate the conditions under which the series converge rapidly.

3.20 Derive the result of Problem 3.11 from the wave equation Green's function, by considering the equation $\nabla^2 v - v_{tt} = \delta(x - y)e^{ikt}$.

PART II

BANACH SPACES AND FIXED POINT THEOREMS

In Parts II and III we introduce and develop our main theme, the theory of various types of function space. Broadly speaking, Part II contains the basic theory, and applications to nonlinear problems, while Part III discusses the special ideas and methods available for linear problems. Part I need not be read before embarking on Part II, though many of the applications in the following chapters require some knowledge of Green's functions.

In Chapter 4 we set up our basic theoretical framework. Chapter 5 contains an easy fixed-point theorem, and some applications. Chapter 6 discusses deeper fixed-point theorems, with applications in theoretical mechanics; it also deals with the idea of compactness, which is essential for the theory of Part III. Readers who are mainly interested in linear problems, and wish to reach Hilbert space as soon as possible, should read the following sections only: 4.1–4.5, 5.1, 6.1, 6.3, 6.4.

Chapter 4

Normed Spaces

The theory of this chapter is the foundation of the rest of the book. The main aim of the theory, as far as we are concerned, is to produce solutions of the equations of applied mathematics. Since exact solutions are not usually available, approximation methods are very often used. When a function f is said to be an approximate solution of an equation, one means that in some sense f is close to an exact solution F of the equation. In this way we are led to consider the 'distance' between the functions f and F, and in particular the question of when that distance is small, rather as if the functions were points in space. This analogy between functions and points in space is one of our main themes.

Section 4.1 describes that part of the theory of vector spaces needed for our purposes. The reader familiar with linear algebra can omit this section; the reader unfamiliar with abstract vector spaces but familiar with vectors in three-dimensional space should read it; anyone completely unfamiliar with vectors should read an elementary book on the subject, or the first chapter or so of an introduction to linear algebra (for example, Kreider *et al.*) before proceeding. Section 4.2 introduces the idea of length, and the rest of the chapter discusses theoretical ideas associated with the problem of closeness of approximation mentioned above.

4.1 Vector Spaces

The reader (unless he has ignored the recommendation just above) knows that a point in space can be represented by a vector, which is a set of three numbers giving the distance of the point from three perpendicular axes. Given a vector $\boldsymbol{a} = (a_1,a_2,a_3)$, we can construct a new vector $k\boldsymbol{a} = (ka_1,ka_2,ka_3)$ which represents a point k times as far from the origin of coordinates as the original point. Again, given two vectors $\boldsymbol{a}$ and $\boldsymbol{b}$, we can construct a new vector $\boldsymbol{a} + \boldsymbol{b}$, representing the point given by applying the parallelogram rule of vector addition to $\boldsymbol{a}$ and $\boldsymbol{b}$. From the algebraic point of view, the main thing about vectors is that they form a set of objects which can be added together, the result being another member of the set, and multiplied by real numbers, the result being another member of the set. The analogy between sets of functions and points in space is based on this property.

Consider, for example, the set S of real functions which are continuous on the interval $[0,1]$. If f and g are two such functions, then $f + g$ is also continuous on

[0,1] (a basic theorem of analysis) and therefore belongs to S; and for any real constant k the function kf (that is, the function whose value at x is $kf(x)$) is continuous and belongs to S. We see that S is a set of objects which can be added together and multiplied by numbers, and thus has the same algebraic properties as the set of vectors in space. Any such set is called a 'vector space', and we shall see that it is useful to think geometrically about the elements of such sets, drawing pictures as if they were points in ordinary space. Summing up this paragraph, we have:

Rough Definition 4.1 A **real vector space** is a set of objects which can be added together and multiplied by real numbers, giving in each case another member of the set. □

A useful extension of this definition is to allow multiplication by complex numbers as well as reals; a set of objects which can be added together and multiplied by complex numbers is called a complex vector space.

The addition and multiplication of ordinary vectors satisfies various algebraic rules, such as $\boldsymbol{a} + \boldsymbol{b} = \boldsymbol{b} + \boldsymbol{a}$ for any vectors $\boldsymbol{a}$ and $\boldsymbol{b}$, $k(\boldsymbol{a} + \boldsymbol{b}) = k\boldsymbol{a} + k\boldsymbol{b}$ for any vectors $\boldsymbol{a}$ and $\boldsymbol{b}$ and any number k, and so forth; indeed, we are hardly entitled to use the words addition and multiplication unless these rules are obeyed. In the rough definition above, we certainly intend that the ordinary rules of algebra shall be satisfied, but in order to make the definition complete we must spell them out in detail. The definition below is therefore rather elaborate; the algebraic rules have been carefully chosen to ensure that if they are satisfied, then the elements of a vector space can be manipulated just as if they were ordinary vectors. But although the rules take up so much space, you should remember that the essence of the definition is contained in 4.1 above. In most of the examples in this book, when we wish to prove that some set is a vector space, there is no difficulty in showing that the algebraic laws are satisfied, indeed it is usually obvious; the difficulty, if any, is in showing that the sum of two members, and the product of a member with a number, are also members of the set.

Definition 4.2 A real (or complex) **vector space** is a set V such that

1. there is a rule which, given any $x,y \in V$, determines an element of V called $x + y$, satisfying
   (a) $x + y = y + x$ for all $x,y \in V$;
   (b) $x + (y + z) = (x + y) + z$ for all $x,y,z \in V$;
   (c) there is an element of V, called 0, such that $0 + x = x$ for all $x \in V$;
   (d) given any $x \in V$ there is an element of V called $-x$ such that $x + (-x) = 0$;
2. there is a rule which, given any $x \in V$ and any real (or complex) number k, determines an element of V called kx, satisfying
   (e) $k(mx) = (km)x$ for any numbers k,m and any $x \in V$;

(f) $1x = x$ for any $x \in V$;
(g) $(k + m)x = kx + mx$ for any $x \in V$ and numbers k,m;
(h) $k(x + y) = kx + ky$ for any $x,y \in V$ and any number k. □

Definition 4.2 is to be understood as defining two types of vector space, a real space in which real numbers multiply the vectors, and a complex space in which complex numbers are used. These are sometimes called 'vector spaces over the reals' and 'vector spaces over the complex numbers' respectively, the idea being that the real or complex numbers form a foundation over which the algebraic structure of vectors is erected.

Example 4.3 The simplest example of a real vector space is the set $\mathbb{R}$ of real numbers: the sum of two real numbers is a real number, as is the product of a real number with a real number, and the rules (a)–(h) are obviously satisfied. Similarly, the set $\mathbb{C}$ of complex numbers is a vector space over the complex numbers. □

The term **scalar** is sometimes used to refer to the numbers, real or complex, used as multipliers in a vector space. Vectors are sometimes distinguished from scalars by the use of bold type or underlining; we shall not do this because it leads to clumsy notation in the case of function spaces, and it should always be clear from the context which symbols stand for vectors and which for scalars – that is, it will be clear provided that the text is read with care.

Example 4.4 The most familiar example of a real vector space is the set $\mathbb{R}^n$ of n-tuples of real numbers $(a_1,a_2,\ldots,a_n)$, with operations defined by

$$(a_1,a_2,\ldots,a_n) + (b_1,\ldots,b_n) = (a_1 + b_1,\ldots,a_n + b_n)$$

and $$k(a_1,\ldots,a_n) = (ka_1,\ldots,ka_n) \quad .$$

This clearly satisfies Definition 4.1, and the algebraic laws (a)–(h) of 4.2 follow immediately from the corresponding results for real numbers; thus $\mathbb{R}^n$ with these operations is a vector space. We stress that whether or not a set is a vector space depends upon how the operations are defined as well as upon the nature of the elements of the set. We would be quite entitled to define multiplication by the (admittedly unnatural) rule $k(a_1,\ldots,a_n) = (k^n a_1,\ldots,k^n a_n)$; but then clause 2(g) of Definition 4.2 would be violated, and the set $\mathbb{R}^n$ with that definition of multiplication would not be a vector space.

Example 4.5 The set $\mathbb{C}^n$ of n-tuples of complex numbers, with the same operations as in Example 4.4, is a complex vector space.

Example 4.6 $C[a,b]$ stands for the set of all continuous real-valued functions on the interval $[a,b]$, with addition and multiplication by real numbers defined in the obvious way – that is, $f + g$ is defined as the function whose value at x is $f(x) + g(x)$, and kf is the function whose value at x is $kf(x)$, k being a real constant. It is a basic theorem of elementary analysis that the sum of two continuous functions is continuous, and a constant multiple of a continuous function is continuous. The rules (a) to (h) follow immediately from the corresponding rules for numbers; hence $C[a,b]$ is a vector space over the reals. There is a corresponding complex vector space consisting of continuous complex-valued functions on a (real) interval. □

The laws (a)–(h) have various elementary consequences. For example, the zero vector defined in (c) is unique, and equals $0x$ (where 0 here denotes the number zero) for any vector $x \in V$; the vector $(-x)$ defined in (d) equals $(-1)x$; and so forth. These results are set out in linear algebra textbooks; they are all obvious and easy to prove, and will be taken for granted here. We shall proceed to develop the geometry of vector spaces. The elementary geometry of ordinary space is based on lines and planes. Their analogues in general vector spaces are subspaces.

Definition 4.7 A **subspace** of a vector space V is a subset of V which is itself a vector space with the operations of addition and multiplication defined as in V. □

The process of verifying that a given subset is a subspace can be simplified by the following result.

Lemma 4.8 (Subspace Criterion) A subset W of a vector space V is a subspace if $ax + by \in W$ for all $x,y \in W$ and all scalars a,b.

Proof Setting $a = b = 1$ shows that the sum of any two members of W belongs to W, and setting $b = 0$ shows that any scalar multiple of a member of W belongs to W. Setting $a = b = 0$ shows that W contains a zero element, and setting $b = 0$, $a = -1$ produces the negative $-x$ of any $x \in W$. All the other conditions of Definition 4.2 follow from the fact that they hold in V. □

Every space V has the following two uninteresting subspaces: V is a subspace of itself, and the set consisting of the zero vector alone is a subspace. When we wish to exclude them we use the following terms.

Definition 4.9 A **proper subspace** of a space V is a subspace other than V. A **non-trivial subspace** is a subspace containing non-zero elements. The **trivial subspace** is the subspace $\{0\}$.

Example 4.10 The subset of $\mathbb{R}^3$ consisting of all vectors of the form $(x,y,0)$ is a subspace, since the criterion 4.8 is obviously satisfied. Geometrically speaking, the xy plane is a subspace of three-dimensional space; it is proper and non-trivial. Similarly, any plane through the origin is a subspace, as is any line through the origin, but planes and lines which do not pass through the origin are not subspaces (because every subspace must contain the zero vector). □

Geometrically it is clear that the intersection of two planes is in general a line, unless the two planes coincide, and the intersection of two lines through the origin is either a single point or a line. Generalising this piece of basic geometry gives the following.

Proposition 4.11 The intersection of any two subspaces of a vector space is a subspace.

Proof If $X = V \cap W$ where V and W are subspaces of a vector space U, then for any $x,y \in X$ and any scalars a,b, we have $ax + by \in V$ because x,y belong to the subspace V, and $ax + by \in W$ because x,y belong to the subspace W. Thus $ax + by \in V \cap W = X$, so X is a subspace by Lemma 4.8. □

Example 4.12 Consider the vector space $C[0,1]$, a particular case of Example 4.6. The set S of all twice differentiable functions u satisfying $u'' + u' + 2u = 0$ on the interval $[0,1]$ is a subspace; this follows immediately from the criterion 4.8. The set T of functions which vanish at 0, the set U of functions which vanish at 1, and the set V of functions whose derivative vanishes at 0, are also subspaces. Applying Lemma 4.11 twice shows that $S \cap T \cap V$ is a subspace, and a fundamental theorem on linear ordinary differential equations (uniqueness of the initial-value problem) says that this subspace is the trivial subspace. The question whether the subspace $S \cap T \cap U$ is trivial arises in the solution of partial differential equations by the method of separation of variables. It is a non-trivial question, and we shall return to it in later chapters, under the heading of eigenvalue problems. □

The next piece of geometry to be taken over to the abstract theory of vector spaces is the idea of the dimension of space. Three-dimensional space is characterised by the fact that one can have three, but no more than three, perpendicular vectors through any point. We have not yet introduced the idea of angle into vector spaces, and shall not do so until Chapter 7. But it is not necessary to use perpendicular lines; the essential point is that one can draw three, but no more than three, 'independent' vectors through a point, where 'independent' means that none of them can be constructed from the other two by addition and multiplication by scalars (geometrically, none of them lies in the plane defined by the other two). In a plane one can draw two, but no more than two, independent vectors through a point, so we say that a plane is two-dimensional. An n-dimensional space is defined to be one in which there are n independent vectors. That is the idea behind the following chain of definitions.

Definition 4.13 A set of vectors $\{x_1, \ldots, x_n\}$ is called **linearly dependent** if there is a set of scalars $c_1, \ldots, c_n$, not all zero, such that

$$c_1x_1 + c_2x_2 + \ldots + c_nx_n = 0 \quad .$$

A **linearly independent** set is one that is not linearly dependent.

Definition 4.14 A **linear combination** of a finite set of vectors $x_1, \ldots, x_n$ is a vector of the form $a_1x_1 + \ldots + a_nx_n$ where the a_i are any scalars. □

Definition 4.13 can now be reworded: a set of vectors is linearly dependent if and only if one of them can be expressed as a linear combination of the others. Note that the zero vector is a linear combination of any given set of vectors (take all $a_i = 0$ in Definition 4.14), and therefore any set of vectors which includes zero is linearly dependent.

Definition 4.15 Let X be any set of vectors in a vector space V. The set of all linear combinations of elements of X is called the **subspace spanned by** X, or **generated** by X. □

To justify this term, we must show that it is indeed a subspace of V. This follows immediately from Lemma 4.8. Notice that X may be a finite or an infinite set, but we only take linear combinations of a finite number of elements of X at a time; infinite series have not yet been defined.

Example 4.16 Let V be the set of all continuous complex-valued functions of a real variable. It is easily seen to be a vector space, by the argument of Example 4.6. Consider the functions x^r (that is, the powers, $x \mapsto x^r$). For any positive integer n, the functions $1 = x^0, x, x^2, \ldots, x^n$ are linearly independent. *Proof:* the zero element of V is the identically zero function, so if the powers were linearly dependent we would have a polynomial $c_0 + c_1x + \ldots + c_n x^n = 0$ for all x, which is impossible unless all the coefficients c_i are zero.

The subspace spanned by $x^0, \ldots, x^n$ is the set of all linear combinations of powers of x up to the n-th, that is, the set of all polynomials of degree less than or equal to n. Notice that the set of all polynomials of degree exactly n is not a subspace of V, indeed, not a vector space at all. *Proof:* x^n is an n-th degree polynomial, but $0x^n$ is not, so not all multiples of n-th degree polynomials are n-th degree polynomials, and clause 2 of Definition 4.2 is violated. □

We can now define the dimension of a space.

Definition 4.17 A set $\{x_1, \ldots, x_n\}$ is called a **finite basis** for a vector space V if it is linearly independent and it spans V (that is, the subspace spanned by the set is V itself).

Definition 4.18 A vector space is said to be ***n*-dimensional** if it has a finite basis consisting of n elements. A vector space with no finite basis is said to be **infinite-dimensional.** □

There are several theorems about bases and dimension which are intuitively obvious and easy to prove. For example, any subspace of an n-dimensional space must have dimension $\leqslant n$; if a space is n-dimensional, then every set of $n + 1$ vectors must be linearly dependent and no set of $n - 1$ vectors can be a basis (this ensures that the dimension of a space is unique). We leave these for the reader to work out or look up in any linear algebra textbook. The only trap for the unwary is the term 'infinite-dimensional'. A finite-dimensional space is, by definition, one with a basis consisting of a finite number of vectors. An infinite-dimensional space is not a space with an infinite basis, it is a space which does not have a finite basis – think of in-finite-dimensional as meaning not-finite-dimensional, rather than having a dimension which is 'infinite'. In Chapter 7 we shall extend the definition of a basis to include infinite bases. We shall then find that an infinite-dimensional space, as defined above, may or may not have an infinite basis,

We now consider some examples of finite and infinite-dimensional spaces.

Example 4.19 In $\mathbb{R}^n$ the n vectors $(1,0,0, \ldots ,0)$, $(0,1,0, \ldots ,0)$, $\ldots$, $(0,0, \ldots ,1)$ are linearly independent (proof?) and form a basis. This means that every vector in $\mathbb{R}^n$ is a linear combination of the given vectors, which is obvious:

$$(a_1, \ldots ,a_n) = a_1(1,0, \ldots ,0) + a_2(0,1,0, \ldots ,0) + \ldots + a_n(0,0, \ldots ,1) \quad .$$

Thus $\mathbb{R}^n$ is n-dimensional, which you might think is so obvious as to be not worth proving. What we have shown, however, is that our abstract definition of dimension agrees with the intuitive meaning of the word. $\mathbb{R}^n$ is the prototype of an n-dimensional space, and in fact every n-dimensional space is, in a certain sense, equivalent to $\mathbb{R}^n$ – see linear algebra books for an explanation of this statement.

Example 4.20 Returning to Example 4.16, we can now see that the space of polynomials of degree less than or equal to n is $(n + 1)$-dimensional. Consider now the set of all polynomials. It is clearly a vector space. It cannot be n-dimensional for any integer n, because the $n + 1$ polynomials $x^0, x, x^2, \ldots, x^n$ are linearly independent, as shown in Example 4.16, and we noted above that no set of $n + 1$ vectors in an n-dimensional space can be linearly independent. Hence the space of all polynomials is infinite-dimensional.

Example 4.21 The space $C[a,b]$ of Example 4.6 is infinite-dimensional. *Proof:* the space of all polynomials is a subspace of $C[a,b]$, and if $C[a,b]$ were finite-dimensional then, by the remarks following 4.18, the space of all polynomials would be finite-dimensional too, which contradicts the result of Example 4.20. □

This completes our account of those parts of linear algebra that are needed as a foundation for functional analysis. We now proceed to introduce the idea of the 'length' of a vector.

4.2 Normed Spaces

When we use approximate methods to solve equations in which the unknown is a member of a vector space, we would like to be able to say when the approximation is close to the exact solution, that is, when the difference between the exact and approximate solutions is small. For a real or complex number z, the appropriate measure of size is the modulus $|z|$; for a vector (x,y,z) in $\mathbb{R}^3$ the appropriate measure is the length of the vector as given by Pythagoras' theorem: $(x^2 + y^2 + z^2)^{\frac{1}{2}}$. Just as we took the essential algebraic idea behind geometrical vectors and expressed it as an abstract definition of a vector space, so we shall take the essential algebraic properties of the length of a vector, and formulate an abstract definition. We do this not for love of abstraction for its own sake, but because stripping away the details and revealing the bones of the algebraic structure makes the theory much clearer and easier to grasp. It also saves effort in the long run, because theorems that have been proved once and for all in the abstract theory can be applied to a wide range of problems in applied mathematics with relatively little effort.

The essential idea behind length is that for every nonzero vector there is a positive number which measures its size; this means that if one vector is twice another, then its length is twice the other's. Another important property of length is expressed by the phrase 'the shortest distance between two points is a straight line'. This implies that any side of a triangle is shorter than the sum of the lengths of the other two sides. In the language of vectors we can say that for any vectors $\boldsymbol{a}$ and $\boldsymbol{b}$, the length of $\boldsymbol{a} + \boldsymbol{b}$ is not greater than the sum of the lengths of $\boldsymbol{a}$ and $\boldsymbol{b}$. Expressing these ideas abstractly gives the following definition, in which the word 'norm' is used for the abstract idea corresponding to length.

Definition 4.22 A **normed space** is a vector space V with a given norm. A **norm** on a vector space V is a rule which, given any $x \in V$, specifies a real number $\|x\|$ (read as 'norm of x'), such that

(a) $\|x\| > 0$ if $x \neq 0$, and $\|0\| = 0$;
(b) $\|ax\| = |a| \cdot \|x\|$ for any $x \in V$ and any scalar a;
(c) $\|x + y\| \leqslant \|x\| + \|y\|$ for any $x,y \in V$ (the triangle inequality).

A normed space is called real or complex according to whether the underlying vector space V is real or complex.

Example 4.23 $\mathbb{R}$ becomes a real normed space if we define the norm to be the modulus, $\|x\| = |x|$; (a) and (b) are obviously satisfied, and $|x+y| \leqslant |x| + |y|$ is trivial if x and y have the same sign and obvious if they have different signs.

Similarly $\mathbb{C}$ becomes a complex normed space if we take $\|z\| = |z|$. □

The next example needs the following preliminary result.

Lemma 4.24 (Cauchy-Schwarz Inequality) For any complex numbers $x_1, \ldots, x_n$, $y_1, \ldots, y_n$,

$$|\Sigma_1^n x_i \bar{y}_i|^2 \leqslant (\Sigma_1^n |x_i|^2)(\Sigma_1^n |y_i|^2)$$

where $\bar{y}$ denotes the complex conjugate of y.

Proof This is one of a small number of places in this book where ingenious and not entirely obvious methods (sometimes referred to as artificial and ad hoc tricks) must be used. We shall not try to explain the origin of these tricks, but refer interested readers to Melzak.

For any number a, the following expression is real and non-negative:

$$\begin{aligned} 0 &\leqslant \Sigma_1^n (x_i - ay_i)(\bar{x}_i - \overline{ay_i}) \\ &= \Sigma |x_i|^2 - a\Sigma y_i \bar{x}_i - \bar{a}\Sigma x_i \bar{y}_i + |a|^2 \Sigma |y_i|^2 \quad . \end{aligned}$$

If $\Sigma |y_i|^2 = 0$ then each y_i must be zero and the lemma is certainly true. If $\Sigma |y_i|^2 \neq 0$ then set

$$a = \Sigma x_i \bar{y}_i / \Sigma |y_i|^2$$

in the above, and the result is the Cauchy-Schwarz inequality. □

Example 4.25 $\mathbb{R}^n$ is a real normed space, and $\mathbb{C}^n$ a complex normed space, if we define

$$\|x\| = (\Sigma_1^n |x_i|^2)^{\frac{1}{2}}$$

for any vector $x = (x_1, \ldots, x_n)$. *Proof:* 4.22(a) and (b) are easily seen to be satisfied, and the triangle inequality is proved by squaring both sides and using Lemma 4.24.

Example 4.26 The vector space $\mathbb{R}^n$ can be turned into a normed space different from that of Example 4.25 by defining

$$\|x\| = \max_i |x_i|$$

for any vector $x = (x_1, \ldots, x_n)$. Again 4.22(a) and (b) are easily verified, and (c) follows from the fact that the maximum value of the sum of two things cannot exceed the sum of their maximum values. □

The last two examples show that the same vector space can be given different norms. This may at first seem a confusing state of affairs – if there is no unique norm for a space, how can one choose the 'right' one? The answer is that the appropriate norm for a space depends on the particular problem concerned. For some purposes it may be that a vector should be considered 'large' if its largest element exceeds some given value, regardless of the sizes of the others, and then the norm of 4.26 above would be the right norm; for other purposes, an overall measure of the sizes of the elements might be a better measure of the magnitude of a vector, and then the norm of 4.25 above would be suitable. One of the characteristics of functional analysis is the ability to adapt the norm to the problem at hand. This will be seen more clearly when we discuss the theory of convergence in the next section. Meanwhile we shall look at some examples of norms on function spaces. We will need the following result.

Lemma 4.27 (Cauchy-Schwarz Inequality for Integrals)

$$\left|\int_a^b f(t)\overline{g(t)}\,\mathrm{d}t\right|^2 \leqslant \left(\int_a^b |f(t)|^2\,\mathrm{d}t\right)\left(\int_a^b |g(t)|^2\,\mathrm{d}t\right)$$

for any functions f and g such that the integrals exist.

Proof: completely analogous to that of 4.24, with sums replaced by integrals. □

Example 4.28 The vector space $C[a,b]$ of continuous functions on the interval $[a,b]$ can be turned into a normed space by defining

$$\|f\| = \left(\int_a^b |f(t)|^2\,\mathrm{d}t\right)^{\frac{1}{2}} \quad . \tag{4.1}$$

This is justified in exactly the same way as in Example 4.25, using 4.27 instead of 4.24 to prove the triangle inequality. As in Example 4.25, there are really two spaces under consideration here, a real space consisting of real continuous functions, and a complex space consisting of complex continuous functions. Strictly speaking the two spaces should be denoted by different symbols, but we shall not do so; the two spaces have essentially the same properties.

One aspect of this example is more subtle than might appear at first sight. In Example 4.25, condition (a) of Definition 4.22 is obviously satisfied. The space $C[a,b]$ with the above norm is in many ways similar to $\mathbb{R}^n$, but to show that it satisfies 4.22(a) is not quite trivial. The zero element of the space is the function which vanishes throughout $[a,b]$ (why?). It is not in general true that nonzero functions f give a nonzero value for $\int |f(t)|^2\,\mathrm{d}t$. Indeed, the function f defined by $f(x) = 0$ for $x \neq 1$, $f(1) = 1$, is not the zero function, yet $\int_0^2 |f(t)|^2\,\mathrm{d}t = 0$. We shall face this difficulty in Example 4.60. But in the case of continuous functions we need not worry. The following lemma shows that $C[a,b]$ with the norm (4.1) satisfies 4.22(a), and thus completes the proof that $C[a,b]$ is a normed space.

Lemma 4.29 If f is a continuous function and $\int_a^b |f(t)|^2 \,dt = 0$, then $f(x) = 0$ for all $x \in [a,b]$.

Proof Suppose that there is an $x_0 \in [a,b]$ such that $|f(x_0)| = M$ with $M > 0$. Since f is continuous, there is an interval around x_0 in which $|f|$ differs from M by less than $M/2$, say the interval (c,d) with $a \leqslant c < d \leqslant b$. Then $|f(x)| > M/2$ for $x \in (c,d)$, hence

$$\int_a^b |f(t)|^2 \,dt \geqslant \int_c^d |f(t)|^2 \,dt > (d-c)M^2/4$$

which contradicts the given vanishing of the integral. Thus there can be no point where $|f(x)|$ is positive. □

Example 4.30 The set of continuous functions on $[a,b]$ can be turned into a normed space different from that of Example 4.28 by defining

$$\|f\| = \sup\{|f(x)| : a \leqslant x \leqslant b\} \tag{4.2}$$

(for the meaning of sup{ }, see Appendix C). This time 4.22(a) really is obviously satisfied; so is 4.22(b), and (c) follows from the obvious fact that the maximum or supremum of the sum of two things cannot be greater than the sum of their maxima, or suprema, considered separately.

Remark 4.31 The spaces of Examples 4.28 and 4.30 are regarded as different normed spaces even though they contain the same elements. The rule defining the norm is an essential part of the definition and identity of a normed space; indeed, some authors define a normed space formally as a pair (V,N) where V is a vector space and N a mapping from V to the non-negative reals satisfying the axioms 4.22(a), (b), (c). This has the advantage of making it quite clear that the same vector space can generate many different normed spaces, but we generally use less formal language. Another formal complication would be to use different symbols for the spaces of Examples 4.28 and 4.30. We shall try to steer a middle course between notation which is unambiguous (but fearsomely complicated) and notation which is simple and streamlined (but inexplicit and confusing to all but the expert). □

We have here another example of the multiplicity of norms, which we discussed after Example 4.26. However, in this case it turns out that one of the two norms defined above for $C[a,b]$ is much more useful and satisfactory than the other; the phrase 'the normed space $C[a,b]$' usually means Example 4.30. The reason for preferring the sup norm to the integral norm for continuous functions is connected with the theory of convergence in normed spaces, which is the subject of the next section.

4.3 Convergence

Our ultimate aim in discussing the theory of normed spaces is to deal with problems of approximation to the solutions of various types of equation. In such problems, one has a procedure for generating a sequence of approximations, which one hopes come closer and closer to the true solution of the problem; one then tries to justify the procedure by showing that the sequence converges to the true solution. We must therefore study the theory of convergence in normed spaces.

The theory is very easy, being essentially identical to the theory of convergence in ordinary analysis, with which we assume the reader to be familiar (a brief summary is given in Appendix B). In ordinary calculus, two numbers are considered to be close together if the modulus of their difference is small; in a normed space, two elements are considered to be close together if the norm of their difference is small. Given a sequence (x_n) of elements of a normed space N, we say x_n converges to an element X of N if the norm of the difference between x_n and X can be made as small as we please by taking n sufficiently large.

Definition 4.32(a) (Convergence) If $X, x_1, x_2, \ldots$ belong to a normed space N, we say $x_n \to X$ as $n \to \infty$ if, given any $\epsilon > 0$, there is a number M such that $\| x_n - X \| < \epsilon$ for all $n > M$. □

The following is an alternative version of this definition.

Definition 4.32(b) (Convergence) If $X, x_1, x_2, \ldots$ belong to a normed space N, we say $x_n \to X$ as $n \to \infty$ if $\| x_n - X \| \to 0$ as $n \to \infty$. □

$\| x_n - X \|$ is a real number, so we know what is meant by saying that it tends to 0; putting the familiar definition of convergence of a sequence of real numbers into 4.32(b) shows that it is equivalent to 4.32(a). The idea behind the definition is clear: $x_n \to X$ if the distance between them tends to 0.

Example 4.33 $\mathbb{R}^k$ is the space of k-component real vectors $x = (x^{(1)}, x^{(2)}, \ldots, x^{(k)})$. We use superscripts to denote the components of a vector to avoid confusion with the subscripts used above to denote members of a sequence of vectors. With the norm defined in Example 4.25, $x_n \to X$ means $\Sigma_{i=1}^k (x_n^{(i)} - X^{(i)})^2 \to 0$. But for each j, $(x_n^{(j)} - X^{(j)})^2 \leqslant \Sigma_i (x_n^{(i)} - X^{(i)})^2 \to 0$; it follows that each component $x_n^{(j)}$ tends to the corresponding component $X^{(j)}$ of X as $n \to \infty$. Thus our definition of convergence in this space means just what you would expect it to mean: a sequence of vectors converges to a vector in $\mathbb{R}^k$ if and only if each component converges to the corresponding component of the limit vector. A special case of this is the normed space $\mathbb{R}$; here the norm is just the modulus, so convergence in the sense of our Definition 4.32 is exactly the same as ordinary convergence. □

All the usual properties of limits hold in normed spaces, and the proofs are essentially the same as in ordinary analysis, with modulus signs replaced by norm signs. Thus the sum of two convergent sequences converges to the sum of their limits; any subsequence of a convergent sequence (that is, any sequence formed by selecting certain members of the original sequence) converges to the same limit; a constant multiple of a convergent sequence converges to the multiple of the limit; and so on. The boring proofs of these results can be found in standard textbooks; we shall take them for granted.

We must also define a convergent series. Just as in ordinary analysis, we define the convergence of a series (that is, an infinite number of things to be added up) in terms of the previous definition of a sequence (that is, an infinite number of things to be considered in turn, with no addition being performed) as follows:

Definition 4.34 Let $X, x_1, x_2, \ldots$ belong to a normed space N. We say the **series** $x_1 + x_2 + \ldots$ **converges** to X, and write $\Sigma_1^\infty x_n = X$, if the sequence (s_n) converges to X, where $s_n = x_1 + \ldots + x_n$. X is called the **sum** of the series $\Sigma_1^\infty x_n$. □

Again, it is easy and boring to prove that the sum of two convergent series converges to the sum of their sums, etc.

Example 4.35 In the space $C[a,b]$, with the norm (4.2), a sequence of functions f_n converges to a function F if $\sup\{|f_n(x) - F(x)| : a \leqslant x \leqslant b\} \to 0$, and this is just the condition for uniform convergence of f_n to F. Thus uniform convergence fits naturally into the framework of normed spaces – or, to put it another way, the norm we are using in $C[a,b]$ is designed to agree with uniform convergence. For this reason, the norm (4.2) is sometimes called the **uniform norm.** □

The reader will be aware that although uniform convergence is useful for many purposes, it is not the only kind of convergence used for sequences of functions. Simple pointwise convergence is also a useful idea, and in Chapter 1 we met a quite different definition of convergence in the space of test functions. Some kinds of convergence arise, like uniform convergence, from a norm in a normed space; others, like convergence in the space of test functions, do not (in fact, a generalisation of the concept of a normed space, to what is called a multinormed space, is needed in order to relate test function convergence to the theory of this chapter; see Zemanian (1968)). In most of the rest of this book, we shall deal with types of convergence associated with a norm by means of Definition 4.32.

The uniform norm of Example 4.35 is one of the two most important examples for our purposes. The other is the following.

Example 4.36 The space of continuous functions can also be given the norm $\|f\| = (\int_a^b |f(x)|^2 \,dx)^{\frac{1}{2}}$, as in Example 4.28. Convergence in this normed space means that the average, or more precisely the root mean square, over the interval

$[a,b]$ of the difference between f_n and F tends to zero as $n \to \infty$. This type of convergence is less stringent, generally speaking, than uniform convergence. It is called 'convergence in the mean', because it is the mean value of $f_n - F$ that tends to zero – the value at particular points need not always do so, as we shall see in the next section. □

The idea of convergence in the mean applies to any $\mathbb{R} \to \mathbb{R}$ or $\mathbb{R} \to \mathbb{C}$ functions, not only those in $C[a,b]$.

Definition 4.37 If $F, f_1, f_2, \ldots$ are $\mathbb{R} \to \mathbb{R}$ or $\mathbb{R} \to \mathbb{C}$ functions, we say $f_n \to F$ **in the mean** on $[a,b]$ as $n \to \infty$ if

$$\int_a^b |f_n(x) - F(x)|^2 \, dx \to 0 \quad .$$ □

We have not specified the class of functions to which 4.37 applies; it applies to any functions sufficiently well-behaved for the integral to exist.

The idea of mean convergence is an important one, particularly in mathematical physics. The partial differential equations which arise there are often solved by series expansions, Fourier series or others. Those expansions often do not converge uniformly, but, as we shall see in later chapters, they generally converge in the mean in the sense of 4.37. The following example illustrates the difference between the various types of convergence.

Example 4.38 For $n = 1, 2, \ldots$, define functions f_n by $f_n(x) = e^{-nx}$. Clearly $f_n \in C[0,1]$. Now, for each $x \in [0,1]$,

$$f_n(x) \to F(x) \quad \text{as} \quad n \to \infty \quad , \tag{4.3}$$

where F is defined by

$$F(x) = 0 \quad \text{for} \quad x \neq 0, \quad F(0) = 1 \quad .$$

Note carefully that (4.3) is a statement about convergence of the sequence of numbers $f_n(x)$, for any $x \in [0,1]$; it is not a statement about the sequence of functions f_n regarded as members of a normed space of functions. It says, in fact, that $f_n \to F$ pointwise. It is also easy to show that $f_n \to F$ in the mean; the integral concerned is just $\int_0^1 e^{-2nx} \, dx$ (the fact that the integrand has a discontinuous jump of 1 at the lower limit makes no difference to the value of the integral), which is easily evaluated and shown to tend to zero.

Now, $F \notin C[0,1]$ because it is discontinuous. Consider therefore the function

$$F_0(x) = 0 \quad \text{for all } x \quad .$$

It is certainly continuous, and belongs to $C[0,1]$. We can therefore ask if $f_n \to F_0$ in the sense of convergence in $C[0,1]$. The answer depends on which normed space $C[0,1]$ we consider: there are two, the spaces of Examples 4.28 and 4.30. If we follow 4.30 and use the uniform norm, then we conclude that $f_n \not\to F_0$

in the sense of the uniform norm; indeed, $|f_n(0) - F_0(0)| = 1$ for all n, so $\sup|f_n(x) - F_0(x)|$ certainly does not tend to zero. On the other hand, if we take the space of 4.28, then convergence in that space is just mean convergence, and $f_n \to F_0$ in the mean by the same argument that we used above to show that $f_n \to F$ in the mean. Limits in the mean are not unique. Because the value of an integral is unaffected by changing the value of the integrand at a number of isolated points, it follows that if $f_n \to F$ in the mean, it also tends to any function which differs from F at a finite number of points only. □

Summing up the above discussion, we see that there are two important types of convergence in function spaces: uniform convergence, which is convergence with respect to the sup norm, and convergence in the mean, which is convergence with respect to the integral norm of Example 4.28. A sequence of continuous functions can converge in the mean to a discontinuous function, although such a sequence cannot converge uniformly to a discontinuous function. On the other hand, any uniformly convergent sequence in $C[a,b]$ can be integrated term by term (Appendix B, Theorem 9), and therefore converges in the mean. Thus convergence in the sup norm implies convergence in the integral norm, but not vice versa. Mean convergence will be discussed further in section 4.5, but first we must pause to develop some more theoretical machinery.

4.4 Open and Closed Sets

This section discusses a technical matter, closely related to the theory of convergence, which is essential to almost everything that follows. It is a generalisation of the definition of open and closed intervals of real numbers.

Definition 4.39 A subset S of a normed space is said to be **open** if for each $x \in S$ there is a $\delta > 0$ such that $y \in S$ whenever $\|x - y\| < \delta$. □

The idea behind this definition is that an open set S is one such that every point in it can be surrounded by a sphere of points lying entirely in S; a sphere here means the set of all points lying within some fixed distance of the first point x. The radius δ of the sphere will depend on the location of the point x: the closer x is to the boundary of S (if a boundary exists), the smaller δ will be.

Example 4.40 In $\mathbb{R}^n$ the sphere $S = \{x \in \mathbb{R}^n : \|x - a\| < 1\}$ (where a is a fixed vector) is an open set, since for any $x \in S$ we can take $\delta = 1 - \|x - a\|$ and 4.39 is then satisfied (this is geometrically obvious, as can be seen from Fig. 6(a), and is easily proved using the triangle inequality). In the case $n = 1$, S is an open interval on the real line, and we see that an open interval in the sense of elementary calculus is an open set in the sense of 4.39; our terminology is thus consistent.

If we consider the set $T = \{x \in \mathbb{R}^n : \|x - a\| \leqslant 1\}$, then if x is such that $\|x - a\| = 1$, that is, if x is on the boundary of the sphere T, then there is no δ satisfying 4.39; any sphere centred on such an x is bound to contain points lying outside T. Again this is geometrically obvious, as can be seen from Fig. 6(b), and is easily proved: if there were such a δ, then the point $y = x + \frac{1}{2}\delta(x - a)$ would be in T, contradicting the fact that $\|y - a\| = (1 + \frac{1}{2}\delta)\|x - a\| = 1 + \frac{1}{2}\delta > 1$. Thus T is not an open set. □

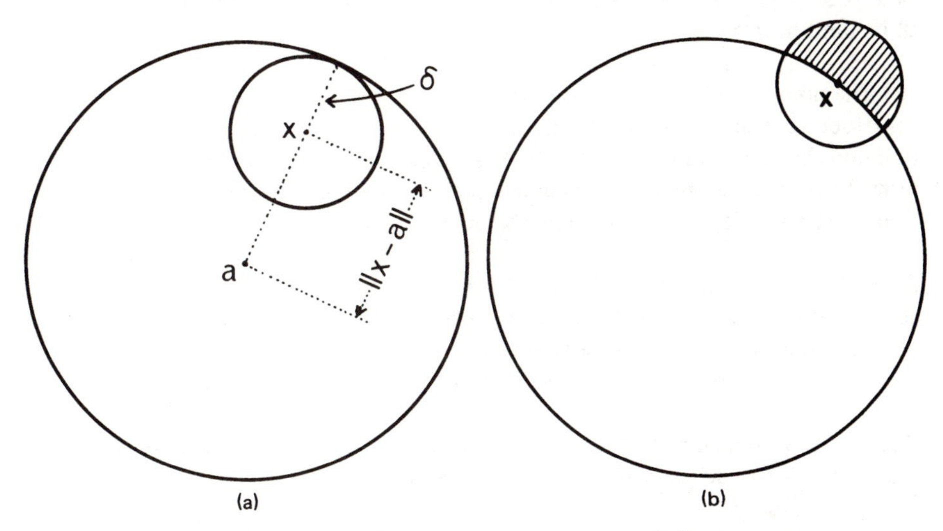

Fig. 6. (a) $\{x : \|x - a\| < 1\}$ is open; (b) $\{x : \|x - a\| \leqslant 1\}$ is not open.

The obvious next step is to define closed sets. We do this by means of the following idea.

Definition 4.41 If S is a subset of a normed space we say x is a **limit point** of S if there is a sequence (y_n) of elements of S such that $y_n \to x$, and for each n, $y_n \neq x$. □

Note that x in this definition is not required to belong to S; a limit point of a set may or may not belong to the set. It can be thought of as a point in whose neighbourhood there are infinitely many elements of S; it is sometimes called a cluster point for this reason. If the condition $y_n \neq x$ were omitted from the definition, this interpretation would not apply, since any member of any set would then be a limit point – the sequence $(x,x, \ldots)$ converges to x. However, this sequence is excluded by Definition 4.41.

Example 4.42 In the space $\mathbb{R}$, consider the half-open interval $I = \{x \in \mathbb{R}: 0 < x \leqslant 1\}$. Every member of I is a limit point of I, since for $x > 0$, the sequence $(x/2, 2x/3, 3x/4, \ldots)$ converges to x. Also, 0 is a limit point of I (*proof:* the sequence $1, 1/2, 1/4, \ldots$ converges to zero), but does not belong to I. Thus I has limit points outside I. However, if we consider the closed interval $J = [0,1]$, the only limit points of J are the members of J. This is the essential property of closed intervals which we shall generalise.

Definition 4.43 A subset S of a normed space is said to be **closed** if it contains all its limit points. □

The point of this idea for our purposes is that we shall often solve equations by constructing sequences of approximations, all of which belong to a set S of functions having certain properties. If S is a closed set, and if the sequence can be proved to converge, then we can deduce that the limit also belongs to the set S, and so possesses the desirable properties characterising S.

Example 4.44 The set T of Example 4.40 is a closed set; it is the n-dimensional generalisation of a closed interval on the real line. More precisely, it is one possible generalisation; for some purposes a rectangular block is a better generalisation than the sphere T. The proof that T is closed follows from the next result.

Theorem 4.45 (Open and Closed Sets) A subset S of a normed space N is closed if and only if its complement $N - S = \{x \in N: x \notin S\}$ is open.

Proof (a) Suppose that S is closed; we shall prove that $N - S$ is open. If it were not open, then there would be an $x \in N - S$ such that no sphere centred on x lies entirely in $N - S$; in other words, each sphere contains an element not in $N - S$; that is, for each $\delta > 0$ there is an element $X(\delta)$, depending on δ, with $\|x - X(\delta)\| < \delta$ but $X(\delta) \notin N - S$, i.e., $X(\delta) \in S$. Taking $\delta = 1, 1/2, 1/3, \ldots$ in succession gives a sequence of elements of S, $X(1/n)$, which converge to x which is not in S, contradicting the hypothesis that S is closed. Thus $N - S$ must be open.

(b) Suppose now that $N - S$ is open, and let (x_n) be any convergent sequence of elements of S, with $x_n \to x$. We shall show that $x \in S$. If not, then x belongs to the open set $N - S$ so that for some $\delta > 0, y \in N - S$ whenever $\|x - y\| < \delta$. But $x_n \to x$, so for any $\delta > 0$ there is an $x_n \in S$ with $\|x_n - x\| < \delta$, contradiction. Hence all convergent sequences in S converge to elements of S, so S is closed. □

We can now complete the discussion of Example 4.44. The complement of T is easily shown to be open by the same method as in 4.40: for any $x \in \mathbb{R}^n - T$, take $\delta = \|x - a\| - 1$, then all points whose distance from x is less than δ belong to $\mathbb{R}^n - T$, so it is open. It follows from 4.45 that T is closed.

There are various properties of open and closed sets which can be found in standard books on analysis and topology (e.g., Kolmogorov & Fomin); for example, the union of a finite number of closed sets is closed, the union of open sets is open, and so on. We shall not need these results. It is important to realise, however, that sets cannot be classified into open and closed sets in the way that, for example, they can be classified as either finite or infinite. It is possible for a set to be both closed and open, and for a set to be neither closed nor open. Indeed, the latter is the most usual case; most sets do not satisfy the rather stringent requirements of 4.39 or 4.43. An easy example of a set which is neither open nor closed is a half-open interval $\{x \in \mathbb{R}: a < x \leqslant b\}$; another example is any set which is the union of a (nonempty) closed set and a (nonempty) open set which does not intersect it. There are also sets which are both closed and open; they are sometimes called clopen. If N is any normed space, then N itself is both open and closed, 4.39 and 4.43 being obviously satisfied. It follows from Theorem 4.44 that the empty set, which is the complement of N, is also both open and closed. In this case 4.39 and 4.43 are vacuously satisfied; that is to say, they are satisfied because any statement about members of an empty set must be admitted to be true because it is impossible to produce a counterexample or contradiction (the reader who does not enjoy this kind of reasoning should take heart: it will not occur again in this book).

We now introduce two more definitions related to the ideas of open and closed sets, and essential for the theory that follows. It is clear from Definition 4.43 that any set can be made closed by adding all its limit points. This operation is called forming the closure of the set.

Definition 4.46 For any set S in a normed space, the **closure** of S is the union of S with the set of all limit points of S. The closure of S is written $\overline{S}$. □

Obviously S is contained in $\overline{S}$, and $S = \overline{S}$ if S is closed. Other elementary properties of S are given below.

Proposition 4.47 (Properties of the Closure)

(a) For any set S, $\overline{S}$ is closed.

(b) If $S \subset T$, then $\overline{S} \subset \overline{T}$.†

(c) $\overline{S}$ is the smallest closed set containing S; that is, if $S \subset T$ and T is closed, then $\overline{S} \subset T$.

Proof (a) Let x be a limit point of $\overline{S}$, with $x_n \to x$ where $x_n \in \overline{S}$ for each n. We must show that $x \in \overline{S}$. For each n, either $x_n \in S$, or x_n is a limit point of S, in which case there is an element of S as close as we please to x_n. In both cases, then, there is a $y_n \in S$ with $\| y_n - x_n \| < 1/n$ (in the case $x_n \in S$, just take $y_n = x_n$). Now,

$$\| x - y_n \| \leqslant \| x - x_n \| + \| x_n - y_n \| < \| x - x_n \| + 1/n \to 0 \quad \text{as} \quad n \to \infty \quad .$$

Hence $y_n \to x$, so x is a limit point of S, and $x \in \overline{S}$ as required.

†The symbol $\subset$ means "is contained in or equal to".

(b) If $x \in \overline{S}$, then $x = \lim(x_n)$ for some sequence (x_n) in S. But $S \subset T$, so (x_n) is a sequence in T, converging to x, hence $x \in \overline{T}$. This proves $\overline{S} \subset \overline{T}$.

(c) This follows immediately from (b), taking $T = \overline{T}$ to be closed. □

When we add points to S to form $\overline{S}$, the new points are all close to points of S; more precisely, for any $x \in \overline{S}$ there is an element of S as close as we please to x. We say S is dense in $\overline{S}$.

Definition 4.48 Let S, T be subsets of a normed space, with $S \subset T$. S is **dense** in T if for each $t \in T$ and each $\epsilon > 0$ there is an $s \in S$ with $\| s - t \| < \epsilon$. □

The idea is that the elements of S are thickly or densely scattered amongst those of T. For example, the set of rational numbers $\{p/q : p, q \text{ integers}, q \neq 0\}$ is dense in $\mathbb{R}$: for every real number, there is a rational as close to it as we please.

The following alternative characterisation of dense subsets is sometimes useful.

Proposition 4.49 S is dense in T if and only if for each $t \in T$ there is a sequence (s_n) of elements of S such that $s_n \to t$.

Proof (a) Suppose that each $t \in T$ is the limit of a sequence (s_n) in S. Then for any $\epsilon > 0$ there is an N such that $\| s_N - t \| < \epsilon$, showing that S is dense in T.

(b) If S is dense in T, then for each $t \in T$ and each n there is an $s_n \in S$ with $\| s_n - t \| < 1/n$. Then $s_n \to t$ as $n \to \infty$ (take $N = \epsilon^{-1}$ in the definition of convergence). □

Corollary 4.50 Every set is dense in its closure. $\overline{S}$ is the largest set in which S is dense.

Proof If $x \in \overline{S}$ then either $x \in S$ or x is a limit point of S. In either case, there is a sequence in S converging to x (if $x \in S$ take the sequence $(x, x, x, \ldots)$). Hence S is dense in $\overline{S}$ by 4.49.

To show that $\overline{S}$ is the largest set in which S is dense, suppose that S is dense in a set T. Then each $t \in T$ is either a member or a limit point of S by 4.49, and therefore belongs to $\overline{S}$. Therefore $T \subset \overline{S}$; $\overline{S}$ contains every set in which S is dense. □

The following examples of dense subsets are useful in the application of the theory to differential equations. The details are rather complicated, however, and 4.51 may be omitted on a first reading of this chapter.

Examples 4.51 In the space $C[a,b]$, consider the following subsets: $C_0[a,b] = \{f \in C[a,b] : f(a) = f(b) = 0\}$; $C^\infty[a,b]$ = the set of smooth functions on $[a,b]$; $C_0^\infty[a,b] = C^\infty[a,b] \cap C_0[a,b]$ = the set of smooth functions which vanish at the endpoints. We shall show that

(i) $C_0[a,b]$ is dense in $C[a,b]$, and $C_0^\infty[a,b]$ is dense in $C^\infty[a,b]$, with the square-integral but not with the sup norm;

(ii) $C^\infty[a,b]$ is dense in $C[a,b]$, and $C_0^\infty[a,b]$ is dense in $C_0[a,b]$, with either the sup or the square-integral norm;

(iii) $C_0^\infty[a,b]$ is dense in $C[a,b]$ with the square-integral but not with the sup norm.

The proof uses two technical devices, one designed to smooth a given function, the other to make it vanish at the endpoints. We define a **mollifier** m_α, as follows. For each positive number α, m_α is a smooth non-negative function, with $m_\alpha(x) = 0$ for $|x| > \alpha$, and $\int_{-\alpha}^{\alpha} m_\alpha(x)\,dx = 1$. There are many such functions; we choose one arbitrarily (for example, take $a = -\alpha$, $b = \alpha$ in 1.4 and multiply by a suitable number to make the integral = 1). Then it is easy to show that for any $f \in C[a,b]$ the function $\int f(y) m_\alpha(x-y)\,dy$ is smooth (for details, see below).

Now, given any number β with $0 < \beta < (b-a)/2$, take a function v_β satisfying

$$\left.\begin{aligned} &v_\beta \in C_0^\infty[a,b],\ 0 \leqslant v_\beta(x) \leqslant 1 \text{ for all } x, \text{ and} \\ &v_\beta(x) = 1 \quad \text{for} \quad a+\beta \leqslant x \leqslant b-\beta \quad . \end{aligned}\right\} \tag{4.4}$$

For example, one could take $v_\beta(x) = \int_a^x m_{\beta/2}(y - a - \beta/2)\,dy$ for $x < a + \beta$, where m is the mollifier defined above, and a similar expression for $x > b - \beta$. Then for any $f \in C[a,b]$, $v_\beta f \in C_0[a,b]$, and for any $f \in C^\infty[a,b]$, $v_\beta f \in C_0^\infty[a,b]$. The function $v_\beta f$ is obtained from f by bringing it smoothly to zero at the endpoints.

We shall now prove the statements (i) (ii) (iii) above. We abbreviate $C[a,b]$ to C, etc., and write $\|f\|_s = \sup|f(x)|$, $\|f\|_2 = [\int |f(x)|^2\,dx]^{\frac{1}{2}}$.

(i) We must show that for any $f \in C$ and any $\epsilon > 0$ there is a $g \in C_0$ with $\|f - g\|_2 < \epsilon$. Since f is continuous, it has a maximum value, M, say. Set $\beta = \epsilon/2M^2$, and take $g(x) = v_\beta(x) f(x)$ where v_β is defined above. Then $g \in C_0$

and
$$\begin{aligned} \|g - f\|_2^2 &= \int_a^{a+\beta} |v_\beta - 1|^2 |f|^2\,dx + \int_{b-\beta}^b |v_\beta - 1|^2 |f|^2\,dx \\ &\leqslant \int_a^{a+\beta} M^2\,dx + \int_{b-\beta}^b M^2\,dx = 2\beta M^2 = \epsilon^2 \quad . \end{aligned}$$

Hence $\|g - f\|_2 < \epsilon$ as required, showing that C_0 is dense in C with the square-integral norm.

To show that it is not dense with the sup norm, consider an $f \in C$ with $f(a) = 1$. Then for any $g \in C_0$, $\|g - f\|_s \geqslant |g(a) - f(a)| = 1$. Thus f is at a distance $\geqslant 1$ from all g in C_0, so C_0 is not dense in C with this norm.

Exactly the same arguments, starting with $f \in C^\infty$ instead of $f \in C$, show that C_0^∞ is dense in C^∞ with $\|\ \|_2$ but not with $\|\ \|_s$.

(ii) Given any $f \in C$ and any $\epsilon > 0$, we must find a $g \in C^\infty$ with $\|g - f\| < \epsilon$. The function f is defined on $[a,b]$. We extend it to all real x by defining $f(x) = f(a)$ for $x < a$ and $f(x) = f(b)$ for $x > b$. Then f is continuous everywhere. Now, for any positive α, define

$$f_\alpha(x) = \int_{-\alpha}^{\alpha} f(x-y) m_\alpha(y)\,dy \quad , \tag{4.5}$$

where m_α is a mollifier. A change of variable gives

$$f_\alpha(x) = \int_{a-\alpha}^{b+\alpha} f(u) m_\alpha(x-u)\,du \quad ; \tag{4.6}$$

note that the integrand in (4.6) is zero over part of the range of integration. Since m_α is smooth, (4.6) shows that f_α is smooth, by standard theorems on differentiating integrals. To show that f_α is close to f, we use (4.5), and the fact that $\int m_\alpha(x)\,dx = 1$, to give

$$f_\alpha(x) - f(x) = \int_{-\alpha}^{\alpha} [f(x-y) - f(x)]\, m_\alpha(y)\,dy \quad .$$

Since f is continuous, we can choose α such that $|f(x-y) - f(x)| < \epsilon$ when $|y| < \alpha$. Then

$$|f_\alpha(x) - f(x)| < \epsilon \int_{-\alpha}^{\alpha} m_\alpha(y)\,dy = \epsilon \quad .$$

Hence $\| f_\alpha - f \|_s \leqslant \epsilon$, and $\| f_\alpha - f \|_2 \leqslant \epsilon\sqrt{b-a}$. This shows that C^∞ is dense in C with either norm. (Careful readers may notice a gap in this argument; the gap is easily filled using 6.37.)

Now we show that C_0^∞ is dense in C_0. For any $f \in C_0$, define f_α by (4.5). Then $f_\alpha \in C^\infty$, and as shown above we can choose α so that $\| f_\alpha - f \|$ is as small as we please, in particular $\| f_\alpha - f \|_2 < \epsilon/2$. With the $\| \,.\, \|_2$ norm, C_0^∞ is dense in C^∞, hence there is a $g \in C_0^\infty$ with $\| g - f_\alpha \|_2 < \epsilon/2$. Then $\| g - f \|_2 \leqslant \| g - f_\alpha \|_2 + \| f_\alpha - f \|_2 < \epsilon$. This proves that C_0^∞ is dense in C_0 with the square-integral norm.

To prove that it is dense with the sup norm, begin as above with (4.5), with α such that $\| f_\alpha - f \|_s < \epsilon/2$. Then $|f_\alpha(a)| < \epsilon/2$, and since f is continuous there is a $\beta_1 > 0$ such that $|f_\alpha(x)| < \epsilon/2$ for $a < x < a + \beta_1$. Similarly there is a β_2 such that $|f_\alpha(x)| < \epsilon/2$ for $b - \beta_2 < x < b$. Take β equal to the smaller of β_1 and β_2, and define $g(x) = \nu_\beta(x) f_\alpha(x)$, where ν_β is defined in (4.4) above. Then $|g(x) - f_\alpha(x)| = 0$ for $a + \beta < x < b - \beta$, and $|g(x) - f_\alpha(x)| < \epsilon/2$ otherwise. Hence $\| g - f_\alpha \|_s < \epsilon/2$. And $\| f_\alpha - f \|_s < \epsilon/2$ by definition of f_α, so $\| g - f \|_s \leqslant \| g - f_\alpha \|_s + \| f_\alpha - f \|_s < \epsilon$, showing that C_0^∞ is dense in C_0 with the sup norm.

(iii) We use the intuitively clear and easily proved fact (Problem 4.15) that if A is dense in B, and B is dense in C, then A is dense in C. From (i) and (ii) above, C_0^∞ is dense in C_0 which is dense in C with the square-integral norm, hence C_0^∞ is dense in C with this norm.

We showed in (i) that C_0^∞ is not dense in C^∞ with the sup norm. The same argument shows that C_0^∞ is not dense in C with the sup norm. □

4.5 Completeness

Example 4.38 showed that in the space $C[0,1]$ with the square-integral norm, convergence has some unexpected properties. In the first place, convergence in the mean is nonunique, in the sense that the same sequence can have two different limits (though only one of them can be continuous). A related fact is that a sequence of elements of the space $C[a,b]$ may converge in the mean to a function which is not in that space. The following example displays a related but more acute difficulty, the resolution of which will lead us to the last important idea in our discussion of normed spaces.

Example 4.52 Consider the sequence of functions $g_n(x) = \tanh(nx)$ for $n = 1, 2, 3, \ldots$. It is easy to show (using Definition 4.37) that g_n converges in the mean to the function sgn (see 1.27) as $n \to \infty$. Mean convergence is, as we have seen, the same as convergence in the square-integral norm in the space of continuous functions. However, $g_n \not\to \text{sgn}$ in the sense of convergence in $C[-1,1]$ because $\text{sgn} \notin C[-1,1]$ – it is discontinuous. In fact, there is no function $G \in C[-1,1]$ such that $g_n \to G$. For a proof of this, see Problem 4.12; it is intuitively reasonable, because $g_n \to \text{sgn}$, and any other limit of g_n may be expected to equal sgn almost everywhere, and therefore also be discontinuous. Thus the sequence g_n is not convergent in the space $C[-1,1]$ with the square-integral norm. We can think of (g_n) as trying to converge, but being foiled by the fact that $\text{sgn} \notin C[-1,1]$. This idea will be clarified below. □

The difference between Examples 4.52 and 4.38 should be noted. The pointwise limit of the sequence f_n of Example 4.38 is discontinuous, but can be made continuous by changing its value at a single point; the resulting function, called F_0 in 4.38, is therefore the limit of f_n in $C[0,1]$, and the sequence is convergent in $C[0,1]$. In Example 4.52, on the other hand, the discontinuous pointwise and mean limit, sgn, cannot be turned into a continuous function by changing its values at a finite number of points; there is no continuous function to which g_n converges.

The trouble is that our definition of convergence in a normed space is in a way too restrictive. It does not allow us to call a sequence convergent unless we can produce a limit element X to insert into Definition 4.32. Yet there are many cases where we can recognise a sequence as convergent even though we cannot write down the limit. The series Σn^{-7} is a simple example; it obviously converges – that is, the sequence of its partial sums converges – but we cannot readily write down its sum to insert into 4.32. It is thus useful to have a criterion for convergence which does not require one to produce a limit for the sequence in advance, in the way that 4.32 does. For sequences of real numbers we have the Cauchy criterion: a sequence (u_n) of real numbers is convergent if and only if for any $\epsilon > 0$ there is a number N such that $|u_n - u_m| < \epsilon$ for all $n,m > N$ (see Appendix B). This statement can be translated into the language of normed spaces by replacing $|\ |$ by $\|\ \|$; but then it is not true in general. Sequences satisfying the condition of Cauchy's criterion are not necessarily convergent, and must therefore be given a different name.

Definition 4.53 A **Cauchy sequence** of elements of a normed space is a sequence (x_n) such that for any $\epsilon > 0$ there is a number N such that $\|x_n - x_m\| < \epsilon$ for all $n,m > N$. □

The criterion for convergence quoted above says that in the space of real numbers, with the usual norm, every Cauchy sequence is convergent and vice versa. It is often easier to prove that a sequence is Cauchy than to prove directly that it converges, because 4.53 involves only the elements of the sequence itself, and does not require one to produce a candidate for the limit.

But apart from its practical value, the idea of a Cauchy sequence enables us to clarify our statement above that the sequence (g_n) of Example 4.52 is 'trying to converge'.

Example 4.54 The sequence (g_n) of Example 4.52 is a Cauchy sequence. *Proof:* suppose $m > n$; standard trigonometric identities give

$$
\begin{aligned}
|\tanh mx - \tanh nx| &= \left| \frac{\sinh(m-n)x}{\cosh mx \cosh nx} \right| \\
&\leqslant \left| \frac{\sinh mx}{\cosh mx \cosh nx} \right| \\
&\leqslant \operatorname{sech} nx \quad .
\end{aligned}
$$

$$\therefore \|g_m - g_n\|^2 \leqslant \int_{-1}^{1} \operatorname{sech}^2 nx \, dx = (2/n)\tanh n \to 0 \quad \text{as} \quad n \to \infty \quad .$$

Hence $\|g_m - g_n\| \to 0$ as $n \to \infty$ with $m > n$; this is equivalent to Definition 4.53. □

This example shows that not all Cauchy sequences are convergent. However, all convergent sequences are Cauchy.

Theorem 4.55 Every convergent sequence in a normed space is a Cauchy sequence.

Proof If (x_n) is a convergent sequence, with $x_n \to X$, then for any $\epsilon > 0$ there is a number M such that

$$\|x_n - X\| < \epsilon/2$$

for all $n > M$. Therefore, by the triangle inequality,

$$
\begin{aligned}
\|x_n - x_r\| &\leqslant \|x_n - X\| + \|X - x_r\| \\
&< \epsilon \quad \text{for all} \quad n,r > M \quad .
\end{aligned}
$$

□

Example 4.54 shows that the converse of Theorem 4.55 is false for the space $C[a,b]$ with the square-integral norm. Normed spaces can be divided into two kinds: those like $C[a,b]$ with the integral norm, in which not all Cauchy sequences have limits in the space, and those, like $\mathbb{R}$, in which all Cauchy sequences are convergent. The first kind is called 'incomplete', the idea being that Cauchy sequences 'really converge' in some sense, but the 'limit' does not belong to the space, just as in

Example 4.52 the sequence g_n converges in the mean to the function sgn $\notin C[-1,1]$. The space has certain essential elements missing. On the other hand, a space which does contain the limits of all Cauchy sequences of its elements is called complete.

Definition 4.56 A normed space is called **complete** if every Cauchy sequence is convergent, and **incomplete** otherwise. A complete normed space is called a **Banach space.** □

Complete spaces are more useful than incomplete spaces, because a typical strategy for solving equations is to construct a sequence of approximations to a solution, and then prove that it is a Cauchy sequence; in a complete space we can then deduce that the sequence converges to a member of the space under consideration. For an example of this procedure, see Theorem 5.15.

Theorem 4.57 The normed spaces $\mathbb{R}^n$ and $\mathbb{C}^n$, with the usual norm as given in Example 4.25, are complete. □

The proof of this theorem follows easily from the statement that all Cauchy sequences of real or complex numbers converge (see Problem 4.20); this statement says that the spaces $\mathbb{R}$ and $\mathbb{C}$ are complete, and its proof requires a careful analysis of what is meant by a real number and a complex number. Completeness is one of the fundamental properties of real and complex numbers; we shall take Theorem 4.57 for granted. A full discussion will be found in books on analysis, such as Whittaker & Watson; many modern books, however, take completeness as an axiom instead of proving it from the definition of real numbers.

The discussion above shows that $C[a,b]$ with the square-integral norm is incomplete. However, with the sup norm it is complete.

Theorem 4.58 The space $C[a,b]$ with norm $\|f\| = \sup\{|f(x)|: a \leqslant x \leqslant b\}$ is complete.

Proof Let (f_n) be a Cauchy sequence in $C[a,b]$. For any $\epsilon > 0$ there is an N such that

$$|f_n(x) - f_r(x)| < \epsilon \tag{4.7}$$

for all $n,r > N$ and all $x \in [a,b]$. Hence for each $x \in [a,b]$ the sequence of numbers $(f_n(x))$ is a Cauchy sequence in $\mathbb{R}$, and therefore converges (using the case $n = 1$ of Theorem 4.57) to a real number, depending on x, which we shall call $F(x)$. We must now show that $F \in C[a,b]$, and that $f_n \to F$ in the sense of convergence in the space $C[a,b]$, that is, uniformly.

In the inequality (4.7) let $r \to \infty$; $f_r \to F$ pointwise, and we deduce that for any $\epsilon > 0$ there is an N such that

$$|f_n(x) - F(x)| \leqslant \epsilon$$

for all $n > N$ and all $x \in [a,b]$. Thus $f_n \to F$ uniformly, and therefore F is continuous (Theorem 7 of Appendix B), and belongs to $C[a,b]$. □

Thus $C[a,b]$ with the sup norm is a Banach space. In the rest of Part II of this book, $C[a,b]$ will be assumed to have the sup norm, unless otherwise specified. In Part III, however, in the context of inner product spaces, the square-integral norm is more natural.

We now reconsider the space of continuous functions with the square-integral norm. Although this is an incomplete space, and therefore unsatisfactory for many purposes, yet the square-integral norm is a useful one, and it would be a pity to abandon it. Fortunately this is not necessary. We can take any incomplete space N and add to it all the 'missing' elements, obtaining a new space Ω which is complete and includes the original space as a subspace. This process is called completing the space N.

Theorem 4.59 (Completion) For any normed space N there is a complete space Ω such that N is a dense subspace of Ω. Ω is called the **completion** of N. □

This theorem is a slightly simplified version of the truth; for a detailed discussion and proof, see Appendix F. For our purposes the details are not important; the main thing is to realise that an incomplete space can always (somehow) be completed. Applying this procedure to $C[a,b]$ gives our last important example of a normed space.

Example 4.60 $L_2[a,b]$ denotes the completion of $C[a,b]$ with the square-integral norm. Thus $L_2[a,b]$ contains all functions which are the limits of continuous functions in the sense of mean convergence. Example 4.52 shows that $L_2[a,b]$ contains discontinuous functions. Problem 4.14 shows that it contains certain functions with infinite discontinuities, provided that the discontinuity is mild enough for the square of the function to be integrable. Thus the function $x^{-\frac{1}{3}}$ belongs to $L_2[a,b]$, but $x^{-\frac{2}{3}}$ does not. In fact every function f such that $\int_a^b |f(x)|^2 \, dx$ exists belongs to $L_2[a,b]$. Such functions are called **square-integrable** over $[a,b]$.

$L_2[a,b]$ also contains functions which are not integrable at all, according to the elementary Riemann definition of integration. Consider, for example, the function F defined by $F(x) = 1$ for rational values of x, $F(x) = 0$ for irrational x. $F \in L_2[a,b]$, but it is so discontinuous that $\int |F|^2 \, dx$ does not exist in the Riemann sense (for further discussion of this function, see Problem 4.19). However, there is a more powerful theory of integration, the Lebesgue theory, in which this function can be integrated, and $L_2[a,b]$ can be shown to contain precisely those functions f for which $\int_a^b |f|^2 \, dx$ exists in the Lebesgue sense. A brief sketch of Lebesgue integration from the point of view of distribution theory is given by Hoskins; for an account of the theory, see Hutson & Pym, Titchmarsh (1939), or Rudin. But from the point of view of this book there is no need to understand Lebesgue integration. $L_2[a,b]$ can be regarded as the Banach space containing ordinary (Riemann) square-integrable functions, together with other functions with severely discontinuous behaviour, such as the function F defined above, which must be included to make the space complete. These functions do not normally appear in the solution of practical problems, and can usually be ignored. (For an alternative point of view, see Chapter 12.)

We note for future use the fact that $C[a,b]$ is a dense subspace of $L_2[a,b]$. This follows from 4.59 since $L_2[a,b]$ is the completion of $C[a,b]$.

We should verify that the set of square-integrable functions is a normed space. It is not even obvious that it is a vector space; given two square-integrable functions with infinite discontinuities, is it obvious that their sum is square-integrable? It is not obvious, but it is true: if f and g are square-integrable, then the inequality $|f+g|^2 \leqslant 2(|f|^2 + |g|^2)$ (easily proved using the fact that $|f-g|^2 \geqslant 0$) shows that $f+g$ is square-integrable. The other properties of a vector space follow immediately. So do the properties of a normed space, except for condition (a) of Definition 4.22. Unfortunately, there are many nonzero functions f such that $\int_a^b |f(x)|^2\,dx = 0$; any function which vanishes except at a finite number of points is an example. So the set of square-integrable functions is not a normed space.

However, there is a standard method of dealing with this problem: one works with equivalence classes of functions, defined as follows (the reader unfamiliar with the concept of equivalence relations should consult Appendix E). Two functions f and g are said to be **equivalent** if $\int_a^b |f-g|^2\,dx = 0$. Roughly speaking, this means that they are equal almost everywhere. It is easy to verify that this is an equivalence relation, and it follows that functions can be divided into equivalence classes, each class consisting of all functions equivalent to a given function. We now redefine $L_2[a,b]$ to be the set of all equivalence classes, with the various operations defined in the obvious way (the sum of two classes is the class of functions equivalent to the sum of representatives of the two classes, and so on). The zero element of the space is now the class of all functions whose square-integral is zero, and this is the only member of the space whose norm is zero. All the normed space axioms are now satisfied.

Regarding $L_2[a,b]$ as a space of equivalence classes is mathematically impeccable, but conceptually rather confusing until one gets used to it. It is simpler to think of $L_2[a,b]$ as a space of functions, and to regard two functions as identical if they are equal almost everywhere, that is, if the integral of the modulus of their difference squared is zero. The justification for this apparently illogical statement is the existence of the more sophisticated (and correct) way of saying the same thing outlined in the last paragraph.

Readers who are familiar with Chapter 1 may have been reminded of distribution theory. In 1.18 we saw that two functions which are equal almost everywhere generate the same regular distribution, just as they correspond to the same element of L_2. There is indeed a close connection between distribution theory and the space L_2, which is explored in Chapter 12.

Example 4.61 The set of square-integrable functions can be made into a Banach space different from that of Example 4.60 by defining

$$\| f \| = (\int_a^b |f(x)|^2 r(x) \, dx)^{\frac{1}{2}} \tag{4.8}$$

where r is a fixed continuous positive function on $[a,b]$. If $\int_a^b |f|^2 \, dx$ exists, then the integral in (4.8) exists. The function r must never be negative, otherwise some functions would have complex norm. r cannot be zero except at isolated points, for if $r(x) = 0$ for $c < x < d$, say, then a function f which is positive in $[c,d]$ and zero elsewhere would have zero norm, which is not allowed because f is not equivalent to the zero function. We need not insist on r being continuous, but we cannot allow it to behave too badly, otherwise the integral in (4.8) may fail to exist. This norm will be useful in our treatment of differential equations in Chapter 9.

Example 4.62 We define $L_2(-\infty,\infty)$ to be the space of functions such that

$$\int_{-\infty}^{\infty} |f(x)|^2 \, dx$$

converges. This condition means that the function cannot be too singular, and also restricts its behaviour at infinity; roughly speaking, functions in this space must die away fairly fast at infinity. Just as for $L_2[a,b]$, it can be shown that it is a Banach space with the norm $\| f \| = (\int_{-\infty}^{\infty} |f(x)|^2 \, dx)^{\frac{1}{2}}$. In an obvious way one can also define spaces $L_2[0,\infty)$ and $L_2(-\infty,0]$.

Example 4.63 Let V be a region in $\mathbb{R}^n$. $L_2[V]$ is the real (or complex) Banach space of $\mathbb{R}^n \to \mathbb{R}$ (or $\mathbb{R}^n \to \mathbb{C}$) functions square-integrable over V, with norm

$$\| f \| = (\int_V |f(x)|^2 \, d^n x)^{\frac{1}{2}} \quad .$$

It is a straightforward extension of the space $L_2[a,b]$ to n dimensions, and is proved to be a normed space in the same way, using the n-dimensional version of the Cauchy-Schwarz inequality.

4.6 Equivalent Norms

The subject of this section will not be used in an essential way until Chapter 9. But the idea of equivalent norms is a generally useful one, and will be applied in some of the problems.

Definition 4.64 Let $\| \cdot \|_1$ and $\| \cdot \|_2$ be two different norms on the same vector space V. We say $\| \cdot \|_1$ is **equivalent** to $\| \cdot \|_2$ if there are positive numbers a and b such that

$$a \| x \|_1 \leqslant \| x \|_2 \leqslant b \| x \|_1$$

for all $x \in V$. □

It is easy to show that if $\|\cdot\|_1$ is equivalent to $\|\cdot\|_2$, then $\|\cdot\|_2$ is equivalent to $\|\cdot\|_1$: 4.64 is then satisfied with a and b replaced by b^{-1} and a^{-1}. In fact, 4.64 defines an equivalence relation in the sense of Appendix E.

Example 4.65 On the vector space $C[0,1]$ the norms $[\int_0^1 |f|^2 \,dx]^{\frac{1}{2}}$ and $[\int_0^1 (1+x)|f|^2 \,dx]^{\frac{1}{2}}$ are equivalent; to show that the first is equivalent to the second, take $a = 1$ and $b = \sqrt{2}$ in 4.64. Similarly the norms $[\int_0^1 |f|^2 \,dx]^{\frac{1}{2}}$ and $[\int_0^1 r(x)|f|^2 \,dx]^{\frac{1}{2}}$ are equivalent for any function r satisfying $0 < A \leqslant r(x) \leqslant B$ for all $x \in [0,1]$, where A and B are numbers independent of x. □

Theorem 4.66 (Convergence with Equivalent Norms) If a sequence converges with respect to one norm, then it converges with respect to any norm equivalent to it.

Proof If (x_n) is a sequence with $\|x_n - X\|_1 \to 0$ as $n \to \infty$, and $\|\cdot\|_2 \leqslant b\|\cdot\|_1$, then $\|x_n - X\|_2 \leqslant b\|x_n - X\|_1 \to 0$ as $n \to \infty$. □

Corollary 4.67 If a space is complete with respect to one norm, then it is complete with respect to any norm equivalent to it.

Proof If (x_n) is a Cauchy sequence with respect to one norm, then it follows at once from the definition that it is a Cauchy sequence with respect to a second equivalent norm. Suppose the space is complete in the first norm; then any Cauchy sequence with respect to the second norm is Cauchy with respect to the first norm, therefore converges with respect to the first norm, therefore converges with respect to the second norm by 4.66. □

These results show that changing from one norm to an equivalent norm does not alter the essential convergence and completeness properties. One norm may be more convenient for a particular problem; in section 9.2, for example, we shall treat the Sturm-Liouville problem using a norm equivalent to the usual L_2 norm, but tailored to fit the particular Sturm-Liouville equation under consideration.

Corollary 4.67 shows that in the space $C[a,b]$ the sup norm and the L_2 norm are not equivalent. In finite-dimensional spaces, however, all norms are equivalent, and therefore every finite-dimensional space, like $\mathbb{R}^n$, is complete.

Theorem 4.68 In a finite-dimensional space, all norms are equivalent. □

The proof of this theorem uses ideas of compactness which are discussed in Chapter 6; see Problem 6.16.

4.7 Summary and References

We began with a brief summary of the theory of vector spaces; Halmos (1958) is a standard textbook, and Kreider *et al.* and many other books can be consulted. In section 4.2 we introduced the idea of the norm or magnitude of a vector, and defined a normed space as a vector space in which every vector is given a norm. In section 4.3 we defined convergence of sequences and series in a normed space, and in section 4.4 we discussed the closely related matter of open and closed sets, and the idea of a dense subset. In section 4.5 we identified a type of sequence, called a Cauchy sequence, whose terms are getting closer together at such a rate that it seems to be trying to converge. We found that in some spaces all Cauchy sequences have limits, and we called those spaces complete; our main examples were $C[a,b]$ and $L_2[a,b]$. Finally, in section 4.6 we defined equivalence of norms, and showed that a space which is complete with respect to one norm is complete with respect to any equivalent norm.

The idea of a complete normed space, or Banach space, is general enough to include all our examples and applications. But from the point of view of pure functional analysis it is quite a special and concrete thing; it is a special example of a more general and abstract 'metric space', which is in turn a special case of the yet more abstract 'topological space'. Many books on functional analysis begin at this more abstract level; on the other hand, some others concentrate on the special type of normed space known as a Hilbert space. Our point of view is broadly similar to that of Hutson & Pym, Luenberger, and Stakgold; see also Kreyszig, Vulikh, and, for a more complete account of the theory, Kantorovich & Akilov. For a beautiful and concise exposition of the theory of metric and topological spaces, see Kolmogorov & Fomin (1957 or 1970). For the history of the various kinds of abstract space, see Kline or Bernkopf.

There is one important branch of the theory of normed spaces which we have not discussed: the theory of the dual space. For the theory and its applications in optimisation and control theory, see Luenberger; for applications in mechanics and engineering, see Milne.

PROBLEMS 4

Section 4.1

4.1 Show that the set of all $m \times n$ matrices with complex elements is a complex vector space, with the obvious definitions of addition and multiplication. What is its dimension? Are the following subsets subspaces?

(a) The set of real $m \times n$ matrices.

(b) The set of matrices with first row $(0,0,\ldots,0)$.

(c) The set of matrices with first row $(1,1,\ldots,1)$.

4.2 $\mathbb{C}$ is a one-dimensional complex vector space. What is its dimension when regarded as a vector space over $\mathbb{R}$?

4.3 Show that $C[a,b]$ is a subspace of the vector space $C[c,d]$ if $[c,d] \subset [a,b]$.

4.4 Let $\{x_1, \ldots, x_n\}$ be a basis for a vector space V. Prove that for any $u \in V$ there is a unique set of scalars $\{u_1, \ldots, u_n\}$ such that $u = \Sigma_1^n u_i x_i$ (these scalars are called the components of u with respect to the basis $\{x_i\}$).

4.5 M and N are subspaces of a vector space V. When is $M \cup N$ a subspace?

Section 4.2

4.6 Show that $| \, \|x\| - \|y\| \, | \leqslant \|x - y\|$ for any x,y in a normed space.

4.7 Consider the space of all complex sequences $x = (x_n)$ with only a finite number of terms nonzero (the number of nonzero terms may be different for different members of the space). Show that it is a normed space if $\|x\|$ is defined as $(\Sigma |x_i|^2)^{\frac{1}{2}}$.

4.8 Prove the following sharper version of the Cauchy-Schwarz inequality, valid for integrals over finite or infinite regions, and for infinite sums: if the integrals or sums on the right hand side of the inequalities in Lemmas 4.27 or 4.24 converge, then the integral or sum on the left converges and satisfies the inequality.

4.9 l_2 denotes the set of all sequences $x = (x_n)$ such that $\Sigma |x_n|^2$ converges. Show, using Problem 4.8, that it is a vector space if $x + y$ is defined as the sequence $(x_n + y_n)$, and a normed space if $\|x\|$ is defined as $(\Sigma |x_n|^2)^{\frac{1}{2}}$.

4.10 Consider the space of $n \times n$ real matrices. For a matrix M, define $\|M\| = \sup\{\|Mx\| : \|x\| \leqslant 1\}$ where $x \in \mathbb{R}^n$ and $\|x\|$ denotes the usual norm in $\mathbb{R}^n$. Show that the space of $n \times n$ matrices becomes a normed space with this definition. Show that $\|Mx\| \leqslant \|M\| \, \|x\|$ for any matrix M and vector x, and that $\|MN\| \leqslant \|M\| \, \|N\|$ for any matrices M,N.

Calculate the norm of a 2×2 matrix explicitly in terms of its coefficients.

Sections 4.3, 4.4

4.11 Show that for any numbers p and q, $\{f \in C[a,b] : p \leqslant f(x) \leqslant q \text{ for } x \in [a,b]\}$ is a closed subset of $C[a,b]$. Similarly for $L_2[a,b]$.

4.12 Show that there is no function $G \in C[-1,1]$ such that $g_n \to G$ in the mean, where g_n are the functions of 4.52.

4.13 Show that every subspace of $\mathbb{R}^n$ is closed.

4.14 Define $f_n(x) = (|x| + 1/n)^{-\frac{1}{3}}$ and $F(x) = |x|^{-\frac{1}{3}}$. Show that $\int_{-1}^{1} [f_n(x) - F(x)]^2 \, dx = 2n^{-\frac{1}{3}} \int_0^n u^{-\frac{2}{3}} [(1 + 1/u)^{-\frac{1}{3}} - 1]^2 \, du$. Deduce that $f_n \to F$ in the mean on $[-1,1]$.

4.15 Show that if A is dense in B, and B is dense in C, then A is dense in C.

4.16 Show that if A is dense in B, then A is dense in the closure of B.

4.17 Let $\|\cdot\|_1$ and $\|\cdot\|_2$ be two norms on a vector space V, and suppose that there is a number a (independent of x) such that $\|x\|_1 \leqslant a\|x\|_2$ for all $x \in V$. Show that if a set A is dense in B with respect to the norm $\|\cdot\|_2$, then it is dense with respect to the norm $\|\cdot\|_1$.

4.18 Prove or disprove: if A is dense in B, then for any set C, $A \cap C$ is dense in $B \cap C$.

Section 4.5

4.19 (i) Construct a sequence of functions f_n such that f_n is piecewise continuous for each n, and $f_n(x) \to F(x)$ for all x, where $F(x) = 1$ for rational x, $F(x) = 0$ for irrational x. This shows that the non-Riemann-integrable function F is the limit of a sequence of integrable functions.

(ii) Consider the set of all rational numbers $p/q \in (0,1)$ with denominator $q \leqslant n$; call them $r_{n1}, r_{n2}, \ldots, r_{nK}$ (where K depends on n). Define a function g_n by $g_n(x) = \sum_{i=1}^{K} \phi_n(x - r_{ni})$, where $\phi_n(u) = 1 - e^{n} \cdot |u|$ for $|u| \leqslant e^{-n}$, $\phi_n(u) = 0$ for $|u| > e^{-n}$. Sketch the graph of g_n. Show that $g_n \in C[0,1]$, $\int |g_n|^2 \, dx \to 0$ as $n \to \infty$, and $g_n(x) \to F(x)$ for rational x, where F is defined in (i) above. (The question whether $g_n(x) \to F(x)$ for irrational x is much more difficult.)

4.20 Assuming that the space $\mathbb{R}$ of real numbers is complete, prove that the spaces $\mathbb{C}$, $\mathbb{R}^n$, and $\mathbb{C}^n$ are complete.

4.21 Is the space of Problem 4.7 complete? Prove it.

4.22 (Harder) Prove that the space of Problem 4.9 is complete.

4.23 Define $C_1^1[a,b]$ to be the space of continuously differentiable functions on $[a,b]$ (cf. Definition 1.2), with norm $\|f\|_1 = (\int_a^b [|f|^2 + |f'|^2] \, dx)^{\frac{1}{2}}$. Show that this is a proper definition of a norm. Is this normed space complete?

4.24 Show that $\|f\| = \int_a^b |f(x)| \, dx$ defines a norm on $C[a,b]$. Is the resulting normed space complete?

4.25 What conditions must the function r satisfy in order that $\|f\| = \sup\{|f(x)r(x)| : a \leqslant x \leqslant b\}$ should define a norm on the vector space $C[a,b]$? Is it then a Banach space?

4.26 A series Σx_n in a normed space is said to converge absolutely if $\Sigma \| x_n \|$ converges. Prove that in a Banach space every absolutely convergent series is convergent in the usual sense. Is this true in every normed space?

4.27 (x_n) is a sequence in a Banach space such that for any $\epsilon > 0$ there is a convergent sequence (y_n) such that $\| y_n - x_n \| < \epsilon$ for all n. In other words, (x_n) can be approximated arbitrarily closely by convergent sequences. Prove that (x_n) is convergent.

Give an example to show that the above statement becomes false if 'Banach space' is replaced by 'normed space'.

Section 4.6

4.28 Show that the space of matrices defined in Problem 4.10 is a Banach space.

4.29 In the vector space of continuous functions on $[a,b]$ which of the following norms are equivalent to each other?

(a) $\sup|f(x)|$;

(b) $\sup|f(x)w(x)|$, where w is a fixed positive continuous function;

(c) $\int_a^b |f(x)|\,dx$; (d) $(\int_a^b |f(x)|^2\,dx)^{\frac{1}{2}}$.

4.30 Let $BC[0,\infty)$ be the set of functions continuous for $x \geqslant 0$ and bounded – that is, for each $f \in BC[0,\infty)$ there is a number M, depending on f, such that $|f(x)| \leqslant M$ for all $x \geqslant 0$. Show that for each $a > 0$, $\|f\|_a = (\int_0^\infty e^{-ax}|f(x)|^2\,dx)^{\frac{1}{2}}$ defines a norm on $BC[0,\infty)$, and $\|\cdot\|_a$ is not equivalent to $\|\cdot\|_b$ if $a > b > 0$. What about the case $a = 0$?

4.31 Theorem 4.68 implies that the norms of Examples 4.25 and 4.26 are equivalent. Show this directly, by finding suitable values to use for a and b in Definition 4.64.

4.32 Show that the norm of Problem 4.23 is not equivalent to the usual L_2 norm.

4.33 In Problem 4.25, the function r need not be strictly positive, but may vanish at isolated points. Is the norm in this case equivalent to the norm $\sup\{|f(x)|\}$?

4.34 Show that the norms of Examples 4.60 and 4.61 are equivalent.

Chapter 5

The Contraction Mapping Theorem

The theory of normed spaces, as described in the last chapter, may not seem very helpful in solving the differential and integral equations of applied mathematics. The link between these equations and normed spaces is the idea of an operator on a space. This is defined in section 5.1, and examples are given. In section 5.2 we consider an iterative method of solving operator equations, and prove a theorem about the existence of solutions of such equations, and how they can be calculated. The rest of the chapter discusses applications to differential, integral, and partial differential equations, and in particular the equations of nonlinear diffusive equilibrium.

5.1 Operators on Vector Spaces

The idea of an operator on a space does not involve the norm, and we therefore define operators on vector spaces in general rather than restricting the definition to normed spaces.

Definition 5.1 Let M and N be vector spaces, and X and Y subsets of M and N respectively. An **operator** or **mapping** or **transformation** $T: X \to Y$ is a rule which, given any $x \in X$, associates with it an element of Y, denoted by Tx or $T(x)$. We write $x \longmapsto Tx$, read as 'x is mapped into Tx'. □

The idea is very simple, but the associated terminology is quite elaborate:

Definition 5.2 The sets X and Y in Definition 5.1 are called the **domain** and **codomain** of T respectively. We say 'T maps X into Y', or 'T maps X into N'. The vector Tx is called the **image** of x under T. The **range** of T is the set $\{Tx : x \in X\}$ of images of all elements of the domain. If A is any subset of the domain, the **image** of A is the set $\{Tx : x \in A\}$. Thus the range of T is the image of its domain. See Figure 7. □

Example 5.3 Take $X = M = \mathbb{R}^3$, $N = \mathbb{R}$, $Y = \mathbb{R}^+$ = the set of positive numbers. Then a map $T: X \to Y$ is simply a positive-valued function of three real variables. The function $T: (x,y,z) \longmapsto x^2 + y^2 + z^2 + 1$ is an example; here the codomain of

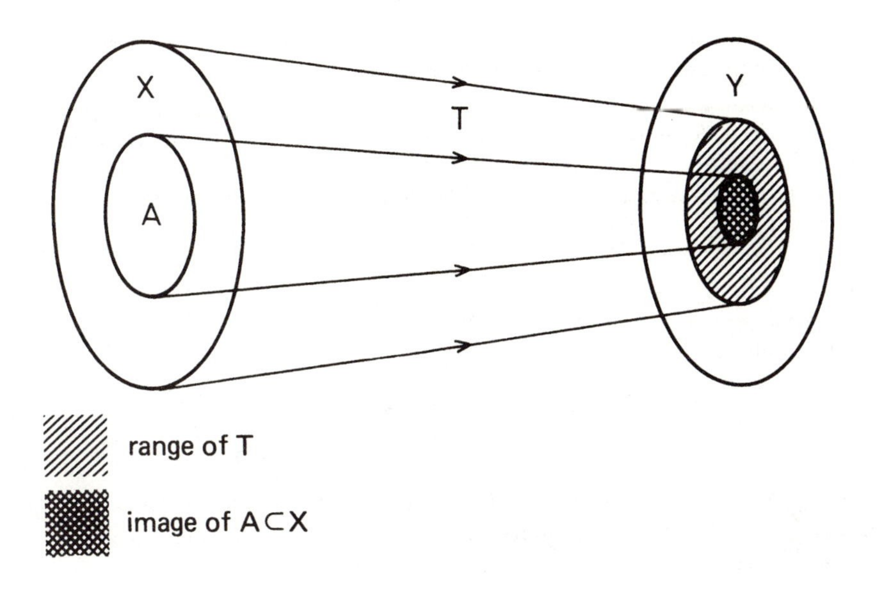

Fig. 7. Range and image of an operator T.

T is $\mathbb{R}^+$, but the range of T is $\mathbb{R}_1 = \{x : x \geqslant 1\}$. Note that once the domain of T is specified, the range is uniquely defined, but what one chooses to call the codomain is to some extent arbitrary: in this example one could regard T as an operator $\mathbb{R}^3 \to \mathbb{R}^+$ or $\mathbb{R}^3 \to \mathbb{R}$ or $\mathbb{R}^3 \to \mathbb{R}_1$, it makes no essential difference.

The image of the vector (0,0,0) is the number 1; the image of the set of vectors of length less than or equal to 1 (the unit sphere) is the interval [1,2]. □

A mapping is simply a generalisation to vector spaces of the familiar idea of a function. In this chapter we shall need only the case where N and M are the same space; mappings from one space to another will appear later.

Example 5.4 Take $M = N = X = Y = C[a,b]$, where $C[a,b]$ denotes either the real or the complex space of continuous functions. For any given continuous function K: $\mathbb{R}^2 \to \mathbb{R}$ or $\mathbb{R}^2 \to \mathbb{C}$, we can define a mapping T: $C[a,b] \to C[a,b]$ by

$$(Tf)(x) = \int_a^b K(x,y)f(y)\,\mathrm{d}y \quad \text{for all} \quad f \in C[a,b] \quad . \tag{5.1}$$

Note the difference in the meaning of the two pairs of parentheses on the left hand side of (5.1): it defines a function (Tf) by the statement that its value at x, $(Tf)(x)$, is the integral on the right of (5.1). It is fairly obvious that Tf is a continuous function if f and K are. The proof, however, is not as easy as one might think (cf. Problem 5.3). It needs the idea of uniform continuity, which is discussed in Chapter 6; see Problem 6.15.

T is called a **linear integral operator** on $C[a,b]$, and K is called its **kernel** (the idea of a linear operator in general will be discussed in Chapter 8). □

Example 5.5 We can define an integral operator on $L_2[a,b]$ by the same formula (5.1) as for $C[a,b]$. If K is continuous and $f \in L_2[a,b]$, then Tf as defined by (5.1) is in $L_2[a,b]$. But we can say more. If K is not continuous but still satisfies

$$\int_a^b \int_a^b |K(x,y)|^2 \,dx\,dy \quad \text{is finite,} \tag{5.2}$$

then the operator defined by (5.1) maps $L_2[a,b]$ into itself. This follows from the Schwarz inequality, Lemma 4.27, as follows.

If (5.2) is satisfied and $f \in L_2[a,b]$, then we have

$$\int_a^b \int_a^b |K(x,y)|^2 \,\|f\|^2 \,dx\,dy = \int_a^b \{\int_a^b |K(x,y)|^2 \,dy \int_a^b |f(y)|^2 \,dy\} \,dx$$
$$\geqslant \int_a^b |\int_a^b K(x,y)f(y)\,dy\,|^2 \,dx$$

by 4.27. Thus the function $\int K(x,y)f(y)\,dy = (Tf)(x)$ is square-integrable, and T maps $L_2[a,b]$ into itself. We note for future reference that the above inequality may be written

$$\|Tf\|^2 \leqslant \int_a^b \int_a^b |K(x,y)|^2 \,dx\,dy \,\|f\|^2 \quad . \tag{5.3}$$

There is a subtle point that should be mentioned here. The space $L_2[a,b]$ was defined as the completion of $C[a,b]$, and it contains ill-behaved functions which are not integrable from the elementary Riemann point of view. It is not obvious that the integral in (5.1) exists for all $f \in L_2[a,b]$. In section 6.1 we shall prove that there is no difficulty here, and the integral operator can indeed be applied to all $f \in L_2[a,b]$. □

Example 5.6 If $K(x,y,z)$ is continuous for all z and for $a \leqslant x,y \leqslant b$, then we can define an operator $T: C[a,b] \to C[a,b]$ by

$$(Tf)(x) = \int_a^b K(x,y,f(y))\,dy \quad . \tag{5.4}$$

This is a nonlinear analogue of the linear operator in Example 5.4.

If K is continuous only for $c \leqslant z \leqslant d$ and $a \leqslant x,y \leqslant b$, then (5.4) defines a mapping from X to $C[a,b]$, where $X = \{f \in C[a,b] : c \leqslant f(x) \leqslant d \text{ for all } x \in [a,b]\}$.

Example 5.7 Let $C^1[a,b]$ be the subset of $C[a,b]$ consisting of all functions with a continuous derivative on $[a,b]$. Then the differentiation operator D maps $C^1[a,b]$ into $C[a,b]$. We can construct various related operators, such as $a\mathrm{D} + b$: $f \mapsto af' + b$, where a and b are constants. □

As Example 5.7 suggests, we can combine operators in various ways. If A and B are two operators with the same domain and codomain, then for any scalars a and b the operator $aA + bB$ is defined in the obvious way as the operator $x \mapsto a(Ax) +$

$b(Bx)$ for all x in the domain of A and B. Another way of combining two operators is to perform them in succession.

Definition 5.8 For operators $A: X \to Y$ and $B: Y \to Z$, where X, Y, Z are subsets of vector spaces, we define the **product** operator $BA: X \to Z$ by $(BA)x = B(Ax)$ for all x in X. □

Notice that with A, B as defined here, AB cannot in general be defined, because for any u in the domain of B, $Bu \in Z$, and A does not generally operate on elements of Z. On the other hand, with operators such as those of Examples 5.4–5.6 which map a space into itself, the product of two operators in either order can be defined.

Example 5.9 Let A, B be linear integral operators on $C[a,b]$ with kernels K and L respectively. Then for any $u \in C[a,b]$ we have

$$(ABu)(x) = \int_a^b K(x,y)\{\int_a^b L(y,z)u(z)\,dz\}\,dy$$
$$= \int_a^b K(x,z)\{\int_a^b L(z,y)u(y)\,dy\}\,dz$$

on relabelling the dummy variables of integration,

$$= \int_a^b \{\int_a^b K(x,z)L(z,y)\,dz\}\,u(y)\,dy$$

on reversing the order of integration. Thus

$$(ABu)(x) = \int_a^b M(x,y)u(y)\,dy$$

where $\quad M(x,y) = \int_a^b K(x,z)L(z,y)\,dz \quad .$

We see that AB is an integral operator with kernel M obtained from K and L by a procedure reminiscent of matrix multiplication. BA is a different integral operator, with kernel $N(x,y) = \int L(x,z)K(z,y)\,dz$. □

For operators which map a space into itself we can define powers by the natural rule $A^2 : x \mapsto A(Ax)$, $A^3 : x \mapsto A(A^2)x$, and so on. Before using the power notation we should verify that operator multiplication is associative, that is, that $A(BC) = (AB)C$ for any operators A,B,C for which the products are well-defined. But that is obvious: both $A(BC)$ and $(AB)C$ correspond to performing the three mappings C,B,A in succession.

Example 5.10 Let $C^n[a,b]$ be the subset of $C[a,b]$ consisting of all n times continuously differentiable functions on $[a,b]$ (cf. Definition 1.2). It is easy to see that $C^n[a,b]$ is a subspace of $C[a,b]$, and in fact of $C^r[a,b]$ for any $r \leqslant n$. For any n the operator D defined in Example 5.7 maps $C^n[a,b]$ into $C^{n-1}[a,b]$. D^2 is the operator which differentiates a function twice, and it maps $C^n[a,b]$ into $C^{n-2}[a,b]$; higher powers of D correspond to higher order derivatives. Now consider the general linear differential equation with constant coefficients, on the interval $[a,b]$.

It can be written in the form $Lu = f$ where $L = a_0 D^n + a_1 D^{n-1} + \ldots + a_n$, u is an unknown element of $C^n[a,b]$, and f a given element of $C[a,b]$. □

From the last example it can be seen that the equations of applied mathematics can often be written as operator equations on a normed space. Such equations often appear in the form

$$x = Tx \tag{5.5}$$

where x is an unknown element of a normed space and T is an operator on that space. The reader familiar with eigenvalue problems will recognise (5.5) as being an eigenvalue equation for $\lambda = 1$, and therefore of great practical importance; and many equations which are not in the form (5.5) can be put into that form by some manipulation. Solutions of (5.5) are called **fixed points** of the transformation T; they are the elements of the space which are unchanged by the action of T. There is a large body of mathematics dealing with the existence of fixed points, and how to find them. The simplest theorem is given in the next section; deeper methods are discussed in Chapter 6.

5.2 The Contraction Mapping Theorem

The form of (5.5) suggests the following method for constructing a sequence of approximations to its solution. An exact solution is an element x of the space which is transformed into itself by T. Suppose we have somehow found a rough first approximation x_0 to the solution. If x_0 were the exact solution, then Tx_0 would equal x_0. If x_0 is reasonably close to the exact solution, then although Tx_0 will not equal either x_0 or the exact solution x, we might perhaps expect Tx_0 to be closer to x than x_0 is, so that if we set $x_1 = Tx_0$, then x_1 is a better approximation than x_0. We now repeat this operation, defining a sequence (x_n) by $x_{n+1} = Tx_n$ for $n = 0, 1, 2, \ldots$, and ask whether the sequence converges to the exact solution.

Example 5.11 Consider the space of real numbers, and let T be the operator which maps any number into its cosine. Then (5.5) is the equation $x = \cos x$. Sketching the graphs of the functions x and $\cos x$ shows that they intersect at a value of x between 0 and $\pi/2$, so there is a solution of the equation in that region. If any number in $[0, \pi/2]$ is chosen as a crude first approximation x_0, and the sequence $x_1 = \cos x_0, x_2 = \cos x_1, \ldots$ is constructed, then a minute or two spent with a calculator, or a table of cosines, will provide convincing evidence that the sequence converges to a solution of the equation. This procedure for solving (5.5) is called **iteration**, and the sequence (x_n) is called an **iteration sequence.** □

The scheme outlined above for solving (5.5) is thus a sensible one. However, the following example shows that it can break down.

Example 5.12 The equation $x = \tan x$ looks very similar to the last example, and sketching the graphs of x and $\tan x$ shows that it has a solution in the interval $[\pi, 3\pi/2]$. But applying the same method gives a sequence which diverges, and is quite useless for finding solutions of the equation. □

The question of when the method works is thus a non-trivial one. We must look for a condition ensuring that the iteration sequence (x_n) constructed by the rule $x_{n+1} = Tx_n$ is convergent. The Cauchy condition (Definition 4.53) might give such a condition, provided that our space is complete. Roughly speaking, a Cauchy sequence is one in which the distance between terms (that is, the norm of their difference) tends to zero as one goes farther out along the sequence. What condition on the operator T will ensure that this happens?

Successive terms in the sequence (x_n) are obtained by applying T to the preceding term. Therefore the distance between two successive terms will be less than the distance between the preceding two terms if the operator T is what is called a 'contraction', that is, an operator which when applied to two elements of the space transforms them into new elements which are closer together than the original pair. The formal definition of a contraction is slightly stronger than this:

Definition 5.13 A mapping $T: X \to X$, where X is a subset of a normed space N, is called a **contraction mapping**, or simply a **contraction**, if there is a positive number $a < 1$ such that

$$\| Tx - Ty \| \leqslant a \| x - y \| \quad \text{for all} \quad x, y \in X \quad . \qquad \square \quad (5.6)$$

The difference between this definition and the statement in the preceding paragraph is that to qualify as a contraction according to Definition 5.13, an operator must shrink the distance between all pairs of elements of X by factors which are not merely all less than 1 but all less than some $a < 1$, that is, all less than 1 by some definite margin, the same for all pairs of elements of X. The following example may help to clarify this point.

Example 5.14 Consider the function $f(x) = x + e^{-x}$, regarded as a mapping of the set $\mathbb{R}^+$ of positive numbers into itself. For any $x, y \in \mathbb{R}^+$, we have $\| f(x) - f(y) \| = |f(x) - f(y)|$ (cf. 4.23) $= |x - y| \cdot |f'(z)| = \| x - y \| \cdot |f'(z)|$ for some number z between x and y, by the Mean Value Theorem. Now, $|f'(z)| < 1$ for all $z > 0$; but f is not a contraction according to Definition 5.13, because the factor by which f shrinks the distance between pairs of points can be made as close to 1 as we please by taking the points far enough out on the x-axis. To prove more formally that the condition of Definition 5.13 is not satisfied, suppose that there exists an $a < 1$ satisfying (5.6); then the argument above shows that $\| f(x) - f(y) \| = \| x - y \| \cdot |f'(z)| = \| x - y \| \cdot (1 - e^{-z}) > a \, \| x - y \|$ if both x and y (and therefore z) are $> -\log(1 - a)$; this contradicts (5.6). Hence no such number a can exist, and f is not a contraction mapping on $\mathbb{R}^+$. □

Definition 5.13 is the 'right' definition of a contraction mapping (as opposed to the rough definition preceding it) for the following reason. We are interested in the convergence of the iteration sequence $x_0, Tx_0, T^2x_0, \ldots$, which depends on the distance between pairs of elements tending to zero as one goes out along the sequence. Going out along the sequence corresponds to multiplying by higher and higher powers of T; each time a pair of elements is multiplied by T, the distance between them is decreased; if T is a contraction then the distance between them is decreased by a factor of at least a^n after multiplication by T^n, and therefore tends to zero, whereas if (5.6) is not satisfied for some $a < 1$, there is no guarantee that multiplying two elements by T^n and letting $n \to \infty$ will make their distance tend to zero, and therefore no guarantee that the iteration sequence is a Cauchy sequence.

To recapitulate: if T is a contraction mapping on a subset X of a normed space, then the argument above suggests that the iteration sequence is a Cauchy sequence. If the space is complete, the Cauchy sequence must converge, but its limit need not in general lie in the domain X of T, and therefore may not be a candidate for a solution of the equation $Tx = x$. However, if X is a closed set, then the limit of the sequence must lie in X, and may then be expected to satisfy the equation. We have now (at last) assembled all the conditions necessary for the success of our method for solving (5.5), and can state the main result of this chapter:

Theorem 5.15 (Contraction Mapping Theorem) If $T: X \to X$ is a contraction mapping of a closed subset X of a Banach space, then there is exactly one $x \in X$ such that $Tx = x$. For any $x_0 \in X$, the sequence (x_n) defined by $x_{n+1} = Tx_n$ converges to x.

Proof For any $x_0 \in X$, set $x_n = T^n x_0$. Let a be as in Definition 5.13, then

$$\begin{aligned} \|x_{n+1} - x_n\| &\leqslant a\|x_n - x_{n-1}\| \leqslant a^2\|x_{n-1} - x_{n-2}\| \leqslant \ldots \\ &\leqslant a^n\|x_1 - x_0\| \quad . \end{aligned}$$

Hence for any $m > n$ the triangle inequality gives

$$\begin{aligned} \|x_m - x_n\| &\leqslant \|x_m - x_{m-1}\| + \|x_{m-1} - x_{m-2}\| + \ldots + \|x_{n+1} - x_n\| \\ &\leqslant (a^{m-1} + a^{m-2} + \ldots + a^n)\|x_1 - x_0\| \\ &\leqslant (a^n/[1-a])\|x_1 - x_0\| \\ &\to 0 \quad \text{as} \quad n \to \infty \quad . \end{aligned}$$

Hence for any $\epsilon > 0$ there is an N such that $\|x_m - x_n\| < \epsilon$ whenever $m > n > N$, which means that (x_n) is a Cauchy sequence. Since the space is complete, (x_n) is convergent; call its limit $\bar{x}$. Since X is closed, $\bar{x} \in X$. We must now show that $T\bar{x} = \bar{x}$.

We have for any n,

$$\begin{aligned} \|T\bar{x} - \bar{x}\| &\leqslant \|T\bar{x} - Tx_n\| + \|Tx_n - \bar{x}\| \\ &\leqslant a\|\bar{x} - x_n\| + \|\bar{x} - x_{n+1}\| \quad . \end{aligned} \tag{5.7}$$

But (5.7) tends to zero as $n \to \infty$; thus $\| T\bar{x} - \bar{x} \|$ is less than every member of a sequence which tends to zero, and must therefore be zero. Hence $T\bar{x} - \bar{x} = 0$, as required.

Finally, to prove uniqueness suppose that $T\bar{x} = \bar{x}$ and $T\bar{y} = \bar{y}$. Then $\| \bar{x} - \bar{y} \| = \| T\bar{x} - T\bar{y} \| \leqslant a \| \bar{x} - \bar{y} \|$, contradiction unless $\| \bar{x} - \bar{y} \| = 0$, that is, $\bar{x} = \bar{y}$. □

This theorem gives simultaneously a pure existence and uniqueness proof and a computational technique (iteration) for finding the solution. It is not hard to extract from the above proof an estimate of the rate of convergence of the iteration; see Problem 5.6. The theorems of the next chapter are in a way less powerful: they give only existence proofs, no uniqueness results or computational techniques. On the other hand, they apply to a much wider class of operators than contractions.

5.3 Application to Differential and Integral Equations

Consider a first-order ordinary differential equation of the general form

$$F(x,y,y') = 0 \quad , \tag{5.8}$$

where F is a given function of three variables, y an unknown function of x, and $y' = \mathrm{d}y/\mathrm{d}x$. It may be possible to solve this equation algebraically for y in terms of x and y', and thus express it in the form $y = Ty$, where T is an operator involving differentiation. If T is a contraction, in some suitable function space, then we can use Theorem 5.15 to obtain solutions of (5.8). Unfortunately, this is not usually the case. The operator of differentiation D is certainly not a contraction, and the operator T obtained by solving (5.8) for y is not likely to be a contraction. To see that D is not a contraction, consider the functions $f(x) = 1$ for all x, and $g(x) = 1 + (1/n)\sin(n^2 x)$. No matter what interval is chosen to construct the function space, and no matter whether the sup norm or the L_2 norm is used, $\| \mathrm{D}f - \mathrm{D}g \|$ can be made as large as we please, while $\| f - g \|$ becomes as small as we please, by choosing n sufficiently large. Thus D is not only not a contraction, it can in fact expand by an indefinitely large factor. This irregular behaviour of D will be considered further in later chapters; for our present purposes it means that the obvious way of applying the contraction mapping theorem to (5.8) does not work.

However, there is another way of putting (5.8) into the form $y = Ty$. The equation (5.8) can be solved for y' as a function of x and y (assuming that F is well-behaved) giving

$$y' = f(x,y)$$

for some function f. This equation can be integrated with respect to x, and if we are given an initial condition of the usual type associated with (5.8), say $y(a) = c$ where a and c are given numbers, integration of the above equation gives

$$y(x) = c + \int_a^x f(t,y(t))\,\mathrm{d}t \quad .$$

This is an integral equation, that is, an equation in which an unknown function appears inside an integral. The operator on the right-hand side can be shown to be a contraction under certain conditions, and thus existence and uniqueness results for (5.8) can be obtained from the contraction mapping theorem. Exactly the same method applies to higher-order differential equations, which can be expressed in the form (5.8) with y a vector. We shall not discuss the details because the full statement and proof of the result, known as Picard's theorem, is rather messy, and can be found in almost every book which deals with the contraction mapping theorem as well as many books on differential equations. We shall just note that it is often useful to turn a differential equation into an integral equation, and then turn our attention to integral equations.

We shall consider an integral equation of the following form:

$$u(x) - \lambda \int_a^b K(x,y,u(y))\,\mathrm{d}y = f(x) \tag{5.9}$$

for an unknown function u; f is a given function, K a given function of three variables, and λ a given constant. The factor λ could of course be absorbed into K and thus removed from the equation, but it is more convenient, as we shall see below, to use the form (5.9).

We shall find that the contraction mapping theorem can be applied to (5.9) provided that K satisfies certain conditions. In particular, it must be a reasonably well-behaved function of its third argument; it must satisfy a condition of the following type.

Definition 5.16 A function $F(x)$ is said to satisfy a **Lipschitz condition** if there is a constant k such that

$$|F(x_1) - F(x_2)| \leqslant k\,|x_1 - x_2| \quad \text{for all} \quad x_1, x_2 \quad . \qquad \square \tag{5.10}$$

If F is differentiable and $|F'(x)| \leqslant k$ for all x, then F satisfies a Lipschitz condition (this follows easily from the Mean Value Theorem). The converse is not true: a function satisfying a Lipschitz condition need not be differentiable. The condition can be thought of as a slight generalisation of differentiability.

We can now state conditions under which (5.9) has a unique solution. There are many such sets of conditions; the following theorem is an example of a whole class of theorems about integral equations. When applying functional-analytic methods to such problems, one must decide what function space to work with. We shall work with $L_2[a,b]$, that is, we shall look for solutions in the class of square-integrable functions, and we shall impose conditions which will make the operator a contraction with respect to the L_2 norm. Different conditions are needed to make it a contraction with respect to the sup norm in $C[a,b]$ – see Problem 5.14.

Theorem 5.17 (Existence and Uniqueness for Integral Equations) The integral equation (5.9) has a unique solution $u \in L_2[a,b]$, provided that

(i) $f \in L_2[a,b]$;

(ii) K satisfies a Lipschitz condition with respect to its third argument,

$$|K(x,y,z_1) - K(x,y,z_2)| \leqslant N(x,y)|z_1 - z_2| \quad \text{for all} \quad z_1, z_2 \tag{5.11}$$

where N is square-integrable, with

$$\int_a^b \int_a^b |N(x,y)|^2 \, dx \, dy = P^2, \quad \text{say;} \tag{5.12}$$

(iii) $K(x,y,0)$ is continuous for $x,y \in [a,b]$;

(iv) $|\lambda| < 1/P$.

Proof Define an operator T by

$$(Tu)(x) = f(x) + \lambda \int_a^b K(x,y,u(y)) \, dy \quad .$$

We must first show that T maps $L_2[a,b]$ into itself. Since $L_2[a,b]$ is a vector space, and $f \in L_2[a,b]$, Tu will belong to $L_2[a,b]$ if $\int_a^b K(x,y,u(y)) \, dy$ does. Now, using the Lipschitz condition,

$$|K(x,y,u) - K(x,y,0)| \leqslant N(x,y)|u| \quad ,$$

$$\therefore |K(x,y,u)| \leqslant |K(x,y,0)| + N(x,y)|u| \quad ,$$

$$\therefore \left|\int_a^b K(x,y,u(y)) \, dy\right| \leqslant \int_a^b |K(x,y,0)| \, dy + \int_a^b N(x,y)|u(y)| \, dy \quad .$$

Each term on the right is square-integrable, the first because it is continuous, the second because it is the result of applying the integral operator with kernel N to the function $|u| \in L_2[a,b]$, and N satisfies the condition (5.2) which we found, in Example 5.5, to be sufficient for the operator to map $L_2[a,b]$ to itself. Hence $\int_a^b K(x,y,u(y)) \, dy$ is square-integrable, and we have shown that T maps $L_2[a,b]$ into itself.

We must now show that T is a contraction. We have, for any u_1 and u_2 in $L_2[a,b]$ and any $x \in [a,b]$,

$$\begin{aligned} |Tu_1(x) - Tu_2(x)| &= \left|\lambda \int_a^b [K(x,y,u_1(y)) - K(x,y,u_2(y))] \, dy\right| \\ &\leqslant |\lambda| \int_a^b N(x,y)|u_1(y) - u_2(y)| \, dy \\ &\leqslant |\lambda| \left\{\int_a^b |N(x,y)|^2 \, dy \int_a^b |u_1(y) - u_2(y)|^2 \, dy\right\}^{\frac{1}{2}} \end{aligned}$$

by the Cauchy-Schwarz inequality. Hence

$$\begin{aligned} \|Tu_1 - Tu_2\| &\leqslant |\lambda| \left\| \left\{\int_a^b |N(x,y)|^2 \, dy \, \|u_1 - u_2\|^2\right\}^{\frac{1}{2}} \right\| \\ &= |\lambda| P \|u_1 - u_2\| \quad . \end{aligned}$$

So condition (iv) ensures that T is a contraction on $L_2[a,b]$, and the equation $Tu = u$, which is just equation (5.9), has exactly one solution. □

Application 5.18 Consider the Sturm-Liouville system

$$u'' + \lambda f(x)u = 0, u(0) = u(1) = 0 \tag{5.13}$$

(readers unfamiliar with Sturm-Liouville systems should consult Appendix G). Theorem 5.17 gives useful information about its eigenvalues. We can use Green's function to turn (5.13) into an equivalent integral equation, as in Example 2.17. To do this we need Green's function for the operator $-D^2$ with boundary conditions $u(0) = u(1) = 0$. This can easily be worked out as in Chapter 2, either from first principles or by taking the limit as $k \to 0$ of formula (2.27); the result is

$$g(x;y) = \begin{cases} x(1-y) & \text{for} \quad x \leqslant y \quad , \\ y(1-x) & \text{for} \quad y \leqslant x \quad . \end{cases} \tag{5.14}$$

Applying this Green's function to the equation $-u'' = \lambda fu$ gives

$$u(x) = \lambda \int_0^1 g(x;y)f(y)u(y)\,dy \quad ,$$

which is an integral equation equivalent to the original differential equation and boundary conditions†. It is of the form (5.9) with $K(x,y,z) = g(x;y)f(y)z$, which satisfies condition (ii) of Theorem 5.17 with

$$N(x,y) = |g(x;y)f(y)| \quad .$$

The theorem says that there is a unique solution if $|\lambda| < P^{-1}$. But the equation always has the trivial (that is, the identically zero) solution. Hence we deduce that the trivial solution is the only solution if $|\lambda| < P^{-1}$, and therefore there is no eigenvalue (that is, no value of λ for which (5.13) has a non-trivial solution) such that $|\lambda| < P^{-1}$. Hence all the eigenvalues satisfy

$$|\lambda| \geqslant \left[\int_0^1 \int_0^1 |g(x;y)f(y)|^2\,dx\,dy\right]^{-\frac{1}{2}} \quad . \tag{5.15}$$

This result gives a useful estimate for the eigenvalues in cases where they cannot be found exactly – that is, for almost all choices of the function f. In later chapters we shall develop more powerful methods of estimating eigenvalues. Meanwhile we can form an idea of the usefulness of (5.15) by applying it to a simple example where the eigenvalues are known exactly. If $f(x) \equiv 1$, then (5.13) is easy to solve exactly, and the eigenvalues are $\pi^2, 4\pi^2, \ldots$. Evaluating (5.15) gives $|\lambda| \geqslant 3\sqrt{10}$; the bound given by (5.15) is thus only 4% away from the exact lowest eigenvalue. □

We have been working in the space $L_2[a,b]$ throughout. One could work in the space $C[a,b]$ instead, and get quite similar results but with differences in detail. We have already seen, in Example 5.6, that the integral operator in (5.9) maps $C[a,b]$ to itself provided that K is continuous. If K is assumed to satisfy a

†A full proof that the differential equation is equivalent to the integral equation is given in section 9.2.

Lipschitz condition as in Theorem 5.17, it is not difficult to work out conditions that will make the operator a contraction on $C[a,b]$; see Problem 5.14.

Application 5.19 (Nonlinear Eigenvalue Problem) The following two-point boundary-value problem is a nonlinear version of the eigenvalue problem in the previous example:

$$u'' + \lambda f(x,u) = 0, u(0) = u(1) = 0 \quad . \tag{5.16}$$

We assume that the given function f is such that $f(x,0) = 0$. This condition will be satisfied when f is approximately proportional to u for small u, which is the most usual case: it means that the system is well approximated by a linear equation for small u.

The classic simple example of (5.16) is the problem of an elastic rod under compression with its ends clamped; the angular displacement u of the rod under a compressive load λ satisfies (5.16) with $f(x,u) = \sin u$:

$$u'' + \lambda \sin u = 0, u(0) = u(1) = 0 \quad . \tag{5.17}$$

The solution $u \equiv 0$ of (5.17) corresponds to the rod being straight. For small loads, that is, for small values of λ, we expect this to be the only solution, but as the load λ is increased, we expect the rod to buckle at some stage, corresponding to the appearance of nonzero solutions of (5.17). In the linear case, where $f(u) = u$, we know that the zero solution is the only solution of (5.16) unless λ is an eigenvalue, in which case there are infinitely many solutions since any multiple of an eigenfunction is an eigenfunction. For the nonlinear eigenvalue problem (5.17) one finds that for small λ the only solution is the zero solution, as in the linear case, but as λ increases it reaches a critical value λ_1 at which a nonzero solution appears, corresponding to buckling of the rod. For $\lambda > \lambda_1$ the nonlinear problem behaves quite differently from the linear problem: for a range of values $\lambda_1 < \lambda < \lambda_2$ there is exactly one nonzero solution of (5.17) for each λ, and when λ exceeds λ_2 a second nonzero solution appears; similarly there is a value λ_3 beyond which there are three nonzero solutions, and so on. This behaviour is a simple example of the phenomenon of **bifurcation** or **branching**, which will be discussed further in Chapter 11; it occurs in many areas of applied mathematics. In mechanics in particular there are many situations where sudden jumps from one kind of behaviour to another (analogous to the buckling of a rod) occur as some parameter (analogous to the compressive load on the rod) is continuously varied; such problems are described by nonlinear eigenvalue equations similar to (5.16) and (5.17).

We now return to the more general equation (5.16), and begin by turning it into the integral equation

$$u(x) = \lambda \int_0^1 g(x;y) f(y,u(y))\, dy \tag{5.18}$$

by means of the Green's function (5.14). We shall use Theorem 5.17 with

$K(x,y,z) = g(x;y)f(y,z)$. To ensure that the theorem applies, let us assume that f_u is bounded,

$$|f_u(x,u)| \leqslant M \quad \text{for} \quad 0 \leqslant x \leqslant 1 \quad \text{and all} \quad u \quad , \tag{5.19}$$

where f_u is the partial derivative of f with respect to its second argument. Then by the Mean Value Theorem,

$$|K(x,y,z_1) - K(x,y,z_2)| = |g(x;y)| f_u(y,\zeta) | z_1 - z_2|$$

for some ζ between z_1 and z_2,

$$\leqslant M|g(x;y)|\,|z_1 - z_2|$$

by (5.19). So, in the notation of Theorem 5.17,

$$P^2 = M^2 \int_0^1 \int_0^1 |g(x;y)|^2 \,\mathrm{d}x\,\mathrm{d}y \quad ,$$

and evaluating the integral gives $P^2 = M^2/90$. Hence the solution of (5.16) is unique if

$$\lambda < 3\sqrt{10}/M \quad . \tag{5.20}$$

But the zero function always satisfies the equation; hence it is the only solution if (5.20) is satisfied. In the column problem, $f(x,u) = \sin u$, so $M = 1$, and we deduce that the rod will not buckle for loads up to $3\sqrt{10}$; we expect that nonzero solutions will appear for some larger values of λ, as in linear eigenvalue problems. These matters are discussed further in Chapter 11.

5.4 Nonlinear Diffusive Equilibrium

The equation

$$\partial u/\partial t = \partial^2 u/\partial x^2 + h(u,x) \quad \text{for} \quad a < x < b,\ t > 0 \tag{5.21}$$

represents many physical processes involving diffusion and nonlinear growth. For example, if a chemically reacting substance is diffusing through a medium, then its concentration $u(r,t)$ satisfies the equation $u_t = K\nabla^2 u + h(u,r)$, where K is the diffusion coefficient and $h(u,r)$ represents the rate of increase of the substance due to the chemical reaction. In the case of diffusion across a slab lying parallel to the yz plane, u depends only on the x-coordinate, and obeys (5.21); the fully three-dimensional problem is discussed in section 5.5. The equation (5.21) also governs the temperature when chemical reactions are generating heat at a rate depending on temperature; other problems modelled by (5.21) include the spread of animal or plant populations, where h represents the net growth rate.

One of the main questions to ask about such growth processes is whether they ever settle down to equilibrium. An equilibrium solution $u(x)$ satisfies the (usually nonlinear) differential equation

$$d^2u/dx^2 = -h(u,x) \quad . \tag{5.22}$$

To determine a solution we need a pair of boundary conditions. We shall assume that (5.22) holds for $0 < x < 1$, and that $u(0) = u(1) = 0$; in the heat flow interpretation this corresponds to the flow of heat in a slab when both faces of the slab are held at temperature zero. Other boundary conditions can be treated in the same way – see Problem 5.17. We shall now ask whether (5.22) has a solution satisfying the boundary conditions. If the answer is yes, then there is a possible steady state for the system; if the answer is no, we conclude that the system never settles down, but u either grows indefinitely or oscillates.

We begin by replacing the differential equation and boundary conditions by the equivalent integral equation

$$u(x) = \int_0^1 g(x;y)h(u(y),y)\,dy \quad , \tag{5.23}$$

where $g(x;y)$ is the Green's function (5.14). We now apply Theorem 5.17, with $f = 0$ and $\lambda = 1$. If h has a bounded derivative, with $|\partial h/\partial u| \leqslant M$ for all u and all x, then we can take $N(x,y) = Mg(x;y)$ in (5.11), and deduce that there exists a unique solution if

$$M = \sup|\partial h/\partial u| < [\int_0^1 \int_0^1 |g(x;y)|^2\,dx\,dy]^{-\frac{1}{2}} \quad .$$

Evaluating the integral gives the criterion

$$\sup|\partial h/\partial u| < 3\sqrt{10} \tag{5.24}$$

for existence of an equilibrium solution for problems where $|\partial h/\partial u|$ is bounded.

As an illustration of the use of (5.24), suppose that the growth rate is greatest when $u = 0$ and tends to zero as $u \to \infty$; such a situation might be represented by $h(u,x) = a(x)/(1 + u^2)$. Then an easy calculation shows that $\sup|\partial h/\partial u| = (\sqrt{3}/2)^3 \sup|a(x)|$, so (5.24) says that there will be an equilibrium if $|a(x)| < 8\sqrt{10/3} = 14.6$ for all x. In the next chapter we shall show, by means of a more powerful fixed-point theorem, that in fact there is a solution for any a (see Problem 6.23).

There are many cases to which (5.24) does not apply. For example, the function

$$h(u,x) = a(x)u(1 - u/b) \tag{5.25}$$

has an unbounded derivative, and so the above theory does not work. But it can be adapted, as we shall see, so as to apply to (5.25).

The function (5.25) represents a growth rate which is proportional to u for small u, but decreases as u increases, and vanishes when $u = b$. It corresponds to the growth of a population u when there is a limit b on the size of population that the environment can support; if $u > b$ then $h < 0$, so the population shrinks whenever u is greater than the limiting value b. This interpretation of (5.25) suggests that for the purpose of studying equilibrium solutions, we may be able to restrict ourselves to values of u less than the limiting value b. Thus we define

$$X = \{u \in L_2[0,1] : 0 \leqslant u(x) \leqslant b \quad \text{for} \quad 0 \leqslant x \leqslant 1\};$$

X is clearly a closed subset of $L_2[0,1]$ (Problem 4.11). Even though the integral operator appearing in (5.23) is not a contraction on $L_2[0,1]$, it may be a contraction on the subset X, since $|\partial h/\partial u|$ is bounded when $0 \leqslant u \leqslant b$. We shall therefore apply the contraction mapping theorem to the operator $T: X \to L_2[0,1]$ defined by

$$(Tu)(x) = \int_0^1 g(x;y)h(u(y),y)\,dy \quad \text{for} \quad u \in X \quad .$$

We first ask if T maps X into itself. We see from (5.14) that $g \geqslant 0$ everywhere, and

$$0 \leqslant h(u,y) \leqslant \tfrac{1}{4}ba(y) \quad ,$$

assuming that $a(y) > 0$ for all y. Hence if $u \in X$, then

$$0 \leqslant (Tu)(x) \leqslant \tfrac{1}{4}b\int_0^1 g(x;y)a(y)\,dy \leqslant \tfrac{1}{4}bAG$$

where $\quad A = \max\{a(x): 0 \leqslant x \leqslant 1\}$

and $\quad G = \max\{\int_0^1 g(x;y)\,dy : 0 \leqslant x \leqslant 1\}$

$$= 1/8$$

after some calculation. Hence $Tu \in X$ if $bAG/4 \leqslant b$, so $T: X \to X$ if

$$A \leqslant 32 \quad . \tag{5.26}$$

Now we ask if $T: X \to X$ is a contraction. We have, for any $u, v \in X$,

$$\begin{aligned} |(Tu)(x) - (Tv)(x)| &= |\int_0^1 g(x;y)[h(u,y) - h(v,y)]\,dy| \\ &\leqslant \sup|\partial h/\partial u| \int_0^1 g(x;y)|u(y) - v(y)|\,dy \\ &\leqslant \sup|\partial h/\partial u| \{\int_0^1 |g(x;y)|^2\,dy\}^{\frac{1}{2}} \|u - v\| \end{aligned}$$

by the Schwarz inequality. Hence

$$\|Tu - Tv\| \leqslant \sup|\partial h/\partial u| \{\int_0^1\int_0^1 |g|^2\,dx\,dy\}^{\frac{1}{2}} \|u - v\| \quad ,$$

and T is a contraction on X if $\sup\{|\partial h/\partial u| : u \in X\} < \{\int_0^1\int_0^1 |g|^2\,dx\,dy\}^{-\frac{1}{2}} = 3\sqrt{10}$. With h given by (5.25) this condition becomes

$$A < 3\sqrt{10} \quad . \tag{5.27}$$

The contraction mapping theorem now says that there is a unique equilibrium solution if (5.27) is satisfied.

Now, the reader will have noticed that the equation $u'' = -au(1 - u/b)$, with $u(0) = u(1) = 0$, always has the solution $u = 0$. What, it may be asked, is the use of an elaborate proof that it has a solution if $A = \sup\{a\} < 3\sqrt{10}$, when it is obvious that it always has a solution? The answer is that the uniqueness clause of the contraction mapping theorem is the useful part. It shows that if (5.27) holds, then there is no equilibrium other than the zero solution. This suggests that if $A < 3\sqrt{10}$ then u will tend towards the unique equilibrium $u \equiv 0$ as $t \to \infty$, but of course we have not proved this; see section 11.5. We might also conjecture that if A is large, then the conclusion of the contraction mapping theorem should break down. Since the existence clause is always true, we expect the uniqueness clause to fail, as in Application 5.19, so we surmise that for some values of A greater than $3\sqrt{10}$ there will exist nonzero equilibrium solutions, representing a balance between diffusion and growth. This is another example of the phenomenon of bifurcation discussed in Application 5.19. For the equation $u'' = au(1 - u/b)$ with a and b constant, the existence of bifurcating nonzero solutions can be verified in detail, by a phase-plane analysis or by an exact solution in terms of elliptic integrals. These methods do not work if a depends on x; for the general theory of bifurcations, see section 11.6.

5.5 Nonlinear Diffusive Equilibrium in Three Dimensions

In this section we illustrate the application of the contraction mapping theorem to partial differential equations. We shall consider the reaction-diffusion equation $u_t = \nabla^2 u + h(u,x)$ discussed at the beginning of section 5.4, in a three-dimensional region V bounded by a surface S, and look for equilibrium solutions $u(x)$ satisfying

$$\nabla^2 u = -h(x,u(x)) \tag{5.28}$$

for $x \in V$, with Dirichlet boundary conditions $u = 0$ for x on S.

We begin as usual by turning (5.28) into an integral equation. In section 3.5 we showed that there is always a Green's function $g(x;y)$ for $-\nabla^2$ with our boundary conditions, so that (5.28) is equivalent to

$$u(x) = \int_V g(x;y)h(y,u(y))\,\mathrm{d}^3y \quad . \tag{5.29}$$

The equivalence of the differential and integral equations is discussed more carefully in section 9.3.

We wish to interpret (5.29) as an operator equation of the form $u = Tu$, where u is a member of a complete normed space. The obvious space to use is the three-dimensional analogue of the space $L_2[a,b]$ used above to deal with ordinary differential equations. This is the Banach space $L_2[V]$ defined in Example 4.63. We define an operator T on $L_2[V]$ by

$$(Tf)(x) = \int_V g(x;y)f(y)\,\mathrm{d}^3y \quad .$$

We must prove that T maps $L_2[V]$ into itself, which is perhaps not entirely obvious, since g has a singularity at $x = y$. However, in Example 5.5 we found a criterion (5.2) which is sufficient for an integral operator to map $L_2[a,b]$ into itself, and the argument extends immediately to $L_2[V]$. It is easy to show that g satisfies this criterion.

Proposition 5.20 If $g(x;y)$ is the Dirichlet-problem Green's function for $-\nabla^2$ in a bounded region $V \subset \mathbb{R}^3$, then $\int_V \mathrm{d}^3y\,|g(x;y)|^2$ is a bounded function of x, and $\int_V \mathrm{d}^3x \int_V \mathrm{d}^3y\,|g(x;y)|^2$ is finite.

Proof In 3.35 we showed that $|g(x;y)| \leqslant 1/4\pi|x-y|$, so our result will follow if we can show that $\int d^3y\,|x-y|^{-2}$ is bounded. Let d be the diameter of V, that is, $d = \sup\{|x-y| : x,y \in V\}$. We shall show that

$$\int_V |x-y|^{-2}\,d^3y \leqslant 4\pi d \tag{5.30}$$

for all $x \in V$. Take any fixed $x \in V$, and let Σ be the sphere of radius d centred on x. Then $V \subset \Sigma$, hence

$$\int_V |x-y|^{-2}\,d^3y \leqslant \int_\Sigma |x-y|^{-2}\,d^3y \quad . \tag{5.31}$$

We evaluate this integral using polar coordinates centred on $x : r = |y-x|$, and two angular variables θ, ϕ. We have

$$\begin{aligned}\int_\Sigma d^3y\,|x-y|^{-2} &= \int_0^d r^2\,dr \int_0^\pi \sin\theta\,d\theta \int_0^{2\pi} d\phi\,r^{-2} \\ &= 4\pi d \quad .\end{aligned}$$

(5.30) now follows immediately from (5.31). Since $|g| \leqslant 1/4\pi|x-y|$, we see that $\int d^3y\,|g|^2 \leqslant d/4\pi$, and therefore $\int d^3x \int d^3y\,|g|^2 \leqslant dV/4\pi$ where V denotes the volume of V. □

Corollary 5.21 The linear integral operator whose kernel is the Green's function of Proposition 5.20 maps $L_2[V]$ into itself.

Proof We must show that if $f \in L_2[V]$, then the integral

$$F(x) = \int_V d^3y\,g(x;y)f(y)$$

converges and gives a square-integrable function of x. We shall use the following three-dimensional version of the Cauchy-Schwarz inequality:

$$\left|\int_V \phi(y)\psi(y)\,d^3y\right|^2 \leqslant \left(\int_V |\phi|^2\,d^3y\right)\left(\int_V |\psi|^2\,d^3y\right) \quad . \tag{5.32}$$

If the integrals on the right of (5.32) converge, then the integral on the left converges and (5.32) is valid; this is proved in the same way as for single integrals or sums (Problem 4.8).

Now, take $\phi(y) = g(x;y)$ and $\psi(y) = f(y)$ in (5.32). The first integral on the right of (5.32) converges by Proposition 5.20; the second converges because $f \in L_2[V]$, so (5.32) gives

$$|F(x)|^2 = \left|\int_V g(x;y)f(y)\,d^3y\right|^2 \leqslant \left(\int |g(x;y)|^2\,d^3y\right)\|f\|^2 \quad ,$$

$$\therefore \int_V |F(x)|^2\,d^3x \leqslant \|f\|^2 \int_V\int_V |g(x;y)|^2\,d^3x\,d^3y \quad , \tag{5.33}$$

which shows that $F \in L_2[V]$. □

We can now deal with equation (5.28).

Theorem 5.22 (Existence and Uniqueness for Nonlinear Dirichlet Problem) The equation $\nabla^2 u = -h(x,u(x))$ for $x \in V$, where V is a three-dimensional region bounded by a closed surface S, has exactly one square-integrable solution satisfying $u = 0$ on S, provided that

(i) h is differentiable with respect to u, with $|h_u| \leqslant M$ for all u and all $x \in V$;

(ii) $M < [\int\int |g(x;y)|^2 \,\mathrm{d}^3x\,\mathrm{d}^3y]^{-\frac{1}{2}}$ where g is the Dirichlet-problem Green's function for V;

(iii) $h(x,0) \in L_2[V]$.

Proof The differential equation and boundary conditions are equivalent to the integral equation (5.29), which we write as $u = Tu$ where the operator T is defined by

$$(Tu)(x) = \int_V g(x;y)h(y,u(y))\,\mathrm{d}^3y \quad . \tag{5.34}$$

A solution thus corresponds to a fixed point of T. We must first show that T maps $L_2[V]$ into itself. By the Mean Value Theorem we have $|h(x,u(x)) - h(x,0)| \leqslant M|u(x)|$, hence

$$|h(x,u(x))| \leqslant |h(x,0)| + M|u(x)|$$

$$\therefore |(Tu)(x)| \leqslant \int_V g(x;y)|h(y,0)|\,\mathrm{d}^3y + M\int_V g(x;y)|u(y)|\,\mathrm{d}^3y \quad .$$

Each term on the right here is the result of applying the operator of Corollary 5.21 to a member of $L_2[V]$, and therefore is itself a member of $L_2[V]$. Hence $Tu \in L_2[V]$.

We must now show that T is a contraction.

$$|(Tu_1)(x) - (Tu_2)(x)| \leqslant \int \mathrm{d}^3y\, g(x;y)|h(y,u_1(y)) - h(y,u_2(y))|$$
$$\leqslant M\int \mathrm{d}^3y\, g(x;y)|u_1(y) - u_2(y)|$$

by the Mean Value Theorem,

$$\leqslant M\{(\int \mathrm{d}^3y\,|g(x;y)|^2)(\int \mathrm{d}^3y\,|u_1 - u_2|^2)\}^{\frac{1}{2}}$$

by the Schwarz inequality. Hence

$$\| Tu_1 - Tu_2 \|^2 \leqslant M^2 \int \mathrm{d}^3x \int \mathrm{d}^3y\,|g(x;y)|^2 \,\| u_1 - u_2 \|^2$$

and T is a contraction by condition (ii) of the theorem. Hence there is exactly one solution in $L_2[V]$. □

For an application of this theorem in combustion theory, see Problem 5.20. As in section 5.3, the theory can be reworked using the space of continuous functions; the condition (ii) of Theorem 5.22 is then replaced by the condition

$$M < [\sup\{|\int g(x;y)\,\mathrm{d}^3y| : x \in V\}]^{-1} \quad .$$

Again as in section 5.3, the theorem can be used to give estimates for the lowest eigenvalues of the operator $-\nabla^2$, that is, the lowest resonant frequencies of bounded regions, once the Green's function for the region has been found.

5.6 Summary and References

In section 5.1 we discussed operators on normed spaces; in section 5.2 we defined a contraction as an operator which decreases the distance between pairs of points, and proved that a contraction has just one fixed point. Most books on functional analysis deal with this; we mention in particular Kolmogorov & Fomin (1957 and 1970), Kreyszig, Sawyer, and Smart. In section 5.3 we discussed existence and uniqueness results for ordinary differential equations and integral equations; see Hochstadt. In section 5.4 we considered some problems of nonlinear diffusive equilibrium in one dimension, and in section 5.5 we proved an analogous result in three dimensions. The partial differential equation of section 5.5 is discussed by Smart. For further information on nonlinear diffusion, see Murray, Fife, Kordylewski, and section 11.7.

There are many other applications of the contraction mapping theorem. For example, Kreyszig discusses applications to linear algebraic equations (see Problem 5.9), and Hale applies the theorem to functional differential equations. For applications to numerical analysis, see Collatz. For an application in fluid mechanics, see Keady & Norbury.

PROBLEMS 5

Section 5.1

5.1 For what values of the constant a does the rule $u \mapsto \int_0^1 x^a u(x)\,dx$ define an operator $C[0,1] \to \mathbb{C}$? For what values of a does it define an operator $L_2[0,1] \to \mathbb{C}$?

5.2 l_2 is the space defined in Problem 4.9. Define an operator A as follows: Ax is the sequence whose n-th term is $\Sigma_m a_{nm} x_m$, where a_{nm} is a given array (double sequence) of numbers. Show that if $\Sigma_n \Sigma_m |a_{nm}|^2$ converges, then A maps l_2 into itself. Give an example to show that if $\Sigma_n \Sigma_m |a_{nm}|^2$ does not converge, then A may not map l_2 into itself.

5.3 Find the mistake in the following argument, which claims to prove that the integral operator T of Example 5.4 maps $C[a,b]$ into itself.

K is continuous, hence for any $\epsilon' > 0$ there is a $\delta > 0$ such that $|K(x_1,y) - K(x_2,y)| < \epsilon'$ whenever $|x_1 - x_2| < \delta$. Hence

$$|(Tf)(x_1) - (Tf)(x_2)| = |\int_a^b [K(x_1,y) - K(x_2,y)] f(y)\,dy|$$
$$< \epsilon' \int_a^b |f(y)|\,dy \quad .$$

Now, given any $\epsilon > 0$, take $\epsilon' = \epsilon / \int_a^b |f(y)|\,dy$ and choose δ as above, then $|(Tf)(x_1) - (Tf)(x_2)| < \epsilon$ whenever $|x_1 - x_2| < \delta$, so Tf is continuous and belongs to $C[a,b]$.

Section 5.2

5.4 Show that the equation $x^3 - x - 1 = 0$ has a root between 1 and 2. There are two obvious ways of putting the equation in the form $x = Tx$ where $T: \mathbb{R} \to \mathbb{R}$. Show that for one of these ways T is a contraction on $[1,2]$, and use the iteration method to find the root to 4 significant figures.

5.5 Define $T: \mathbb{R}^2 \to \mathbb{R}^2$ by $T(x,y) = (y^{\frac{1}{3}}, x^{\frac{1}{3}})$. What are the fixed points of T? What happens when you iterate, starting from various places in $\mathbb{R}^2$ (find out by numerical experiments)? In what regions of $\mathbb{R}^2$ is T a contraction?

5.6 Let $E_n = \| x_n - x \|$ be the error at the n-th stage of the iteration process for finding solutions of $Tx = x$. Use the proof of the contraction map theorem to show that

$$E_n \leqslant (a^n/[1-a])\| x_1 - x_0 \|$$

where a is as in Definition 5.13.

Use this result to estimate how many iterations would be needed to solve the equation of Problem 5.4 to six significant figures.

5.7 X is a closed subset of a Banach space. Show that if T^n is a contraction for some n, then $T: X \to X$ has a unique fixed point. (For an application of this result, see Problem 5.10).

5.8 Completeness of the space is essential in the contraction mapping theorem; give an example showing that Theorem 5.15 becomes false if "Banach" is replaced by "normed".

5.9 Consider the equation $Ax = b$ where A is an $n \times n$ matrix, b a given n-vector and x an unknown n-vector. If all the diagonal entries of A are non-zero, then the equation $Ax = b$ can be scaled so that the diagonal entries of A all become 1. We suppose this has been done. If the entries a_{ij} of A then satisfy

$$\sum_{\substack{j=1 \\ j \neq i}}^{n} |a_{ij}| < |a_{ii}| = 1 \quad \text{for each} \quad i,$$

A is said to have 'strict diagonal dominance'.

Now, the equation $Ax = b$ can be written $x = T(x)$ where the operator $T: \mathbb{R}^n \to \mathbb{R}^n$ is defined by $T(x) = (I - A)x + b$ where I is the unit matrix. Show that if A has strict diagonal dominance, then T is a contraction on $\mathbb{R}^n$ with the norm $\| x \| =$

$\max_i |x_i|$. Invent an example using 3 X 3 matrices, and use the iteration method to solve the equation to 3 significant figures.

(For further developments of this idea, see Kreysig section 5.2, or books on numerical analysis.)

Section 5.3

5.10 If K is a given function of two real variables, the operator L defined by

$$(L\phi)(x) = \int_0^x K(x,y)\phi(y)\,dy$$

is called the Volterra integral operator with kernel K; note the difference from Example 5.4. The equation $\phi = f + L\phi$ is called a Volterra integral equation. Write $T\phi = f + L\phi$.

Show that L^n is the Volterra operator with kernel K_n, where the sequence of 'iterated kernels' K_n is defined by $K_1 = K$ and

$$K_{n+1}(x,y) = \int_y^x K(x,z)K_n(z,y)\,dz \quad .$$

Deduce that $T^n\phi = f + Lf + \ldots + L^{n-1}f + L^n\phi$. Now suppose that K is continuous and bounded in the unit square: $|K(x,y)| \leqslant M$ for $0 \leqslant x,y \leqslant 1$. Show (by induction) that $|K_n(x,y)| \leqslant M^n|x - y|^{n-1}/(n-1)!$ Deduce that there is a number r such that T^r is a contraction on $C[0,1]$. It follows from Problem 5.7 that the Volterra equation with a continuous kernel has a unique solution for any continuous f. Deduce that the Volterra operator L has no nonzero eigenvalues corresponding to continuous eigenfunctions. Can it have discontinuous eigenfunctions?

5.11 Use iteration to obtain the first few terms in a sequence of approximations to the solution of the (easy) equation $u(x) = 1 + \lambda\int_0^x u(y)\,dy$. Hence guess the exact solution, and verify your guess.

5.12 Consider the Sturm-Liouville system $u'' + u' + \lambda x^2 u = 0$, $u(0) = u(1) = 0$. Use Green's function for the operator $D^2 + D$ to obtain an equivalent integral equation. Use the method of Application 5.18 to obtain a lower bound on the eigenvalues, in the form of a double integral. If you have experience in numerical computation, evaluate the integral numerically.

5.13 This problem shows how a clever choice of norm can pay dividends.

(a) A nonlinear version of the Volterra operator of Problem 5.10 is defined as follows: $(Lx)(t) = \int_0^t K(t,s)f(s,x(s))\,ds$ where K and f are continuous functions, and $|f(s,u) - f(s,v)| \leqslant N|u - v|$ for all u,v,s where N is a constant. Then L maps $C[0,T]$ into itself for any $T > 0$. Give an example to show that L is not a contraction on $C[0,T]$ with the usual norm $\|x\| = \sup|x(t)|$.

(b) Show that for any $a > 0$, $\| x \|_a = \sup\{ e^{-at} |x(t)| : 0 \leqslant t \leqslant T\}$ defines a norm on $C[0,T]$ which is equivalent to the usual norm, in the sense of section 4.6. Deduce that $C[0,T]$ with the norm $\| \ \|_a$ is a Banach space.

(c) Set $M = \max\{|K(s,t)| : 0 \leqslant s,t \leqslant T\}$. Show that $\| Lx - Ly \|_a \leqslant (MN/a)(1 - e^{-aT}) \| x - y \|_a$, for all $x,y \in C[0,T]$. Deduce that for any $T > 0$ the integral equation $x = Lx + g$, where g is a given continuous function, has a unique solution.

5.14 Consider the integral equation (5.9) where K is a continuous function satisfying the Lipschitz condition $|K(x,y,z_1) - K(x,y,z_2)| \leqslant N(x,y)|z_1 - z_2|$ for all z_1, z_2. Prove that the equation has a unique solution in $C[a,b]$ if $|\lambda| < [\sup\{\int_a^b N(x,y)\,dy : a \leqslant x \leqslant b\}]^{-1}$. For the simple eigenvalue problem considered at the end of Application 5.18, compare the estimate given by this result with that given by Theorem 5.17; which is better?

5.15 Consider the boundary-value problem $u''(x) = f(x,u(x))$ for $a \leqslant x \leqslant b$, $u(a) = u(b) = 0$, where f has a bounded u-derivative. Use the equivalent integral equation to show that there is a unique solution provided that the interval is short enough, that is, if $b - a < \sqrt{8/A}$ where $A = \sup|\partial f/\partial u|$. Can you extend this result to the case with boundary conditions $u(a) = p$, $u(b) = q$ where p and q are given numbers?

5.16 Reconsider Problem 5.14 in the light of Problem 5.13. Is it possible to improve the result of 5.14 by using the norm of Problem 5.13(b)? Use the simple eigenvalue problem as a test case to answer this question.

Sections 5.4, 5.5

5.17 The equation $u'' = -h(u,x)$ with $u'(0) = 0$, $u(1) = 0$ represents diffusive equilibrium when there is a reflecting boundary at $x = 0$, and zero temperature or concentration at $x = 1$. Calculate Green's function for D^2 with these boundary conditions, and derive a condition for the existence of an equilibrium solution analogous to (5.24).

5.18 Rework the theory of section 5.4 using the sup norm in $C[0,1]$, as in Problem 5.14.

5.19 Condition (ii) of Theorem 5.22 is not very easy to apply because Green's functions for partial differential equations are usually quite complicated. Use the results in the proof of Proposition 5.20 to give a simpler (but weaker) condition to replace 5.22(ii). Hence show that for a sphere of radius R there is a unique solution if $|h_u| < \sqrt{3/2}R^{-2}$.

Use this result to give a crude bound on the eigenvalues of $-\nabla^2$ inside a sphere of radius R (by the method of Application 5.18). Use it also to find an estimate for the critical radius for a sphere containing reacting matter governed by the equation $u_t = \nabla^2 u + a(1+u^2)^{-1}$, with $u = 0$ on the boundary. The critical radius is defined as a number R_c such that if $R < R_c$ then there is equilibrium but if $R > R_c$ there is no equilibrium.

5.20 The equation $u_t = \nabla^2 u + ae^{-\frac{1}{u}}$ describes temperature growth in a heat-producing chemical reaction. The function $e^{-\frac{1}{u}}$ has a severe singularity at $u = 0$, and therefore does not satisfy the conditions of Theorem 5.22. However, if we define a function h by $h(u) = ae^{-\frac{1}{u}}$ for $u > 0$ and $h(0) = 0$, then h is well-behaved for all $u \geqslant 0$, and has a bounded derivative; verify this and find $\max |dh/du|$. Now consider the operator T of equation (5.34) acting on non-negative functions. Show that T maps G into G, where $G = \{f \in L_2[V] : f(x) \geqslant 0 \text{ for all } x \in V\}$. Now use the contraction mapping theorem to show that T has a fixed point, and hence that there is an equilibrium temperature distribution, if $a < 1.8[\iint |g(x;y)|^2 d^3x\, d^3y]^{-\frac{1}{2}}$.

More problems will be found on page 376.

Chapter 6

Compactness and Schauder's Theorem

In this chapter we discuss fixed-point theorems which are deeper than the contraction mapping theorem; even to state them requires more subtle ideas, and their proofs are difficult (and will not be given here). In section 6.1 we discuss continuous operators, a straightforward generalisation of continuous $\mathbb{R} \to \mathbb{R}$ functions. In section 6.2 we state a fixed-point theorem for continuous operators on $\mathbb{R}^n$, with a sketch of the proof, and give an application to matrix theory.

In order to extend the theorem to function spaces we must make a long digression on compactness, occupying sections 6.3–6.5. These sections are quite theoretical in character; an understanding of compactness is vital for the applications later in the chapter, and for the theory of linear operators in Chapters 8 and 9 and its application to differential and integral equations.

6.1 Continuous Operators

The contraction mapping theorem applies to operators satisfying the very strong condition that any two points are mapped into points closer together than the original pair. The fixed-point theorems of this chapter apply to operators which map any pair of points into points 'not too much further apart' than the original pair, in the following sense.

Definition 6.1 Let N, M be normed spaces, and X a subset of N. An operator $T: X \to M$ is **continuous at a point** $x \in X$ if for any $\epsilon > 0$ there is a $\delta > 0$ such that $\| Tx - Ty \| < \epsilon$ for all $y \in X$ such that $\| x - y \| < \delta$. T is **continuous on** X, or simply **continuous**, if it is continuous at all points of X. □

This means that if two points are close together, then their images are close together. This is the same as the idea of continuity in elementary analysis. Continuous $\mathbb{R} \to \mathbb{R}$ functions in the usual sense are continuous mappings in the sense of Definition 6.1.

Example 6.2 The 'translation' operator on $\mathbb{R}^2$ defined by $(x_1,x_2) \mapsto (x_1,x_2+1)$ is a very simple example of a continuous operator. Since the distance between a pair of points is unchanged by the application of this operator, which simply shifts all points a unit distance in the x_2 direction, Definition 6.1 is satisfied with $\delta = \epsilon$. □

The operator in 6.2 has a property stronger than simple continuity at all points. For a given ϵ, the value of δ required in 6.1 may in general depend on x; but in 6.2 we found a value of δ, independent of x, which works for all x. This is reminiscent of the distinction between pointwise and uniform convergence; we have here a similar distinction between pointwise and uniform continuity.

Definition 6.3 Let N,M be normed spaces and X a subset of N. An operator $T: X \to M$ is **uniformly continuous on** X if for each $\epsilon > 0$ there is a $\delta > 0$, depending only on ϵ, such that $\| Tx - Ty \| < \epsilon$ whenever $\| x - y \| < \delta$. □

The operator of Example 6.2 is uniformly continuous on $\mathbb{R}^2$: we can take $\delta = \epsilon$. Most continuous operators that we shall deal with are uniformly continuous; the following is an example of a nonuniformly continuous operator.

Example 6.4 Take $M = N = \mathbb{R}$, $X = (a,1)$, where $a \geqslant 0$, and set $Tx = x^{-1}$ for $a < x < 1$. If $a > 0$, then $\| Tx - Ty \| = |y - x|/xy < |y - x|/a^2$. Thus we can take $\delta = \epsilon a^2$ in 6.3, and $T: (a,1) \to \mathbb{R}$ is uniformly continuous if $a > 0$. But for $a = 0$ the above argument breaks down. T is still continuous at x for each $x \in (0,1)$; it is easy to verify that 6.1 is satisfied with $\delta = \epsilon x^2/2$ (unless $\epsilon > 1$, in which case $\delta = x^2/2$ will do). To prove that T is not uniformly continuous, we must show for some $\epsilon > 0$ that no value of δ will suffice for all x; that is, that no matter what $\delta > 0$ we take, $\| Tx - Ty \|$ will be greater than ϵ for some x,y with $\| x - y \| < \delta$. We take $\epsilon = 1$ for simplicity. If $\delta \geqslant 1$, take $x = 1, y = \frac{1}{2}$; if $0 < \delta < 1$, take $x = \delta, y = \delta/2$; in both cases we have $|x - y| < \delta$ and $|x^{-1} - y^{-1}| > 1$. Hence we can never find $\delta > 0$ such that $|x^{-1} - y^{-1}| < 1$ for all x,y such that $|x - y| < \delta$. Thus $T: (0,1) \to \mathbb{R}$ is not uniformly continuous. □

Uniform continuity is needed in Theorem 6.14, and in section 6.5, but for many purposes, including the fixed-point theorems of section 6.2 and section 6.6, ordinary continuity is sufficient.

Example 6.5 Given an $n \times r$ matrix A we can define a mapping $T: \mathbb{C}^r \to \mathbb{C}^n$ by the obvious rule $T(x) = Ax$ for $x \in \mathbb{C}^r$. The Cauchy-Schwarz inequality 4.24 shows that for any i

$$\left| \sum_j A_{ij}(x_j - y_j) \right| \leqslant \sqrt{\sum_j |A_{ij}|^2} \, \| x - y \| \quad ,$$

and therefore

$$\|Ax - Ay\| = \|A(x - y)\| \leqslant \sqrt{\sum_{ij}|A_{ij}|^2}\, \|x - y\|$$
$$\leqslant \sqrt{nr}\, M\|x - y\| \quad ,$$

where M is the modulus of the largest element of A. Hence for any ϵ the continuity condition is satisfied by taking $\delta = \epsilon/M\sqrt{nr}$ for any x, and the operator is uniformly continuous.

Example 6.6 In any normed space N, the norm gives a continuous mapping $x \mapsto \|x\|$ from N to $\mathbb{R}$. To prove this we must show that for any $\epsilon > 0$ and any x there is a $\delta > 0$ such that $|\|x\| - \|y\|| < \epsilon$ for all y such that $\|x - y\| < \delta$. But $|\|x\| - \|y\|| \leqslant \|x - y\|$ by Problem 4.6. Hence we can take $\delta = \epsilon$ and the uniform continuity condition is satisfied.

Example 6.7 Consider the integral operator T with kernel K defined in Example 5.4. We must distinguish carefully between the statements (i) K is a continuous function, and (ii) T is a continuous operator on $C[a,b]$. They mean different things, though (i) implies (ii), as we shall now show.

$K(x,y)$ is continuous on a closed region, and therefore bounded (see Theorem 6.25), $|K(x,y)| \leqslant M$, say, for $x,y \in [a,b]$. Hence

$$|(Tf)(x) - (Tg)(x)| = |\textstyle\int_a^b K(x,y)[f(y) - g(y)]\,dy|$$
$$\leqslant M(b - a)\sup|f(y) - g(y)|,$$
$$\therefore \|Tf - Tg\| = \sup|Tf(x) - Tg(x)| \leqslant M(b - a)\|f - g\|,$$

where $\|\cdot\|$ denotes the sup norm on $C[a,b]$. So for any $\epsilon > 0$ the condition 6.3 is satisfied by taking $\delta = \epsilon/M(b - a)$, and the operator is uniformly continuous on $C[a,b]$.

Example 6.8 The integral operator of the last example can also be applied to $L_2[a,b]$, as in Example 5.5. If

$$\textstyle\int_a^b\int_a^b |K(x,y)|^2\,dx\,dy = M$$

we have

$$\|Tf - Tg\|^2 = \|T(f - g)\|^2 \leqslant M\|f - g\|^2$$

by (5.3). Hence T is uniformly continuous (take $\delta = \epsilon/\sqrt{M}$ in Definition 6.3) if the kernel K is square-integrable. □

The following simple result says that two continuous mappings applied in succession give a continuous mapping.

Theorem 6.9 (Product of Continuous Operators) If the operator $A : X \to Y$ is continuous at $x_0 \in X$, and $B : Y \to Z$ is continuous at $y_0 = Ax_0$, then the operator $BA : X \to Z$ is continuous at x_0.

Proof Since B is continuous at y_0, for any $\epsilon > 0$ there is a $\delta > 0$ such that

$$\| By_1 - By_0 \| < \epsilon \quad \text{whenever} \quad \| y_1 - y_0 \| < \delta \quad . \tag{6.1}$$

Since A is continuous at x_0, there is a $\gamma > 0$ such that

$$\| Ax_1 - Ax_0 \| = \| Ax_1 - y_0 \| < \delta \quad \text{whenever} \quad \| x_1 - x_0 \| < \gamma \quad . \tag{6.2}$$

Combining (6.1) and (6.2) gives

$$\| BAx_1 - BAx_0 \| < \epsilon \quad \text{whenever} \quad \| x_1 - x_0 \| < \gamma \quad ,$$

which shows that BA is continuous at x_0. □

In distribution theory we defined a continuous functional as one that maps convergent sequences of test functions into convergent sequences of numbers. Our definitions of convergence and continuity in normed spaces are different from the definitions used in distribution theory, but they share the property that continuous operators map convergent sequences into convergent sequences.

Theorem 6.10 (Continuous Maps of Convergent Sequences) Let $T : X \to M$ be continuous at $x \in X$, where $X \subset N$, and M and N are normed spaces. If (x_n) is a sequence in X with $x_n \to x$ in N, then $Tx_n \to Tx$ in M.

Proof Since T is continuous, for any ϵ there is a δ such that $\| Tx_n - Tx \| < \epsilon$ whenever $\| x_n - x \| < \delta$; and since $x_n \to x$ there is an n_0 such that $\| x_n - x \| < \delta$ when $n > n_0$. Hence $\| Tx_n - Tx \| < \epsilon$ for $n > n_0$, so $Tx_n \to Tx$. □

Example 6.11 The operator $\mathrm{D} : C^1[0,1] \to C[0,1]$ defined by $\mathrm{D}f = f'$ (Example 5.7) is not continuous anywhere. *Proof:* for any $f \in C^1[0,1]$ set

$$f_n(x) = f(x) + \mathrm{e}^{-nx}/n \quad ;$$

then $f_n \to f$ in $C^1[0,1]$ (that is, $f_n \to f$ uniformly), but $f'_n \not\to f'$ (indeed, $f'_n(0) \to f'(0) - 1$), and this would contradict Theorem 6.10 if D were continuous at f. This is another manifestation of the general principle that differential operators are nasty while integral operators are nice (however, see Problem 6.3).

Example 6.12 Contraction mappings are continuous everywhere: take $\delta = \epsilon/a$ in the definition of continuity, where a is as in Definition 5.13. □

Operators are sometimes defined on a subspace of a Banach space rather than the whole space. In Example 6.11, for instance, D is defined on the subspace

$C^1[0,1]$ of the space $C[0,1]$. Such an operator, defined initially on a restricted domain, can sometimes be extended to a larger domain. The extension procedure will be used in the study of differential equations in Chapter 9; the details are a little complicated, and the rest of this section can be omitted on a first reading.

Definition 6.13 Let T be an operator with domain X, and T' an operator with domain $X' \supset X$. Then T' is called an **extension** of T if $Tx = T'x$ for all $x \in X$. □

Thus an extension T' of T is any operator which agrees with T when applied to elements of the domain of T. In general T can be extended in many different ways: $T'x$ can be given arbitrary values for $x \notin X$. Thus the idea of an extension in general contains so much arbitrariness that it is not very useful. However, if T is a continuous operator, then it is natural to require its extension to be continuous too. This reduces the arbitrariness, and in the commonly occurring case that X is dense in X', the extension is unique.

Theorem 6.14 (Extension of a Continuous Operator) Let T be a uniformly continuous operator $X \to M$, where X is a subset of a normed space and M is a Banach space. If X is a dense subset of X', then T has exactly one continuous extension $T': X' \to M$.

Proof We construct T' as follows. By 4.49, for any $x' \in X'$ there is a sequence (x_n) in X with $x_n \to x'$ as $n \to \infty$. We shall show that (Tx_n) is a Cauchy sequence, and define its limit to be $T'x'$.

Because T is uniformly continuous, for any $\epsilon > 0$ there is a $\delta > 0$ such that $\| Tx_n - Tx_m \| < \epsilon$ whenever $\| x_n - x_m \| < \delta$. Because (x_n) is convergent, and therefore a Cauchy sequence, there is an n_0 such that $\| x_n - x_m \| < \delta$ for $n,m > n_0$. Hence

$$\| Tx_n - Tx_m \| < \epsilon \quad \text{when} \quad n,m > n_0 \quad ,$$

showing that (Tx_n) is a Cauchy sequence. It belongs to the Banach space M, therefore it converges, say

$$Tx_n \to u \quad \text{as} \quad n \to \infty \quad .$$

We now define $T'x' = u$. That is, for each $x' \in X'$ we choose a sequence (x_n) in X with $x_n \to x'$, and define $T'x' = \lim(Tx_n)$. We shall show that although (x_n) is chosen arbitrarily from the many sequences tending to x', yet the operator T' is uniquely defined. But first we show that T' is continuous.

Take $x', y' \in X'$, and corresponding sequences (x_n), (y_n) in X with $x_n \to x'$, $y_n \to y'$. Then

$$\begin{aligned} \| T'x' - T'y' \| &= \| \lim(Tx_n - Ty_n) \| \\ &= \lim \| Tx_n - Ty_n \| \end{aligned}$$

using 6.6 and 6.10. Now, because T is uniformly continuous, given any $\epsilon > 0$ there is a $\delta > 0$ such that

$$\| Tx_n - Ty_n \| < \epsilon/2 \quad \text{whenever} \quad \| x_n - y_n \| < \delta \quad .$$

There is an n_0 such that $\| x_n - x' \| < \delta/3$ and $\| y_n - y' \| < \delta/3$ for $n > n_0$. If we take $\| x' - y' \| < \delta/3$, then

$$\| x_n - y_n \| \leqslant \| x_n - x' \| + \| x' - y' \| + \| y' - y_n \| < \delta \quad ,$$

$$\therefore \| Tx_n - Ty_n \| < \epsilon/2 \quad \text{for} \quad n > n_0 \quad ,$$

and so $\| T'x' - T'y' \| < \epsilon$. Thus T' is continuous on X'.

T' is clearly an extension of T, for if $x \in X$, then $T'x = \lim(Tx_n) = Tx$ by 6.10. We now prove that there is only one continuous extension.

Let T', T'' be two continuous extensions of T. For any $x \in X'$, take a sequence (x_n) in X with $x_n \to x$. Then by 6.10 we have $T'x = \lim(T'x_n) = \lim(Tx_n) = \lim(T''x_n) = T''x$. Thus $T' = T''$. □

According to this theorem, a continuous operator on X can be extended to $\bar{X}$: the closure of X is the largest set in which X is dense. If X is closed, there is in general no natural way of extending an operator beyond X.

We now return to the integral operator of Example 5.5. We pointed out there that $L_2[a,b]$ contains functions which are not Riemann-integrable, and therefore it is not clear that one can apply integral operators to them. This difficulty can be resolved by using the Lebesgue theory; but we now have an alternative approach. We can consider the integral operator T to be defined initially on continuous functions only, where there is no doubt about the existence of the integral. Then the calculation of Example 6.8 shows that T is uniformly continuous on $C[a,b]$. Since $C[a,b]$ is a dense subset of $L_2[a,b]$, 6.14 shows that T has a unique continuous extension to the whole of $L_2[a,b]$, and this extension is what is meant by Tf when f is not Riemann-integrable. We can thus use continuous integral operators on L_2 without qualms.

6.2 Brouwer's Theorem

Since continuous operators include contractions as a special case, it is natural to ask whether the contraction mapping theorem generalises to continuous operators. The answer is no, as the following trivial example shows: the translation operator of Example 6.2 is continuous, but it clearly has no fixed point. We can exclude such cases by considering only mappings of finite regions into themselves, or more precisely, of bounded regions as defined below.

Definition 6.15 A subset S of a normed space N is called **bounded** if there is a number M such that $\| x \| \leqslant M$ for all $x \in S$. □

This is a natural generalisation of the idea of a finite region in $\mathbb{R}^n$. We avoid using the word 'finite' because it suggests the idea of a finite volume, and the volume of sets in infinite dimensional spaces is hard to define.

The simple translation operator does not map a bounded set into itself. So perhaps the following generalisation of the contraction mapping theorem may work: does every continuous mapping of a bounded set into itself have a fixed point? The answer is still no, as the following example shows. Consider the annular region between two concentric circles in a plane. The operator which rotates each point through 1° (say) about the centre of the circles is continuous, because it does not change the distance between pairs of points, but it obviously has no fixed point. We must exclude such cases by considering only regions which have no holes. A full discussion of the meaning of 'regions without holes' would take us deep into topology, which is a delightful subject, but is not the subject of this book. We shall accordingly avoid topological definitions, and exclude sets with holes by the following idea.

Definition 6.16 A set S in a vector space is called **convex** if, for any $x,y \in S$, $ax + (1-a)y \in S$ for all $a \in [0,1]$. □

The meaning of this is that the finite line-segment joining any two members of the set lies entirely in the set (see Fig. 8). A region with a hole is obviously not convex, so if we restrict ourselves to mappings of bounded convex sets, we avoid both the above counterexamples to the fixed-point theorem, and in fact the following is true.

Theorem 6.17 (Brouwer's Fixed-Point Theorem) Every continuous mapping of a closed bounded convex set in $\mathbb{R}^n$ into itself has a fixed point. □

We have discussed the need for the boundedness and convexity conditions. The region must be closed: $x \longmapsto x/2$ is an example of a continuous mapping of the open interval $(0,1)$ into itself which has no fixed point. The restriction to finite-dimensional spaces is essential also, as we shall see in Example 6.19.

Sketch of Proof For a proof using 'elementary methods', that is, calculus and analysis but not topology, see Hochstadt; this proof contains no difficulties of principle, but is not at all simple. We shall briefly outline a proof which is more elegant and transparent, but uses a theorem of topology whose proof is quite beyond our scope.

Let G be the region and B its boundary. We shall suppose that $T: G \to G$ is a continuous map with no fixed point, and obtain a contradiction. Thus for all $x \in G$, $Tx \neq x$; hence there is a unique line joining Tx to x, which, when continued to the boundary (in the sense from Tx to x), hits it exactly once (see Fig. 9). We thus have a mapping from G to B, with the property that all points on B are

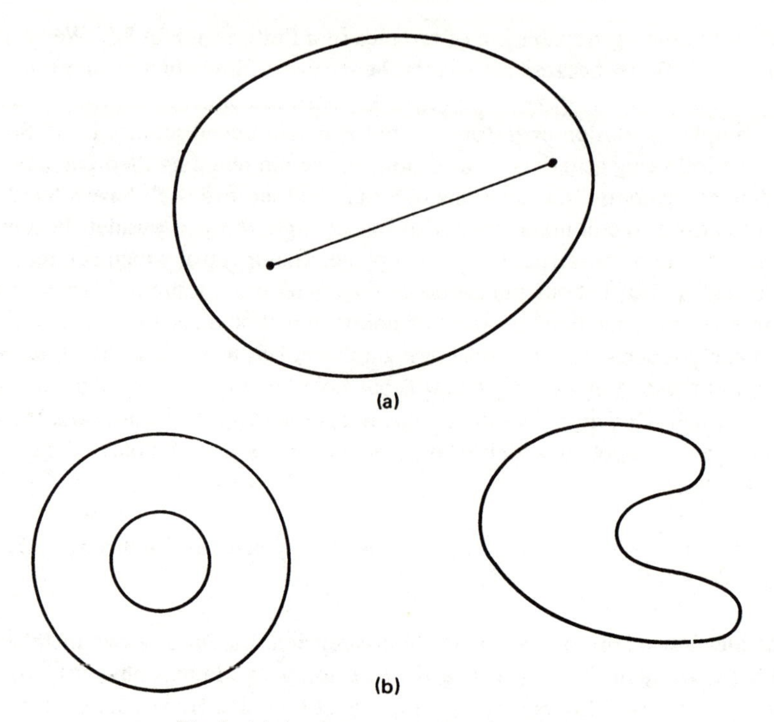

Fig. 8. (a) A convex set. (b) Non-convex sets.

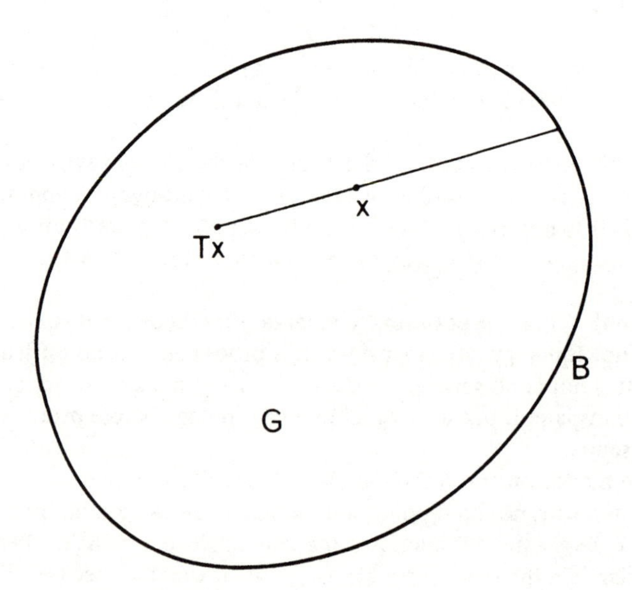

Fig. 9. Sketch for the proof of Brouwer's theorem.

mapped into themselves. This mapping is easily shown to be continuous if T is continuous.

But it is a theorem of topology that there is no continuous mapping of the interior G into B which leaves B itself fixed. This can be made plausible in two dimensions by imagining a mapping of a plane region G as a distortion of a rubber sheet covering the region; a continuous mapping is one in which nearby points are transformed into nearby points, and is represented by a distortion which stretches and contracts the rubber sheet but does not tear it. Now, a continuous mapping of G into B which leaves B fixed would correspond to taking a closed curve B made of wire, say, covering it with a rubber sheet like a drumhead, gluing the sheet where it touches the wire, and then distorting the sheet so that every point of it is brought into contact with the wire. It is clearly impossible to do this without tearing the sheet. A similar picture can be envisaged for three-dimensional space, and the whole argument can be made precise and rigorous (see Smart). This gives a contradiction, and proves the theorem. □

One of the most obvious continuous operators on $\mathbb{R}^n$ is given by a matrix, as in Example 6.5. The following result is an application of Brouwer's theorem to the theory of positive matrices, which are important in economics, probability theory, and in other fields; see Bellman (1960).

Theorem 6.18 (Perron's Theorem) A matrix whose elements are all positive has at least one positive eigenvalue, and the elements of the corresponding eigenvector are all non-negative.

Proof A positive matrix transforms vectors with positive components into vectors with positive components, so that it is reasonable, when looking for eigenvectors, to consider only vectors with positive components; geometrically, this means working in the n-dimensional analogue of the first quadrant. The eigenvalue equation concerns vectors mapped into multiples of themselves. If we could somehow disregard the magnitude of the vector and concentrate on its direction, then an eigenvector could be regarded as a 'fixed' vector under the transformation, in the sense that its direction is unchanged, and Brouwer's theorem applied. One way of doing this would be to normalise all vectors, that is, consider $x/\|x\|$ in place of x, which is equivalent to working on the unit sphere. In view of the remarks above, we would then apply our operator to the n-dimensional analogue of the positive quadrant (or rather, octant) of the surface of the unit sphere.

The surface of a sphere, however, is not convex. We shall therefore use a plane rather than a spherical surface, and define

$$S = \{x \in \mathbb{R}^n : x_i \geqslant 0 \quad \text{for each} \quad i \quad \text{and} \quad \sum_1^n x_i = 1\} \quad .$$

This is an n-dimensional analogue of the line-segment in two dimensions joining (0,1) and (1,0). It is easily shown to be closed, bounded, and convex. The condition defining S can be expressed neatly by means of the notation

$$\| x \|_1 = \sum_1^n |x_i| \quad . \tag{6.3}$$

As the notation suggests, this has all the properties of a norm. S can then be expressed as

$$S = \{x \in \mathbb{R}^n : \| x \|_1 = 1 \quad \text{and} \quad x_i \geqslant 0 \quad \text{for each} \quad i\} \quad ;$$

it is a section of the unit sphere with respect to the 1-norm (6.3). It is clear how to project vectors on to S: for any x in the region of $\mathbb{R}^n$ where all components are non-negative, $x/\| x \|_1 \in S$.

Now we define an operator $B: S \to S$ by

$$Bx = Ax/\| Ax \|_1 \quad \text{for} \quad x \in S \;;$$

since the elements of A are positive, $\| Ax \|_1 > 0$ for all $x \in S$. It can easily be shown (see Problem 6.7) that the continuity of A implies that of B. So Brouwer's theorem implies that there is an $x \in S$ such that $Ax/\| Ax \|_1 = x$; x is thus an eigenvector of A, with $\| Ax \|_1$ as its eigenvalue, and the elements of x are non-negative because $x \in S$. □

For a detailed discussion of this result and some generalisations, see Gantmacher. For a similar theorem on integral operators, see Problem 6.22.

Brouwer's theorem can be applied to ordinary differential equations; see Smart, for example. But we are more interested in infinite-dimensional spaces, to which it does not apply, as Example 6.19 shows. (For another example, see Problem 6.27.)

Example 6.19 Consider the space N of sequences which are infinite in both directions, $\ldots, x_{-2}, x_{-1}, x_0, x_1, x_2, \ldots$ and such that $\sum_{-\infty}^{\infty} |x_n|^2$ converges. This space is very similar to that of Problem 4.9; $\| x \|^2 = \sum_{-\infty}^{\infty} |x_n|^2$ is easily shown to define a norm. The set

$$B = \{x \in N : \| x \| \leqslant 1\} \quad ,$$

the unit ball, or solid sphere, is clearly bounded, convex, and closed (the fact that it is closed is proved exactly as in the n-dimensional case, Example 4.44). We shall give a continuous operator $T: B \to B$ with no fixed point.

First we define an operator U which shifts the sequence along one place:

$$Ux = y \quad \text{where} \quad y_n = x_{n-1} \quad .$$

Clearly $\| Ux \| = \| x \|$. Now define T by

$$Tx = Ux + (1 - \| x \|)z$$

where z is the sequence with $z_0 = 1$ and $z_r = 0$ for $r \neq 0$. T is easily shown to be continuous, and it maps B into B since if $\| x \| \leqslant 1$,

$$\| Tx \| \leqslant \| Ux \| + (1 - \| x \|) \| z \|$$

$$= \| x \| + 1 - \| x \| = 1 \quad .$$

Now we show that T has no fixed point. Suppose it has, and call it x. Then

$$x_n = (Tx)_n \quad ,$$

$$\therefore x_n = \begin{cases} x_{n-1} & \text{if} \quad n \neq 0 \\ x_{n-1} + 1 - \| x \| & \text{if} \quad n = 0 \end{cases} \tag{6.4}$$

by the definition of T. Hence $x_{-1} = x_{-2} = \ldots$, and $x_0 = x_1 = \ldots$. But the convergence of $\Sigma_{-\infty}^{\infty} |x_n|^2$ implies that $x_n \to 0$ as $n \to \infty$ and as $n \to -\infty$. So the only possibility is $x_{-1} = x_{-2} = \ldots = 0$ and $x_0 = x_1 = \ldots = 0$. But then $\| x \| = 0$, and the case $n = 0$ of equation (6.4) cannot be satisfied. This contradiction shows that T has no fixed point. Brouwer's theorem cannot be applied unchanged to infinite-dimensional spaces. □

This is our first example of infinite-dimensional spaces behaving in a way essentially different from finite-dimensional spaces; we shall meet others later. The point here is roughly as follows. A bounded region in a finite-dimensional space has a finite volume. Applying an operator T to a point many times in succession, in the manner of section 5.2, gives an infinite sequence, and if an infinite number of points lie in a bounded volume, there must be an infinite number of them concentrated in the neighbourhood of some point. This point is a limit point of the sequence, and may be a fixed point of the transformation T. But in an infinite-dimensional space, even a bounded region is so large, so to speak, by virtue of the infinite number of dimensions into which it extends, that a sequence can wander through the region indefinitely, each point being far away from the preceding ones, so that the sequence never converges and never reaches a fixed point. The condition of Definition 6.15 is not enough to give a region the intuitive properties of a finite region, in the infinite-dimensional case, and a new idea is needed. This is one of the most important ideas in analysis, and must be discussed at some length before we can formulate an infinite-dimensional analogue of Brouwer's theorem.

6.3 Compactness

The property needed to replace the condition of boundedness in infinite-dimensional spaces is the following, stemming directly from the ideas outlined at the end of the last section.

Definition 6.20 A subset S of a normed space B is **compact** if every infinite sequence of elements of S has a subsequence which converges to an element of S.

Example 6.21 The archetypal example of a compact set is a closed interval in $\mathbb{R}$. It is a theorem of classical analysis that every infinite sequence of real numbers in a closed interval has a convergent subsequence; we shall prove it below (Corollary 6.34). The archetypal example of a noncompact set is an open interval. In (0,1) for example, the sequence 1/2, 1/3, . . . converges to 0; therefore every subsequence converges to zero, and no subsequence can converge to a member of (0,1). Another example of a noncompact set is the set of all positive numbers; the sequence 1,2,3, . . . gives an obvious counterexample to Definition 6.20. □

In finite-dimensional spaces, the condition of Definition 6.20 is satisfied by closed and bounded sets. Compact sets can be thought of as a kind of generalisation of closed and bounded sets to infinite-dimensional spaces, – that is to say, compact sets in general have the nice properties associated with closed and bounded sets in finite-dimensional spaces (see Theorem 6.25, for example), while closed and bounded sets in infinite-dimensional spaces do not always share those properties.

Theorem 6.22 Compact sets are closed and bounded, but not vice versa, in general.

Proof A compact set S is closed, for if x is a limit point of S, then S contains a sequence converging to x; all its subsequences converge to x, but one of its subsequences must converge to an element of S if S is compact, hence x must be that element of S; thus S contains all its limit points and is closed. S must be bounded; for if not, then it must contain elements of norm greater than any given number, and in particular, for any integer n there must be an element x_n with $\|x_n\| > n$; but (x_n) is a sequence no subsequence of which can possibly converge, contradicting the compactness of S. This proves the first part of the theorem.

To show that the converse is false, we must display a non-compact closed and bounded set. Consider the unit ball (solid sphere) in $L_2[0,1]$, that is, the set of functions f such that $\|f\| = [\int_0^1 |f(x)|^2 \, dx]^{\frac{1}{2}} \leqslant 1$. It is obviously bounded, and almost as obviously closed: the limit of any convergent sequence of elements with norm $\leqslant 1$ will also have norm $\leqslant 1$, and therefore belong to the set. Now consider the sequence (f_n) where

$$f_n(x) = \sqrt{2} \sin(n\pi x) \quad .$$

It belongs to the unit ball, in fact $\|f_n\| = 1$ for all n. But it does not converge, and no subsequence of it can converge, because $\|f_n - f_r\| = \sqrt{2}$ for all $n \neq r$ (as an easy calculation shows), and therefore the Cauchy convergence criterion cannot be satisfied. Thus the unit ball is closed and bounded but not compact. □

Remark 6.23 In finite-dimensional spaces, closed and bounded sets are the same as compact sets. Half of this statement follows at once from Theorem 6.22; the other half is proved below (Corollary 6.34). □

Before proceeding towards the infinite-dimensional generalisation of the Brouwer fixed-point theorem, we shall develop the idea of compactness by showing one of the ways in which compact sets behave like closed bounded sets in $\mathbb{R}^n$. A continuous real-valued function on a closed interval has a maximum and a minimum value; a continuous function on an open interval or an infinite region need not have a maximum or a minimum. This obviously important distinction extends to $\mathbb{R}^n$ in the following form: a function $f: \mathbb{R}^n \to \mathbb{R}$ which is continuous on a closed bounded region of $\mathbb{R}^n$ has a maximum and a minimum value. Now, the literal extension of this result to all Banach spaces is false: a real-valued function on a closed bounded region of an infinite-dimensional space need not have a maximum value. But if 'closed and bounded' is replaced by the stronger condition 'compact', then the result is true for all normed spaces. We shall prove it as a consequence of the following theorem, which is interesting in its own right as indicating a natural connection between the ideas of continuity and compactness.

Theorem 6.24 (Continuous Images of Compact Sets) Continuous mappings take compact sets to compact sets. In other words, if M,N are normed spaces, $X \subset M$ is compact, and $T: X \to N$ is continuous, then the set $T(X) = \{Tx : x \in X\}$, the image of X under T, is compact.

Proof Take any sequence (y_n) of elements of $T(X)$; we shall show that it contains a convergent subsequence. For each y_n there is an element x_n of X such that $y_n = Tx_n$ (there may be more than one, in which case one must be chosen arbitrarily). The sequence (x_n) in the compact set X must contain a convergent subsequence, say $x_n' \to x' \in X$. By Theorem 6.10, the corresponding subsequence (y_n') of (y_n) converges to $y' = Tx' \in T(X)$, which proves the theorem. □

The significance of this result will become clearer in Chapter 8, when we discuss the property of continuity of mappings further and introduce the idea of complete continuity. Meanwhile we shall use it to prove

Theorem 6.25 A continuous real-valued function f on a compact set S in a normed space is bounded and attains its upper and lower bounds; that is, there exist $x_{\max}$ and $x_{\min} \in S$ such that $f(x_{\min}) \leqslant f(x) \leqslant f(x_{\max})$ for all $x \in S$.

Proof Using the notation and the result of the previous theorem, $f(S)$ is a compact and therefore (by 6.22) closed and bounded set of real numbers. Let M be its supremum; then for any integer n there is a $y_n \in f(S)$ such that $M - 1/n < y_n \leqslant M$ (Appendix C, Proposition 7). Clearly $y_n \to M$ as $n \to \infty$, hence $M \in f(S)$ because a closed set contains its limit points, so there is an $x_{\max} \in S$ such that $f(x_{\max}) = M \geqslant f(x)$ for all $x \in S$. Similarly for the lower bound. □

When, in later sections of this chapter, we apply fixed-point theorems to differential and integral equations, we shall need to be able to identify compact sets in $C[a,b]$. Theorem 6.41 gives a criterion allowing us to do this easily. The next two sections develop the machinery needed to state and prove this theorem.

6.4 Relative Compactness

The last section discussed an infinite-dimensional analogue of the condition 'closed and bounded'. If we drop the condition of being closed, we obtain the following weaker version of compactness, called relative compactness.

Definition 6.26 A subset S of a normed space N is **relatively compact** if every sequence in S has a subsequence converging to an element of N. □

The crucial difference between Definitions 6.26 and 6.20 is that in 6.26 the limit is not required to belong to the set S. Definition 6.26 is obviously weaker than 6.20; every compact set is relatively compact but not vice versa.

Proposition 6.27 Every subset of a compact or relatively compact set is relatively compact.

Proof Any sequence in the subset is a sequence in the relatively compact set containing it and therefore contains a convergent subsequence. □

Example 6.28 Any bounded set in $\mathbb{R}^n$ is by definition contained in some sphere, which is compact according to Remark 6.23. Therefore bounded sets in $\mathbb{R}^n$ are relatively compact by Proposition 6.27. An open interval in $\mathbb{R}$ is the standard example of a relatively compact set which is not compact (cf. Example 6.21). □

Example 6.28 suggests that the essential difference between compact and relatively compact sets is that compact sets are closed and relatively compact sets need not be closed. This suggests that a relatively compact set can be turned into a compact set by making it closed, that is, forming the closure of S, as defined in 4.46.

Theorem 6.29 The closure of a relatively compact set is compact.

Proof Let (x_n) be a sequence in $\overline{S}$, where S is relatively compact. Since S is dense in $\overline{S}$, for any n there is a y_n in S with $\| x_n - y_n \| < 1/n$. Since S is relatively compact, (y_n) has a convergent subsequence (y_r) with $y_r \to y$, say, and $y \in \overline{S}$. Now

$$\begin{aligned} \| x_r - y \| &\leqslant \| x_r - y_r \| + \| y_r - y \| \\ &< \tfrac{1}{r} + \| y_r - y \| \\ &\to 0 \quad \text{as} \quad r \to \infty \end{aligned}$$

Hence $x_r \to y \in \bar{S}$; thus we have found a subsequence of (x_n) which converges to an element of $\bar{S}$, so $\bar{S}$ is compact. □

Corollary 6.30 If a set is closed and relatively compact, then it is compact.

Proof Follows immediately from 6.29, since $S = \bar{S}$. □

The story so far can be summed up by saying that relatively compact sets are rather like bounded sets, and compact sets are rather like closed and bounded sets. The difficulty with these concepts, as we have developed them so far, is that it is not easy to decide in practice whether a given set is compact, that is, whether every sequence has a convergent subsequence. One would like to have a criterion for compactness which is easier to apply. Theorem 6.33 gives such a criterion. It is based on the following definition, which is another way of characterising sets which have the kind of finiteness property that we need.

Definition 6.31 Let S be a subset of a normed space. An **ϵ-net** for S is a set of elements of S such that each $x \in S$ is within a distance ϵ of some member of the net. A **finite ϵ-net** is an ϵ-net consisting of a finite number of elements.

Example 6.32 The set of all points in $\mathbb{R}^n$ whose coordinates are integer multiples of ϵ is an ϵ-net for $\mathbb{R}^n$. This set is infinite, and there is clearly no finite ϵ-net for $\mathbb{R}^n$. On the other hand, any bounded set in $\mathbb{R}^n$ has a finite ϵ-net consisting of less than V/ϵ^n points, where V is the volume of a box-shaped region containing the bounded set. □

This example suggests a connection between boundedness and the property of having a finite ϵ-net in finite-dimensional spaces. Relative compactness is also closely connected with boundedness in finite-dimensional spaces. The following theorem should therefore cause little surprise. Its proof is not easy, however; the essence of the idea of compactness lies in the statement that compactness is the same as having a finite ϵ-net, and it is quite non-trivial.

Theorem 6.33 (Compactness) A subset S of a Banach space B is relatively compact if and only if for each $\epsilon > 0$ it has a finite ϵ-net.

Proof (a) Suppose that S is relatively compact, and for some $\epsilon > 0$ there is no finite ϵ-net; we shall obtain a contradiction. For any $x_1 \in S$ there is an $x_2 \in S$ with $\|x_1 - x_2\| \geqslant \epsilon$ (otherwise $\{x_1\}$ would be an ϵ-net). There is an $x_3 \in S$ with $\|x_1 - x_3\| \geqslant \epsilon$ and $\|x_2 - x_3\| \geqslant \epsilon$ (otherwise $\{x_1, x_2\}$ would be an ϵ-net); and so on. Thus we obtain a sequence (x_n) with $\|x_r - x_s\| \geqslant \epsilon$ for all $r \neq s$. No subsequence of (x_n) can converge, because the Cauchy condition is violated; this contradicts the compactness of S. Hence there must be a finite ϵ-net for any $\epsilon > 0$.

(b) Suppose now that there is a finite ϵ-net for any $\epsilon > 0$; we shall prove that S is relatively compact. For any integer n, let $\{a_1^{(n)}, a_2^{(n)}, \ldots\}$ be a $(1/n)$-net, and Σ_i^n a sphere of radius $1/n$ centred on $a_i^{(n)}$, that is

$$\Sigma_i^n = \{x \in B : \| x - a_i^{(n)} \| \leqslant 1/n\} \quad .$$

Let (x_r) be any sequence in S. The finitely many spheres $\Sigma_1^1, \Sigma_2^1, \ldots$ together cover S (that is, every point of S lies in at least one of the spheres), so at least one of them must contain infinitely many of the x_r; let $(x_r^{(1)})$ be a subsequence of (x_r) which is contained in one of the Σ_i^1. Again, the finitely many spheres Σ_i^2 cover S, so at least one of them must contain infinitely many of the $x_r^{(1)}$; let $(x_r^{(2)})$ be a subsequence of $(x_r^{(1)})$ contained in one of the Σ_i^2. Repeating this procedure gives an infinite sequence of sequences, the n-th sequence being a subsequence of the $(n-1)$-th and contained in a sphere of radius $1/n$. Consider the 'diagonal' sequence $x_1^{(1)}, x_2^{(2)}, \ldots$. For any $\epsilon > 0$, all its terms after the $[1/\epsilon]$-th are inside a sphere of radius ϵ ($[c]$ denotes the largest integer $\leqslant c$). Hence $\| x_n^{(n)} - x_r^{(r)} \| < \epsilon$ for $n, r > 1/\epsilon$. Thus from any sequence (x_n) we have extracted a subsequence which is Cauchy, hence convergent since B is a Banach space. Therefore S is relatively compact. □

The point of this theorem is that if one wants to prove that a set is compact, it is easier to produce a finite ϵ-net than to show directly that every sequence has a convergent subsequence. We shall use this in the important Theorem 6.41 below, as well as in the following easy corollary.

Corollary 6.34 A bounded set in $\mathbb{R}^n$ is relatively compact; a closed bounded set in $\mathbb{R}^n$ is compact.

Proof The points whose coordinates are integer multiples of ϵ form an ϵ-net for $\mathbb{R}^n$ (in fact we could use more widely-spaced points, but in this kind of proof there is no point in trying to economise, the question is whether the number of points is finite or infinite, not how large or small it is). If S is a bounded set with $\| x \| < L$ for all $x \in S$, then the $(2L/\epsilon)^n$ points in a cube of side $2L$ centred on the origin form an ϵ-net for S. Therefore S is relatively compact by 6.33. The rest follows from 6.30. □

This result, together with Theorem 6.22, shows that in finite-dimensional spaces the notion of compactness is identical with that of being closed and bounded. For the infinite-dimensional function spaces in which we are more interested, this is not so. Since compactness is a key property in most of the rest of this book, it will be useful to have a criterion for deciding when a set of functions is compact. This involves a digression into the theory of continuous functions, and we shall therefore pause for breath.

6.5 Arzelà's Theorem

In this section we prove a theorem giving conditions for a subset of $C[a,b]$ to be compact. We need some new definitions.

Definition 6.35 A function $f: \mathbb{R} \to \mathbb{C}$ is **bounded** on an interval $I \subset \mathbb{R}$ if there is an M such that $|f(x)| \leqslant M$ for all $x \in I$. A family F of functions is **uniformly bounded** on I if there is an M such that $|f(x)| \leqslant M$ for all $x \in I$ and all $f \in F$. □

The force of the word 'uniformly' here is that the constant M is the same for all f in the set F. We follow traditional usage in sometimes using the word 'family' for a set of functions, but 'set' would do just as well.

Example 6.36 For any number a, the function $f_a(x) = ax(1 - x)$ is bounded on $[0,1]$. The family $\{f_a : 0 \leqslant a \leqslant 1\}$ is uniformly bounded on $[0,1]$, since $|f_a(x)| \leqslant 1/4$ for $0 \leqslant a \leqslant 1$. But $\{f_a : a \geqslant 0\}$ is not uniformly bounded, since $f_a(1/4)$ can be made as large as we please by a suitable choice of a. □

In Definition 6.3 we introduced the idea of uniform continuity of a mapping. Here we are mainly interested in $\mathbb{R} \to \mathbb{C}$ mappings, but it is useful to have the following theorem in its general form.

Theorem 6.37 (Uniform Continuity) A continuous mapping of a compact set is uniformly continuous.

Proof Suppose T is continuous but not uniformly continuous on a compact set X. Then the condition of Definition 6.3 is violated, and with some careful thought you will see that this means that there is an $\epsilon > 0$ such that for any n there exist $x_n, y_n \in X$ such that $\| x_n - y_n \| < 1/n$ but $\| Tx_n - Ty_n \| \geqslant \epsilon$. Since X is compact, (x_n) has a subsequence (x_r) converging to $x \in X$. If (y_r) is the corresponding subsequence of (y_n),

$$\| y_r - x \| \leqslant \| x_r - x \| + \tfrac{1}{r} \to 0$$

as $r \to \infty$, hence $y_r \to x$. Now

$$\| Tx_r - Ty_r \| \leqslant \| Tx_r - Tx \| + \| Tx - Ty_r \| \quad ,$$

and since T is continuous at x, each of these two terms can be made less than $\epsilon/2$ by taking r large enough, which contradicts the second sentence of the proof. □

Here we are mainly interested in the $\mathbb{R} \to \mathbb{C}$ case, where 6.37 says that a function continuous on a closed interval is uniformly continuous. Compactness is essential in this theorem; Example 6.4 shows that a function continuous on an open interval (a noncompact set) need not be uniformly continuous.

Definition 6.38 A family F of functions is **equicontinuous** on an interval $I \subset \mathbb{R}$ if for every $\epsilon > 0$ there is a $\delta > 0$ such that $|f(x) - f(y)| < \epsilon$ whenever $|x - y| < \delta$, $x,y \in I$, and $f \in F$.

Proposition 6.39 Every finite set of continuous functions on a closed interval is equicontinuous.

Proof If $f_1, f_2, \ldots, f_n$ are continuous on $[a,b]$, then by Theorem 6.37, for any $\epsilon > 0$ there exist $\delta_1, \delta_2, \ldots, \delta_n > 0$ such that for each i, $|f_i(x) - f_i(y)| < \epsilon$ for all x,y with $|x - y| < \delta_i$. Take δ equal to the smallest of $\delta_1, \ldots, \delta_n$, and the equicontinuity condition is satisfied. □

Remark 6.40 Although 6.35–6.39 have been expressed in terms of complex-valued functions on an interval in $\mathbb{R}$, they generalise immediately to continuous functions in any normed space; $[a,b]$ is then replaced by a compact set. In fact, essentially all of the calculus can be generalised to arbitrary normed spaces; in Chapter 11 we shall study differentiation in Banach spaces, with various applications. □

We have now, at last, assembled the ideas needed for stating a criterion for a set of continuous functions to be compact. The following theorem is useful in the application of functional analysis to concrete problems, as we shall see later in the chapter. Its proof, while not really difficult, is rather long, and may be omitted by readers uninterested in the minutiae of mathematical analysis.

Theorem 6.41 (Arzelà-Ascoli Theorem) A set of functions in $C[a,b]$, with the sup norm, is relatively compact if and only if it is uniformly bounded and equicontinuous on $[a,b]$.

Proof We use the characterisation of relative compactness in terms of finite ϵ-nets given in Theorem 6.33.

(a) Suppose that $S \subset C[a,b]$ is relatively compact. Then it has a 1-net consisting of a finite number of elements. Let M be the largest of their norms; then for any $f \in S$, $\|f\| < 1 + M$ since f is at a distance less than 1 from some member of the net. Since $\|f\| = \sup|f(x)|$, we have $|f(x)| < 1 + M$ for all x and all $f \in S$, so S is uniformly bounded.

Now we shall show that S is equicontinuous, because the elements of S can be approximated by a finite net, which is equicontinuous by Proposition 6.39, leading to equicontinuity of S itself. For any $\epsilon > 0$, let $g_1, g_2, \ldots, g_N$ be an $\epsilon/3$-net. Then for any $f \in S$ there is a k such that $\|f - g_k\| < \epsilon/3$. Hence for any $x_1, x_2 \in [a,b]$,

$$|f(x_1) - f(x_2)| \leqslant |f(x_1) - g_k(x_1)| + |g_k(x_1) - g_k(x_2)| + |g_k(x_2) - f(x_2)|$$
$$< \epsilon/3 + |g_k(x_1) - g_k(x_2)| + \epsilon/3 \quad . \tag{6.5}$$

But by 6.39 there is a $\delta > 0$ such that for all $k \leqslant N$, $|g_k(x_1) - g_k(x_2)| < \epsilon/3$ whenever $|x_1 - x_2| < \delta$. For this δ we therefore have, from (6.5),

$$|f(x_1) - f(x_2)| < \epsilon \quad \text{for all} \quad x_1, x_2 \in [a,b] \quad \text{with} \quad |x_1 - x_2| < \delta \quad . \tag{6.6}$$

But δ is independent of f, because it is independent of k, so (6.6) shows that S is equicontinuous. This completes the proof of the 'only if' part of the theorem.

(b) Suppose now that S is equicontinuous and uniformly bounded; we shall construct a finite ϵ-net. For any $\epsilon > 0$ there is a $\delta > 0$ such that

$$|f(x) - f(\xi)| < \epsilon/5 \quad \text{whenever} \quad |x - \xi| < \delta \quad \text{and} \quad f \in S \quad . \tag{6.7}$$

Choose a set of N numbers $x_1 < x_2 < \ldots < x_N$ with $x_1 = a$, $x_N = b$, and $x_{i+1} - x_i < \delta$ for all i; N is a number $> (b - a)/\delta$ so that the conditions on the x's can be satisfied. Our ϵ-net will consist of functions linear in each interval (x_i, x_{i+1}). To construct these functions, choose numbers $y_1 < y_2 < \ldots < y_R$ with $y_1 = -M$, $y_R = M$, and

$$y_{i+1} - y_i < \epsilon/5 \quad \text{for} \quad i = 1, \ldots, R \quad .$$

M here is the bound on S (that is, $\|f\| \leqslant M$ for all $f \in S$), and R is a number as large as is needed to satisfy the conditions on the y's. We have a finite number of points (x_i, y_j), which we shall call lattice points. Let L be the finite set of all functions on $[a,b]$ whose graphs are polygons with vertices at lattice points; we shall show that L is an ϵ-net for S.

Take any $f \in S$; we must show that L contains a function differing from f by less than ϵ at all points. We choose the function $g \in L$ whose graph passes through the lattice points closest to the graph of f; more precisely, for each i, $g(x_i)$ equals the member of the set $\{y_j\}$ closest to $f(x_i)$. Since the y_j are at most $\epsilon/5$ apart,

$$|f(x_i) - g(x_i)| < \epsilon/5 \quad \text{for each} \quad i \quad . \tag{6.8}$$

Now, for any $x \in [a,b]$, let x_k be the member of the set $\{x_i\}$ closest to x. Then

$$|f(x) - g(x)| \leqslant |f(x) - f(x_k)| + |f(x_k) - g(x_k)| + |g(x_k) - g(x)|$$
$$< \epsilon/5 + \epsilon/5 + |g(x_k) - g(x)| \quad , \tag{6.9}$$

using (6.8) and (6.7), plus the fact that $|x - x_k| < \delta$. Now, because g is monotonic between adjacent x_i's,

$$|g(x_k) - g(x)| \leqslant |g(x_k) - g(x_n)|$$

where $n = k \pm 1$ according as $x \gtrless x_k$. Hence

$$|g(x_k)-g(x)| \leqslant |g(x_k)-f(x_k)| + |f(x_k)-f(x_n)| + |f(x_n)-g(x_n)|$$
$$< \epsilon/5 + \epsilon/5 + \epsilon/5$$

using (6.7) and (6.8). Combining this with (6.9) shows that $\|f-g\| < \epsilon$, and this proves that L is an ϵ-net. Thus S is relatively compact by Theorem 6.33. □

This completes our discussion of compactness, and we can now return to Brouwer's fixed-point theorem, and state its infinite-dimensional analogue.

6.6 Schauder's Theorems

Theorem 6.42 (Schauder's First Theorem) If S is a convex compact subset of a normed space, every continuous mapping of S into itself has a fixed point. □

Remarks 6.43 (a) As in the case of Brouwer's theorem, the condition that S be non-empty should strictly speaking be added. But the case when S is empty is so trivial that it seems pedantic to insist on including 'non-empty' in the statement.

(b) In view of all the song and dance in section 4.5 about the importance of completeness, the reader may be surprised to see that Schauder's theorem applies to any space, whether complete or not. However, when applying the theorem one must prove that S is compact, and this generally involves completeness. In Theorem 6.45, for example, we prove compactness of S by means of the Arzelà-Ascoli theorem, and this involves the completeness of $C[a,b]$ because its proof uses Theorem 6.33. □

The proof of Theorem 6.42 is fairly tedious. The deep topological idea behind the fixed-point theorems is contained in Brouwer's theorem; the proof of Schauder's theorem consists of approximating the infinite-dimensional set S by a finite-dimensional set, applying Brouwer's theorem to deduce the existence of a fixed point of the finite-dimensional approximation, and then taking the limit as the dimension of the approximating space tends to infinity. See Hochstadt or Courant-Hilbert vol. 2 p. 403, for example, for the details.

The following slight generalisation of Theorem 6.42 is often useful.

Theorem 6.44 (Schauder's Second Theorem) If S is a convex closed subset of a normed space and R a relatively compact subset of S, then every continuous mapping of S into R has a fixed point.

Proof This theorem can be deduced from Theorem 6.42 as follows. Let U be the convex hull of R, that is, the convex set obtained by adding to R just enough points to make it convex (more precisely, U is the intersection of all convex sets containing R). Then $U \subset S$ (see Problem 6.5), and U is relatively compact (Problem 6.17). By 6.29, $\bar{U}$ is compact, and by 4.47(c) $\bar{U} \subset S$. Hence the operator maps $\bar{U}$ into itself, and therefore has a fixed point by 6.42. □

We now apply these theorems to integral equations. We consider nonlinear equations of the form

$$u(x) = \int_a^b K(x,y)f(y,u(y))\,dy \quad , \tag{6.10}$$

known as Hammerstein equations; here K and f are given functions and u is unknown. We studied equations of this type in Chapter 5; the contraction mapping theorem gives existence and uniqueness results provided that K is not too large. Schauder's theorem gives no information on uniqueness, and is in this respect less powerful than the contraction mapping theorem, but it involves no restriction on the size of K and is in this respect more powerful. The following is an example of the kind of result one can prove.

Theorem 6.45 (Existence Theorem for Hammerstein Equations) If $K(x,y)$ is continuous for $a \leqslant x,y \leqslant b$, and $f(y,z)$ is continuous and bounded for $a \leqslant y \leqslant b$ and all z, then the equation (6.10) has a continuous solution.

Proof We work in the Banach space $C[a,b]$ and use Schauder's second theorem. We must find a closed convex set in $C[a,b]$ which is mapped into a relatively compact subset of itself by the integral operator in (6.10), and we must prove that the operator is continuous. The easiest convex set to use is a sphere; we shall choose its radius so that it is mapped into itself.

Define an operator T: $C[a,b] \to C[a,b]$ by

$$(Tu)(x) = \int_a^b K(x,y)f(y,u(y))\,dy \quad . \tag{6.11}$$

Now, f is given to be bounded, say $|f(y,z)| \leqslant B$, and K is continuous on a compact region and therefore bounded, say $|K(x,y)| \leqslant A$, by Theorem 6.25. Hence

$$\begin{aligned} (Tu)(x) &\leqslant \int_a^b |K(x,y)|\,|f(y,u(y))|\,dy \\ &\leqslant AB \int_a^b dy \end{aligned}$$

$$\therefore \|Tu\| \leqslant AB(b-a) = D, \quad \text{say.}$$

Thus T maps the whole space into the sphere of radius D, so if we write

$$S = \{u \in C[a,b] : \|u\| \leqslant D\}$$

then $T{:}S \to S$. S is not compact, so we cannot apply Theorem 6.42. We must show that the set

$$R = \{Tu : u \in S\}$$

is relatively compact, and use Theorem 6.44.

We prove the relative compactness of R by the Arzelà-Ascoli theorem. R is uniformly bounded because $\|Tu\| \leqslant D$ for any u. K is continuous and therefore uniformly continuous on the closed square $a \leqslant x,y \leqslant b$, by Theorem 6.37. Hence for any $\epsilon > 0$ there is a $\delta > 0$ such that

$$|K(x_1,y) - K(x_2,y)| < \epsilon/[B(b-a)] \quad \text{whenever} \quad |x_1 - x_2| < \delta \quad .$$

For any u,

$$|(Tu)(x_1) - (Tu)(x_2)| = |\textstyle\int_a^b [K(x_1,y) - K(x_2,y)] f(y,u(y))\,dy\,|$$
$$< (b-a)[\epsilon/B(b-a)]B = \epsilon \quad .$$

Thus R is equicontinuous, and hence relatively compact by Theorem 6.41.

Finally we must show that T is a continuous operator, that is, that for each $v \in C[a,b]$, $Tu \to Tv$ as $u \to v$. We have, for each $x \in [a,b]$,

$$\|Tu - Tv\| \leqslant |\textstyle\int_a^b K(x,y)[f(y,u(y)) - f(y,v(y))]\,dy|$$
$$\leqslant \textstyle\int_a^b |K(x,y)|\,|f(y,u(y)) - f(y,v(y))|dy.$$

Now, $f(y,z)$ is continuous on the compact set $\{(y,z) \in \mathbb{R}^2 : a \leqslant y \leqslant b, |z| \leqslant 2\,\|v\|\}$, and therefore uniformly continuous there by 6.37. Hence for every $\epsilon > 0$ there is a $\delta > 0$ such that $|f(y,u(y)) - f(y,v(y))| < \epsilon/A(b-a)$ when $|u(y) - v(y)| < \delta$, provided that $\|u\| < 2\,\|v\|$, which will certainly be the case when we let $u \to v$. So, if $\|u - v\| < \delta$ we have $\|Tu - Tv\| \leqslant \epsilon \int_a^b |K(x,y)|dy/[A(b-a)] \leqslant [\epsilon/A(b-a)] \int_a^b A dx = \epsilon$. Thus T is a continuous operator. We now apply Theorem 6.44 and the proof is complete.

□

Although this theorem has no restriction on the size of the kernel, and is therefore stronger than the results of Chapter 5, it is powerless to deal with linear equations or equations like (5.17) which reduce to linear homogeneous equations when u is small. Such equations are always satisfied by $u \equiv 0$, so we know already that they have a solution, and results based on the Schauder theorem give us no new information.

6.7 Forced Nonlinear Oscillations

In this section we apply the Schauder and Contraction Mapping Theorems to a problem in the theory of vibrations. One of the simplest vibrating systems is a mass on the end of a spring, possibly subject to an external force as well as the force exerted by the spring. This system is described by the equation

$$d^2x/dt^2 + \alpha^2 x = F(t) \quad , \tag{6.12}$$

where α^2 is proportional to the strength of the spring and $F(t)$ to the external force at time t. If the force F is zero for all t, then the solution is $x = a\sin(\alpha t + \delta)$ where a and δ are constant; this is the natural vibration of the system, with (angular) frequency α and period $2\pi/\alpha$. If there is an external force $F(t)$ which is periodic, with angular frequency ω, then in general (6.12) has a periodic solution with the same frequency ω; physically this means that when an oscillatory force acts on the system it responds by vibrating steadily in sympathy with that force. However, there is an exception to that rule: if $\omega = \alpha$ so that the force has the same frequency as the natural vibration of the system, then there is no periodic solution of (6.12); all solutions grow indefinitely as t increases. The system is said to 'resonate' at frequency α, a force of finite magnitude producing, as it were, an infinite response. As the frequency of the force F approaches α, the response grows without bound; see Fig. 10.

(6.12) is in two respects an idealisation of reality. Firstly, it takes no account of frictional forces. If they are included, then the infinite resonance is tamed, and the equation has a periodic solution, representing steady vibration, for any periodic $F(t)$. This may be explained physically by saying that in the presence of friction there is a limit on the amplitude of the motion that a given force can produce: the faster the system moves in response to the force, the more energy is dissipated by friction, until a balance is reached between energy input from the external force and energy dissipated by friction. This argument suggests that for a steady periodic force there will always be a steady periodic response, and no infinite resonance phenomenon. If there is little friction, then the amplitude of the response is large when $\omega = \alpha$, as shown in Fig 11. The effect of friction is thus to turn the 'infinite peak' of Fig. 10 into the large but finite peak of Fig. 11.

Now, there is a second respect in which (6.12) is an idealisation: it is a linear equation, whereas oscillators in nature are nonlinear. The classic example of an oscillator is a pendulum, described by the equation

$$d^2x/dt^2 + \alpha^2 \sin x = F(t) \tag{6.13}$$

where $x(t)$ is the angle of the pendulum to the vertical, α is related to its length, and $F(t)$ is the external torque applied. (6.12) is a reasonable approximation to (6.13) for small angles, but is clearly inaccurate for the large values of x involved in the resonance phenomenon. It is natural to ask whether or not (6.13) has a periodic solution for all periodic $F(t)$, that is, whether nonlinearity can quench an infinite resonance in the way that friction can. There is no obvious physical mechanism for quenching here, and there is no reason to suppose that (6.13) will not resonate in the way that (6.12) does. However, by means of Schauder's theorem we shall show that (6.13), unlike (6.12), always has periodic solutions, so that nonlinearity is capable of quenching a resonance even in the absence of friction. This general conclusion applies to a wide class of physical systems, though we shall deal only with the relatively simple equation (6.13).

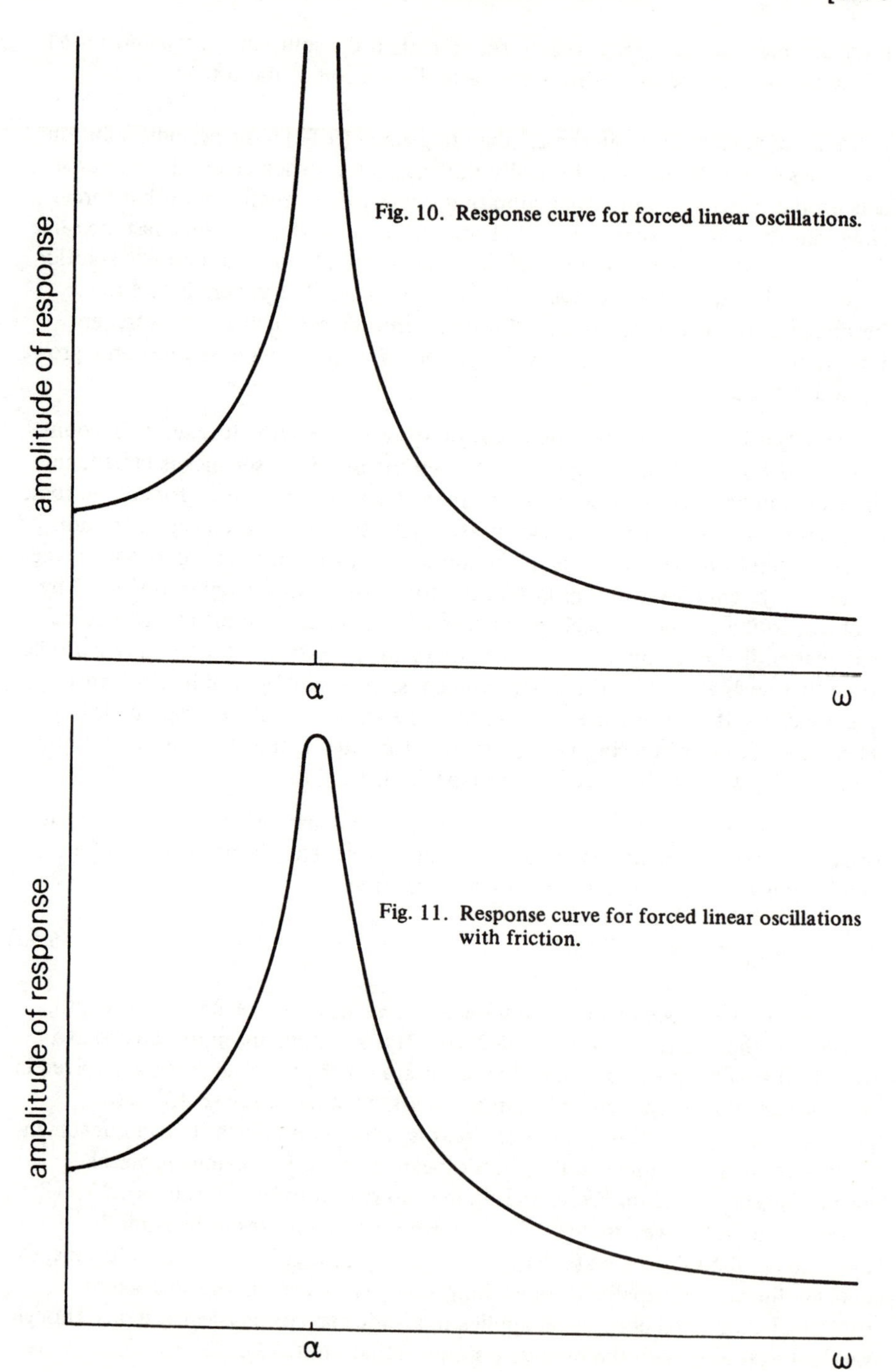

Fig. 10. Response curve for forced linear oscillations.

Fig. 11. Response curve for forced linear oscillations with friction.

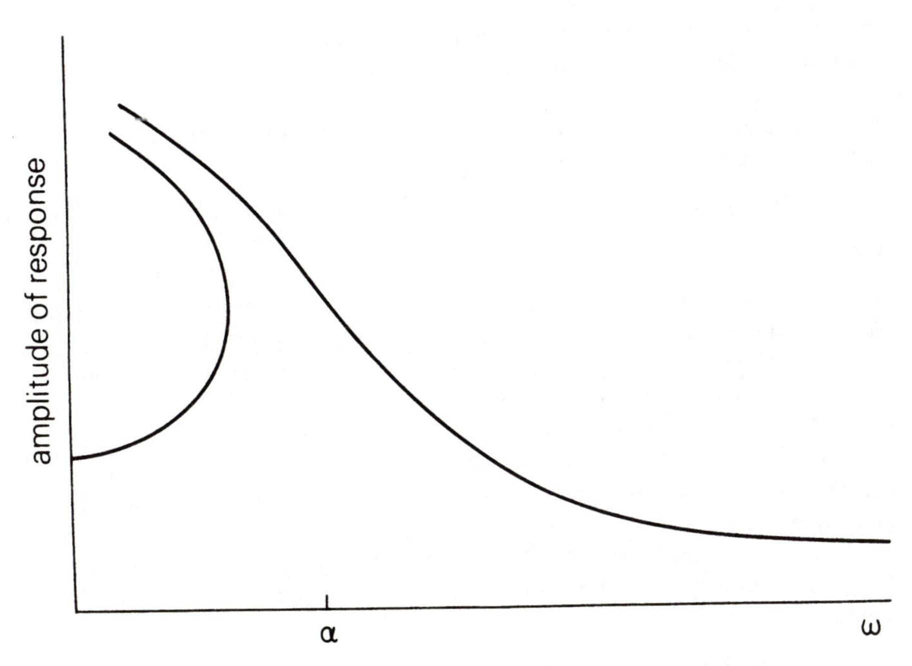

Fig. 12. Response curve for forced nonlinear oscillations.

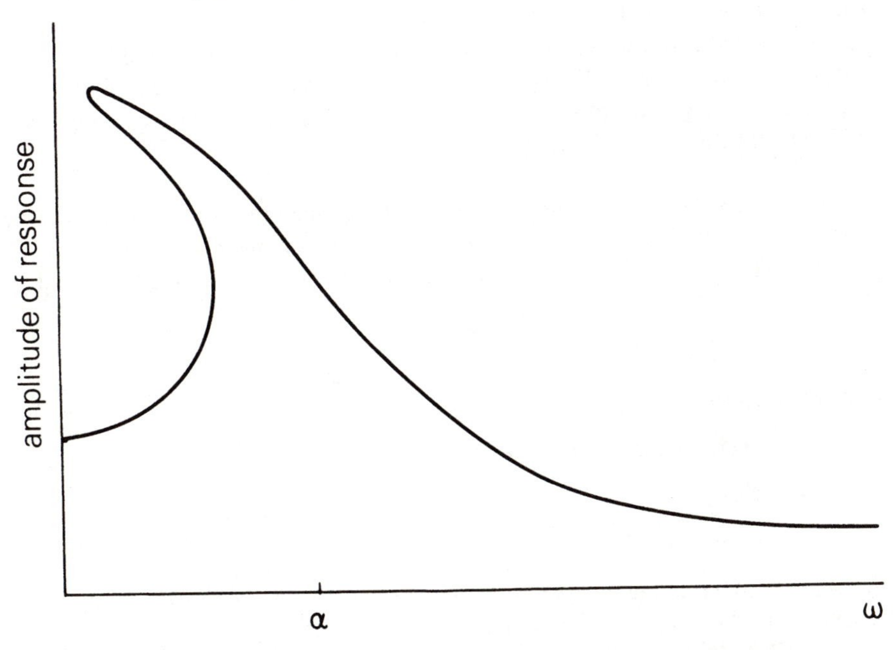

Fig. 13. Response curve for forced nonlinear oscillations with friction.

Theorem 6.46 If the function F is continuous, odd, and periodic, then equation (6.13) has a periodic solution of the same period as F.

Proof We first simplify the problem by rescaling time so that the force has period 2. Define $s = \omega t/\pi$, then (6.13) becomes

$$d^2x/ds^2 + \beta^2 \sin x = f(s) \tag{6.14}$$

where $$\beta = \alpha\pi/\omega$$

and $f = \pi^2 F/\omega^2$. The period of f is 2; we shall prove that there is an odd periodic solution $x(s)$ such that $x(0) = 0$ (as for any odd function) and $x(-1) = x(1)$.

The problem of finding a periodic solution can be reduced to a two-point boundary-value problem on $[0,1]$. For suppose that we have found a solution $x(s)$ of (6.14) on $[0,1]$ satisfying $x(0) = x(1) = 0$. We can extend it to $[-1,1]$ by defining $x(-s) = -x(s)$ for $0 \leqslant s \leqslant 1$, and then extend it to all s by making it periodic with period 2. It then satisfies (6.14) everywhere, and is the required periodic solution.

Thus our problem is to solve (6.14) on the interval $[0,1]$ with homogeneous two-point boundary conditions. We use Green's function to turn it into the integral equation

$$x(s) = \int_0^1 g(s;u)\,[\beta^2 \sin x(u) - f(u)]\,du \tag{6.15}$$

where g is Green's function for the operator $-D^2$, given by equation (5.14). But equation (6.15) is of the type (6.10) to which Theorem 6.45 applies; g is a continuous function, and $\beta^2 \sin x - f(u)$ is continuous and bounded, and has a bounded derivative with respect to x. We conclude that the integral equation has a solution, and this proves the theorem. □

Schauder's theorem tells us nothing about how many solutions there are. Some uniqueness results can be obtained from the contraction mapping theorem, by means of Theorem 5.17. This says that the solution of (6.15) is unique if $P < 1$, where

$$P^2 = \int_0^1 ds \int_0^1 du\, |N(s,u)|^2 \quad ,$$

and N is such that

$$\beta^2 |g(s;u)[\sin x - \sin y]| \leqslant N(s,u)\,|x - y|$$

for all x,y. Now,

$$\sin x - \sin y = (x - y)\cos z$$

for some z, so we can take $N(s,u) = \beta^2 g(s;u)$, giving $P = \beta^2/\sqrt{90}$ after some calculation. The uniqueness condition $P < 1$ thus becomes $\beta < \sqrt[4]{90}$, or

$$\omega > (\pi/3{\cdot}08)\alpha \quad .$$

Thus the nonlinear equation has a unique periodic solution when the forcing frequency is greater than the natural frequency, roughly speaking. We might expect something peculiar to happen in the neighbourhood of the linear resonance frequency, and in fact one finds that there are three possible solutions for some values of ω, one of which is unstable; see Fig. 12. Experiment shows that as ω is steadily increased, the system can jump suddenly from one solution branch to another.

If friction is included as well as nonlinearity, the response diagram is as shown in Fig. 13. It is now possible for the solution to change discontinuously when ω is decreased (the system falls over the edge of the upper branch of the curve) as well as when ω is increased. For a general discussion and analysis of these phenomena, see Jordan & Smith, or Stoker.

6.8 Swirling Flow

In this section we discuss a class of problems in fluid mechanics in which both physical reasoning and numerical computation have been on occasion misleading, and where the methods of this chapter have given useful information about the dynamics. These problems concern the flow of a fluid, such as water or air, in contact with a large rotating disc. We begin with a brief and simple-minded discussion of the physics of the situation; a thorough account will be found in Batchelor (1967). We then sketch some of the recent history of the subject, and finally we show how Schauder's theorem is applied to these problems.

It is a general principle of fluid mechanics that when a fluid is in contact with a solid boundary, the layer of fluid adjacent to the boundary has the same velocity as the boundary. Friction at the fluid-solid interface prevents the fluid from slipping sideways relative to the boundary. Thus, for example, when you stir a cup of tea, there is a thin layer of tea at the side of the cup which never moves. We shall return to the dynamics of tea later.

Consider now a large disc spinning with angular velocity Ω about an axis through its centre perpendicular to its plane. We idealise it as having an infinite radius, and suppose it to be in a horizontal plane, with fluid above it. The layer of fluid in contact with the disc has the same velocity as the disc, as explained above, and thus spins with angular velocity Ω. It will experience a centrifugal force trying to push it outwards, which is balanced by the friction force which keeps it at rest relative to the disc. Now, the layer of fluid just above the first layer will be set into rotation by the first layer, and will therefore also feel a centrifugal force. Since it is not in contact with a solid, the friction force on it is limited, and it will therefore yield to the centrifugal force, to some degree. We thus expect the fluid in a region near the disc to move outwards, away from the axis of rotation; its place must be taken by fluid drawn towards the disc, parallel to the axis of rotation. We are thus led to the picture shown in Fig. 14; the spinning disc acts as a fan, drawing air inwards along the axis and expelling it sideways. Fig. 14 has been simplified by showing only the velocity components in a plane through the axis; there is a swirl around the axis combined with this, so that the fluid really moves in complicated spiral paths. But the pumping effect in axial planes, illustrated in Fig. 14, is the main feature of the motion.

Now consider a different situation, with the disc at rest, but the fluid at some distance from the disc being forced to rotate (perhaps by being stirred by a teaspoon). There must be an inwards pressure to counteract the centrifugal tendency of the spinning fluid, and keep it rotating at a constant distance from the axis. However, there is a layer of fluid near the disc which is spinning more slowly,

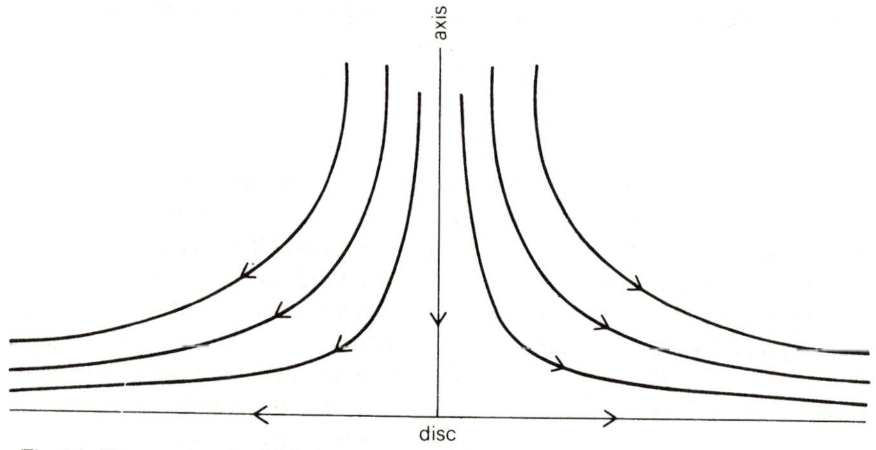

Fig. 14. Flow pattern in axial planes produced by a rotating disc with no swirl at infinity.

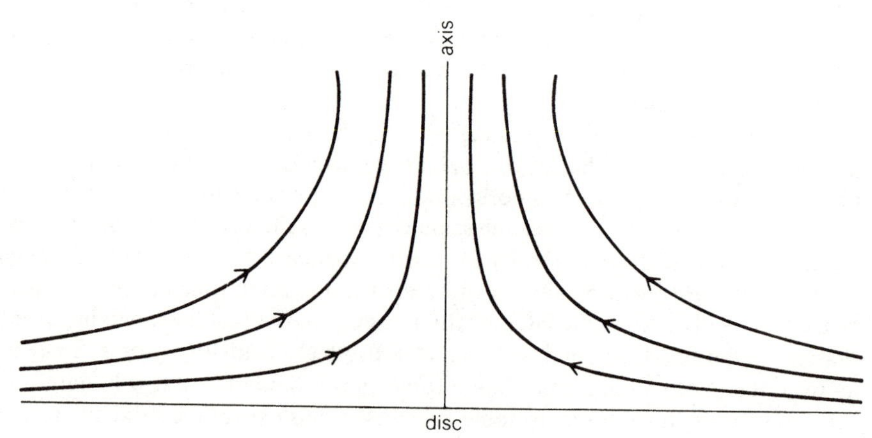

Fig. 15. Flow pattern in axial planes for swirling fluid above a stationary disc.

Figures 14, 15 and 16 are taken, with permission, from: G. K. Batchelor, *Q. J. Mech. Appl. Math.* 4 (1951) 29–41 (Oxford U.P.).

the rotation rate tending to zero as the surface of the stationary disc is approached. Here there is less centrifugal force, so the inwards pressure dominates, and fluid is driven inwards near the disc, and consequently pumped outwards along the axis, as shown in Fig. 15. A flow of this kind is set up when a cup of tea is stirred. The inwards pressure here is exerted by the sides of the cup, and the inflow along the bottom can be clearly seen by the tendency of tealeaves, sugar crystals, etc. to collect at the centre. The circulation is completed by outflow at the surface and downflow at the sides of the cup, but that does not concern us here.

The two cases considered above are fairly straightforward. It is interesting to ask what happens when the disc rotates with angular velocity Ω_0 and the fluid at infinity rotates with angular velocity Ω_∞. If $\Omega_\infty = \Omega_0$ there is an obvious solution in which the whole of the fluid rotates as if it were a solid body. If $0 < \Omega_0 < \Omega_\infty$ it is reasonable to expect a solution resembling the $\Omega_0 = 0$ case, and if $0 < \Omega_\infty < \Omega_0$ it is reasonable to expect a solution resembling the $\Omega_\infty = 0$ case. When applied mathematicians first considered this problem, it was naturally conjectured that there is a solution of the equations for each pair of values of Ω_0, Ω_∞ and that if Ω_0 and Ω_∞ have different signs, then the flow pattern combines the centrifugal outflow along the disc with axial outflow at large distances as shown in Fig. 16

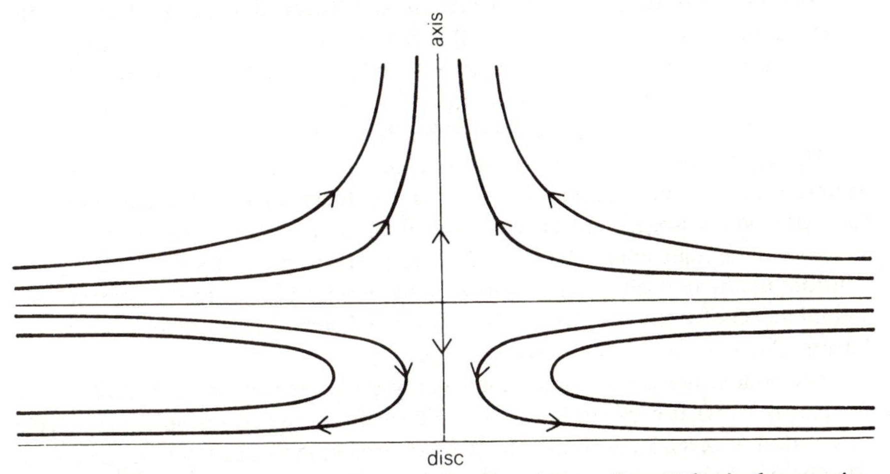

Fig. 16. A conjectured flow pattern for swirling flow above a disc rotating in the opposite direction.

(see Batchelor (1951)). In another paper published two years later, however, a quite different flow pattern was suggested for the case $\Omega_\infty/\Omega_0 < 0$ (Stewartson (1953)). Section III of the review article by Moore (1956) gives a clear impression of the confusion reigning at this time.

An attempt was made by Rogers & Lance (1960) to clarify the position by numerical computation. They obtained satisfactory solutions for $\Omega_\infty/\Omega_0 > 0$. But for $-0.2 < \Omega_\infty/\Omega_0 < 0$ they described their solutions as having anomalous features, and for $\Omega_\infty/\Omega_0 < -0.2$ were unable to find any solutions at all.

The plot thickens further when we consider a very closely related problem: the flow between two discs, in parallel planes a distance d apart, and rotating with angular velocities Ω_0, Ω_1 about the same axis. The one-disc problem described above may be regarded as the limit of the two-disc problem as $d \to \infty$. Batchelor (1951) and Stewartson (1953) considered the two-disc problem as well as the one-disc problem, and when the discs rotate in opposite directions, Stewartson's proposed flow pattern was different from Batchelor's. Again attempts were made to resolve the difficulty by numerical computation. Calculations by Lance & Rogers (1962) supported Stewartson's picture; calculations by Greenspan (1972), on the other hand, supported Batchelor's picture. This state of affairs was, to say the least, unsatisfactory.

It is at this point that Schauder's theorem enters the story, giving a clear resolution of the conflict between the two pictures. McLeod & Parter (1974) studied the case $\Omega_0 = -\Omega_1$ in great mathematical detail, and they used Schauder's theorem to prove the existence of a solution of the type suggested by Stewartson; they showed also that Greenspan's calculations must be incorrect. Their results are confirmed by theoretical work along quite different lines (Matkowsky & Siegmann (1976)).

We now return to the one-disc problem, and Rogers & Lance's failure to find numerical solutions when $\Omega_\infty/\Omega_0 < -0.2$. McLeod (1970) gave a rigorous proof that the equations have no solution when $\Omega_\infty/\Omega_0 = -1$, which is a step in the right direction. But he offered the opinion that a solution does exist for all other values of Ω_∞/Ω_0, despite the difficulties found by Rogers & Lance and other workers.

The key to the puzzle was found by Zandbergen & Dijkstra (1977). The trouble is not that there is no solution, it is that there is more than one solution in the region where Rogers and Lance had difficulties, a situation with which their numerical technique could not deal. Dijkstra (1980) has discussed the multiple solutions mathematically, using approximate methods; Elcrat (1980) has shown nonuniqueness for the one-disc problem from a different point of view, using Schauder's theorem. See also Kreiss and Parter.

The nonexistence and nonuniqueness results quoted above raise some interesting questions. What happens if one tries physically to set up a problem which has no solution or several solutions? One way of attacking this question is the following. The nonexistence and nonuniqueness results apply to the equations of steady flow, in which all variables are assumed independent of time. One can represent the physical problem better by allowing the flow pattern to vary with time. One can then investigate a particular boundary-value problem by starting the flow from some initial state and calculating its time-development. If the steady-flow equations have a unique solution, then we expect that the time-varying flow will approach that unique steady state (though this may not always be true: see Bodonyi (1978) for an exception). If the steady-flow equations have no solution, there are various possibilities; the flow may fluctuate indefinitely, for example, never settling down to a steady state.

Bodonyi & Stewartson (1977) and Banks & Zaturska (1981) have used this approach to study the one-disc problem with $\Omega_0 = -\Omega_\infty$, for which McLeod proved that there is no steady solution. They started from an initial state in which the disc and the fluid all rotate with uniform angular velocity Ω_∞. At time $t = 0$ the rotation of the disc is suddenly reversed, so that $\Omega_0 = -\Omega_\infty$. Numerical calculations show that the resulting disturbance in the flow field grows rapidly, and the velocity of the fluid tends to infinity as $t \to t^* \simeq 2.58/\Omega_\infty$. This is a very remarkable result. Before the disc has had time to make half a turn, the flow field has exploded. Banks and Zaturska have confirmed the numerical results by an analytical approximation to the equations, showing that the velocity components behave like $(t^* - t)^{-1}$ and $(t^* - t)^{-2}$ as $t \to t^*$. There is clearly much more to be done before these swirling flows are fully understood.

We end this section by describing the application of Schauder's theorem in more detail. Much of the work is far too complicated to be given here, but we shall try to give some of its flavour by summarising the earliest, and simplest, application of Schauder's theorem to swirling flow problems. We have seen that the one-disc problem presents difficulties when Ω_0 and Ω_∞ have opposite signs. The case $\Omega_\infty = 0$ is on the border of this difficult region, and is not easy to treat mathematically. But it can be mathematically stabilised and made more amenable to analysis by changing the boundary condition on the disc so as to allow fluid to be sucked into the surface, as if the disc were porous. One of the first existence proofs for swirling flow was for this modified problem; Howard (1961) proved existence of a solution for sufficiently large suction. We shall outline Howard's analysis.

We use cylindrical polar coordinates r, θ, z, with the disc on the plane $z = 0$, spinning about the z-axis with angular velocity Ω. The degree of suction applied at the disc is measured by a parameter $c = -(\nu\Omega)^{-\frac{1}{2}} w_\infty$, where ν is the kinematic viscosity of the fluid (a measure of how much friction it has: ν is relatively small for water and large for treacle), and w_∞ is the axial velocity of the fluid at large distances. It is convenient to map the region $z > 0$ occupied by the fluid to the interval $[0,1]$ by the transformation $x = e^{-cz\sqrt{\Omega/\nu}}$. Now, the motion is governed by a set of partial differential equations for the velocity components u_r, u_θ, u_z as functions of the four variables r, θ, z, t. However, the problem is reduced to a set of ordinary differential equations as follows. One looks for solutions which are independent of t, representing steady rotation, and independent of θ, corresponding to symmetry about the axis of rotation; one then has equations in r and x. The form of these equations is such that there are solutions, called similarity solutions, whose r-dependence is trivial: u_z is independent of r, and u_r and u_θ are proportional to r (for a general discussion of these similarity solutions, see Batchelor (1967)). The differential equations of motion then reduce to the system

$$\left.\begin{aligned} &\mathrm{d}\phi/\mathrm{d}x = -[f^2 - g^2 - k\phi]/c^2, \quad \mathrm{d}\psi/\mathrm{d}x = -[2fg + k\psi]/c^2 \\ &\mathrm{d}(xf)/\mathrm{d}x = -\phi, \quad \mathrm{d}(xg)/\mathrm{d}x = -\psi, \quad \mathrm{d}(xk)/\mathrm{d}x = 2f \end{aligned}\right\} \qquad (6.16)$$

where $f = u_r/\Omega rx$, $g = u_\theta/\Omega rx$, $k = (c/x)[c + u_z(\nu\Omega)^{-\frac{1}{2}}]$, and ϕ and ψ are defined by the third and fourth equations in (6.16).

We have a two-point boundary-value problem for (6.16) on the interval $[0,1]$, with ϕ, ψ, f, g, k bounded as $x \to 0$ and $f(1) = 0$, $g(1) = 1$. After some manipulation, the system is reduced to the pair of nonlinear integral equations

$$\left.\begin{aligned} f(x) &= \textstyle\int_0^1 K(x,y)[g^2(y) - f^2(y)]\,\mathrm{d}y \\ g(x) &= S(x)/xS(1) - \textstyle\int_0^1 K(x,y)2f(y)g(y)\,\mathrm{d}y \end{aligned}\right\} \tag{6.17}$$

where

$$K(x,y) = \begin{cases} [S(y)/c^2xR(y)]\,[1 - S(x)/S(1)] & \text{if} \quad x > y \;, \\ [S(x)/c^2xR(y)]\,[1 - S(y)/S(1)] & \text{if} \quad x < y \;, \end{cases}$$

$$S(x) = \textstyle\int_0^x R(y)\,\mathrm{d}y \quad ,$$

$$R(x) = \exp[-\textstyle\int_0^x k(y)\,\mathrm{d}y/c^2] \quad ,$$

and $$k(x) = (2/x)\textstyle\int_0^x f(y)\,\mathrm{d}y \quad .$$

Let $\mathscr{C}$ be the Banach space whose elements are pairs of continuous functions on $[0,1]$, with norm $\|(f,g)\| = \max\{\|f\|, \|g\|\}$ where $\|f\|$ denotes the usual sup norm on $C[0,1]$. Let $T\colon \mathscr{C} \to \mathscr{C}$ be the operator mapping a pair of functions (f,g) into the pair of functions given by the right hand sides of (6.17). Then a fixed point of T corresponds to a solution of the flow problem.

It is easy to show that T is a continuous operator. To use Schauder's theorem we must find a convex set which is mapped into a compact subset of itself by T. It is not obvious how to choose such a set. A study of approximate solutions of the problem led Howard to consider the set

$$\Sigma_{AB} = \{(f,g) : 0 \leqslant f(x) \leqslant A(1-x) \quad \text{and} \quad g(x) \leqslant B\} \quad ,$$

where A and B are positive constants. Some straightforward analysis shows that if $c \geqslant (3\mathrm{e})^{\frac{1}{4}}$ then A and B can be chosen so that T maps $\mathscr{C}$ into itself (the values are $A = c^2/6$ and $B = \mathrm{e}^{\frac{1}{4}}$). Using the Arzelà-Ascoli theorem, one can show that the image of $\mathscr{C}$ under T is compact. It now follows from Schauder's theorem that there exists a solution if $c \geqslant (3\mathrm{e})^{\frac{1}{4}}$, that is, if the suction at the disc is great enough. This completes our sketch of the first, and simplest, application of fixed-point theory to the swirling-flow problem. Much stronger results have since been obtained; see McLeod (1969), Hastings (1970), and Elcrat (1980).

6.9 Summary and References

We began by defining continuous operators on normed spaces, and then stated Brouwer's fixed-point theorem. The proof sketched in section 6.2 is given in full in Massey. For the two-dimensional case a simple and elegant proof will be found in Courant & Robbins, and the n-dimensional version of this proof (not simple) is in

Cronin; for other approaches see the references listed below under Schauder's theorem.

We then gave an example (due to Kakutani and described by Smart) showing that Brouwer's theorem does not extend in a simple-minded way to all Banach spaces. Sections 6.3 and 6.4 were therefore devoted to the idea of compactness. This subject is discussed in all books on functional analysis, but often in a more abstract setting (general topological or metric spaces), and sometimes with slightly different terminology – for example, 'sequentially compact' is sometimes used to mean what we call 'compact', while Kolmogorov & Fomin (1957) use 'compact' to mean what we and Kolmogorov & Fomin (1970) call 'relatively compact'. Great care is therefore needed when consulting books on this subject.

In section 6.5 we stated and proved a criterion for compactness in $C[a,b]$, sometimes called Arzelà's theorem and sometimes the Arzelà-Ascoli theorem. Its history is discussed in Dunford & Schwarz, a very full treatment of the theory of operators on normed spaces.

In section 6.6 we stated two versions of the fixed-point theorem for general normed spaces. For the proof, see Smart (who gives a survey of the literature), Kantorovich & Akilov, Courant-Hilbert vol. II, Hutson & Pym, Cronin, or Hochstadt. For other fixed-point theorems and their applications, see Smart or Krasnoselskii.

In section 6.7 we applied the Schauder theorem to nonlinear oscillations of a pendulum; see Tricomi. This problem is discussed, using quite different methods, by Chester. In section 6.8 we discussed some recent research in fluid mechanics. The relevant papers are cited in the text; for an account of the fluid-mechanical background, see Batchelor (1967). Applications of Schauder's theorem to quantum and statistical physics are described by Reed & Simon. Applications in numerical analysis are given in Collatz; see also Problem 6.9.

The theorems of this chapter do not tell us how many solutions the equation $x = Tx$ has. This question can be approached using the idea of the 'degree' of a mapping; see Hutson & Pym for a brief introduction, or Krasnoselskii, Cronin, or Lloyd for more extensive treatments.

PROBLEMS 6

Sections 6.1, 6.2

6.1 Consider the operator $A: l_2 \to l_2$ defined in Problem 5.2, where $\Sigma_{n,m} |a_{nm}|^2$ converges. Is it continuous? Prove it.

6.2 Under what conditions is the operator T defined by $(Tu)(x) = \int_0^1 K(x,y,u(y))\,dy$ a continuous mapping of $C[0,1]$ into itself?

6.3 Show that D, the differentiation operator, is a continuous operator $C_1^1[a,b] \to C[a,b]$, where $C_1^1[a,b]$ is the space defined in Problem 4.23 and $C[a,b]$ has the L_2 norm. Compare this result with Example 6.11.

6.4 Show that any sphere $S = \{x \in N: \|x - a\| \leqslant r\}$ in a normed space N (where $a \in N$ and $r > 0$) is convex.

6.5 The **convex hull** $\hat{S}$ of a set S is defined as the intersection of all convex sets containing S.

(a) Show that $\hat{S}$ is convex.

(b) If $S \subset R$ and R is convex, show that $\hat{S} \subset R$.

(c) Sketch the convex hulls of the sets illustrated in Fig. 8.

(d) A **convex combination** of elements $x_1, \ldots x_r$ of a vector space is a linear combination $\Sigma a_i x_i$ with $a_i \geqslant 0$ for each i and $\Sigma a_i = 1$. If R is a convex set, show that any convex combination of a finite number of elements of R belongs to R.

(e) Show that for any set S, $\hat{S}$ equals the set of all convex combinations of finitely many elements of S.

6.6 We define a (partial) ordering on $\mathbb{R}^n$ as follows: given $u = (u_1, \ldots, u_n)$ and $v = (v_1, \ldots, v_n)$, we say $u < v$ if $u_i < v_i$ for $i = 1, \ldots n$, and $u \leqslant v$ if $u_i \leqslant v_i$ for $i = 1, \ldots, n$. (This relation is called a 'partial' ordering because for some pairs of elements u,v of $\mathbb{R}^n$, none of the relations $u < v$, $u = v$, $v < u$ holds.) Verify the usual properties: if $u < v$ and $v < w$, then $u < w$; if $u \leqslant v$ and $v \leqslant u$, then $u = v$; if $u < v$ and $k \in \mathbb{R}$, then $ku < kv$ if $k > 0$, and $ku > kv$ if $k < 0$.

Write $[u,v] = \{x \in \mathbb{R}^n : u \leqslant x \leqslant v\}$. Show that if $u < v$, the set $[u,v]$ is convex. Sketch $[u,v]$ for $n = 2$ and $n = 3$.

6.7 If A is a continuous operator on a set X, and $Ax \neq 0$ for all $x \in X$, show that the operator $B: x \mapsto Ax/\|Ax\|$ is continuous on X.

6.8 Can you find a stronger version of Perron's theorem in which not all elements of the matrix need be strictly positive?

6.9 This problem shows how Brouwer's theorem can give not just existence results but numerical information on the solution. Let A be an $n \times n$ matrix with $A_{ij} \geqslant 0$ for all i,j. In the notation of Problem 6.6, show that if $u \leqslant v$ then $Au \leqslant Av$. If b is a given vector, define an operator T: $\mathbb{R}^n \to \mathbb{R}^n$ by $T(x) = Ax + b$. Show that if $x \in [u,v]$, then $T(x) \in [T(u), T(v)]$. Deduce that if u and v are vectors such that $u \leqslant T(u) \leqslant T(v) \leqslant v$, then the equation $x = Ax + b$ has a solution x with $u_i \leqslant x_i \leqslant v_i$ for all i. Invent a concrete example (with $n = 3$) to try this method out. (Alternatively, look up the example in Collatz, who discusses further developments of this idea, including generalisations to function spaces and application to integral equations. See also Moore (1985) for a systematic treatment.)

6.10 An operator $T: X \to M$ (where X is a subset of a normed space N) is said to be linear if for all $x,y \in X$, $A(\alpha x + \beta y) = \alpha Ax + \beta Ay$ for all scalars α,β such that $\alpha x + \beta y \in X$.

(a) Let X' be the subspace spanned by X (Definition 4.15). Show that T has exactly one linear extension $T': X' \to M$.

(b) If X is a vector subspace of N, and the linear operator T is continuous at the point $a \in X$, show that T is uniformly continuous on X.

Sections 6.3–6.5

6.11 Show that every finite set is compact.

6.12 Let M be a bounded set in $C[a,b]$ (not necessarily compact). Show that the set of all functions $F(x) = \int_a^x f(t)\, dt$ with $f \in M$ is relatively compact.

6.13 X and Y are subsets of a Banach space. Prove that if $T: X \to Y$ with $Y \subset X$ and Y compact, and if $\| Tx - Ty \| < \| x - y \|$ for all $x,y \in X$ with $x \neq y$, then T has a fixed point. (Hint: Apply Theorem 6.25 to the function $f(x) = \| Tx - x \|$.)

Show that the fixed point is unique. (Compare this result with Example 5.14.)

6.14 Construct a set in $\mathbb{R}^2$ which has finite area but is not relatively compact. Generalise to $\mathbb{R}^n$.

6.15 Use Theorem 6.37 to fill the gap in the argument given in Problem 5.3.

6.16 This problem outlines a proof that any two norms on a finite-dimensional space are equivalent (Theorem 4.68). Let V be an n-dimensional space, and $\| \;\|_a$ any given norm on V. Take a basis for V, so that x has components $x_1, \ldots, x_n$, and define another norm on V by $\| x \|_1 = \Sigma_1^n |x_r|$. Write V_1 for the normed space obtained by applying the norm $\| \;\|_1$ to V.

(a) Prove that there is an $M > 0$ such that $\| x \|_a < M \| x \|_1$ for all $x \in V$.

(b) Prove that $x \mapsto \| x \|_a$ is a continuous function $V_1 \to \mathbb{R}$.

(c) Deduce that there is an $m > 0$ such that $\| x \|_a \geqslant m$ for all x such that $\| x \|_1 = 1$.

(d) Deduce that $\| x \|_a \geqslant m \| x \|_1$ for all $x \in V$, hence that $\| \;\|_a$ is equivalent to $\| \;\|_1$, hence that any two norms are equivalent.

6.17 If S is a relatively compact set, prove that its convex hull (Problem 6.5) is relatively compact.

Sections 6.6 and 6.7

6.18 Let $S = \{ u \in C[0,1] : \| u \| \leqslant A \}$, and let R be the image of S under the transformation T of Problem 6.2, assuming that K is continuous and has a continuous derivative with respect to u. Show that R is relatively compact. Can you use Schauder's theorem to deduce anything about the equation $u(t) = \int_0^1 \exp[s^2 t + su(s)]\,ds$?

6.19 Dig out of the proof of Theorem 6.45 a bound on the size of the solution.

6.20 Consider the Hammerstein equation (6.10) when f is not bounded. Assume that K and f are continuous; set $C = \max|K(x,y)|$.
Show that for any $M > 0$ it has a solution u with $\|u\| \leqslant M$ provided that $M \geqslant BC$, where $B = \sup\{\int_a^b |f(y,u(y))|\,dy : \|u\| \leqslant M\}$.

6.21 Apply the previous problem to the question of the existence of periodic solutions of the equation $\ddot{u} + u + au^3 = b\sin(\beta t)$. (This equation is familiar to students of nonlinear oscillations under the name of Duffing's equation; it is important as an approximation to equations like the pendulum equation (6.13) when the amplitude is fairly small yet large enough for nonlinear effects to be significant.)

6.22 A linear integral operator with a positive kernel is a natural analogue of the positive matrix of Perron's theorem. Use Schauder's theorem to prove that an integral operator with a positive continuous kernel has a positive eigenvalue.

6.23 Consider the nonlinear diffusion problem of section 5.4, $u'' = -h(u,x)$, $u(0) = u(1) = 0$. Use Theorem 6.45 to show that if h is bounded and continuous, then there is a solution. Deduce that the inequality (5.24) is not necessary for existence (though if it is not satisfied, uniqueness may fail). Obtain also a bound on the size $\|u\|$ of the equilibrium solution, in terms of the maximum value of h.

6.24 Extend the theory of Problem 6.23 to unbounded functions h as follows. Define a function $B : \mathbb{R} \to \mathbb{R}$ by $B(\alpha) = \sup\{h(u,x) : 0 \leqslant x \leqslant 1, |u| \leqslant \alpha\}$. Assume that h is continuous and differentiable. Show that the integral operator $T : u \mapsto \int g(x;y)h(u(y),y)\,dy$ is continuous (where g is Green's function for $-D^2$), and maps any sphere in $C[a,b]$ into a relatively compact set. Set $G = \max|g(x;y)|$. Show that if α satisfies $\alpha \geqslant GB(\alpha)$, then there exists a solution u with $\|u\| \leqslant \alpha$. If h is monotonic increasing, show that the condition becomes $\alpha \geqslant Gh(\alpha,x)$ for all $x \in [0,1]$.

For the case $h = a + b\sqrt{u}$ (corresponding to slow growth as u increases), show that there is always a solution. For the case $h = a + bu^2$ (corresponding to fairly rapid growth as u increases), obtain conditions on a and b which will ensure the existence of an equilibrium.

More problems will be found on page 376.

PART III

OPERATORS IN HILBERT SPACE

In Part II we developed ways of dealing with nonlinear differential and integral equations. Part III is concerned with linear equations, for which much more powerful methods are available, including techniques for constructing solutions as well as existence theory. Chapters 7 and 8 set out the basic theory of Hilbert space and linear operators. These chapters are very theoretical; they may be regarded as an excitingly rigorous and exhilarating chain of logical development, or an arid desert of definitions and theorems, according to one's point of view. Chapter 9 deals with the main theorem of this subject, the spectral theorem, with applications to ordinary and partial differential equations. In Chapter 10 the theory is applied to the variational methods for approximating the solutions of linear equations. The theory of the earlier chapters is applied to prove that successive approximations converge to the exact solution.

A good deal of Part II is needed as a prerequisite for reading Part III; see the introduction to Part II. Readers who are not interested in theoretical subtleties, and do not require complete proofs, may omit the following without seriously impairing their understanding of the rest: 7.35, all section 7.8, 8.55, 9.1–9.3, 9.21–9.25, all section 10.4, and the proofs of 7.36, 7.60, 8.14, 8.34, 8.51, 8.54, 9.4, 9.11, 9.26, 9.27, 9.44.

Chapter 7

Hilbert Space

In Part II nonlinear equations were discussed in terms of operators on Banach spaces. For the more detailed analysis which is possible for linear equations, we introduce more structure into the space. We define an 'inner product', which is an abstract version of the scalar product of elementary vector algebra, and allows us to define angles, and particularly right angles, in a vector space. In section 7.1 we define the inner product, and the associated idea of a pair of vectors being perpendicular, or orthogonal. The next four sections discuss how sets of perpendicular vectors can form a basis for an infinite-dimensional space, in roughly the way that the x,y,z-axes form a basis for $\mathbb{R}^3$; the theory is of course more elaborate than for finite-dimensional spaces, because questions of convergence must be considered as well as the purely algebraic side of the theory.

In sections 7.6 to 7.8 we discuss ideas and theorems which are needed in order to prove the useful theorems of Chapters 9 and 10. These sections should be skimmed rather lightly by those readers interested in results but not in proofs.

7.1 Inner Product Spaces

Definition 7.1 A (real or complex) **inner product space** is a (real or complex) vector space V with an inner product specified. An **inner product** is a rule which, given any $x,y \in V$, specifies a (real or complex) number (x,y), called the inner product of x and y, such that

(a) (x,x) is real and positive for all $x \neq 0$, and $(0,0) = 0$;
(b) $(x,y) = \overline{(y,x)}$ for all x,y;
(c) $(ax,y) = a(x,y)$ for any scalar a and $x,y \in V$;
(d) $(x,y + z) = (x,y) + (x,z)$ for all $x,y,z \in V$.

Remarks 7.2 (i) For a real space, the bar indicating complex conjugation in (b) is redundant, since the inner products are all real.

(ii) Properties (a)–(d) imply that various other algebraic rules hold. For example, (b) and (d) imply that $(x + y,z) = (x,z) + (y,z)$; to prove this, apply (b), then (d), then (b). These algebraic properties are generally the same as those governing the scalar product in ordinary vector algebra, with which the reader should be familiar. The only rule that is not obvious is the following: in a complex space the inner product is not linear but 'conjugate linear' with respect to the second factor, that is

$$(x,ay) = \bar{a}(x,y) \quad \text{for any scalar} \quad a \quad \text{and} \quad x,y \in V \ . \tag{7.1}$$

This follows from 7.1(b) and (c): $(x,ay) = \overline{(ay,x)} = \overline{a(y,x)} = \bar{a}(x,y)$.

Example 7.3 The complex vector space $\mathbb{C}^n$ (Example 4.5) is an inner product space if we define $(x,y) = \Sigma_{i=1}^n x_i \bar{y}_i$, where x_i are the components of the vector x. The complex conjugation in this definition is needed to satisfy Condition 7.1(a). The corresponding real space $\mathbb{R}^n$ is an inner product space with $(x,y) = \Sigma_1^n x_i y_i$.

Example 7.4 The set of all infinite sequences of numbers (x_i) such that $\Sigma_{i=1}^\infty |x_i|^2$ converges is a vector space (see Problem 4.9), called l_2. By analogy with Example 7.3, we define an inner product by $(x,y) = \Sigma_1^\infty x_i \bar{y}_i$; it is not obvious that this series converges, but it follows from the Cauchy-Schwarz inequality (see Problem 4.8).

Example 7.5 In the vector space $L_2[a,b]$ we define an inner product by $(f,g) = \int_a^b f(x)\overline{g(x)}\,dx$. Lemma 4.27 assures the existence of this integral, and the properties 7.1(a)–(d) of the inner product are easy to verify. Henceforth we shall always assume that $L_2[a,b]$ is given this inner product unless otherwise specified.

Other inner products are possible. For any well-behaved positive function w, $(f,g) = \int_a^b f(x)\overline{g(x)}w(x)\,dx$ defines an inner product, and we shall find this generalisation useful in Chapter 9. Note that if w is negative or zero or complex in some interval, then 7.1(a) is violated.

The n-dimensional analogue of $L_2[a,b]$ is the space $L_2[V]$ of functions square-integrable over $V \subset \mathbb{R}^n$ (see 4.63). We define an inner product by $(f,g) = \int_V f(x)\overline{g(x)}\,d^n x$, or more generally, $(f,g) = \int_V f(x)\overline{g(x)}w(x)\,d^n x$ for a well-behaved positive w.

Example 7.6 In the space $C[a,b]$, $(f,g) = \int_a^b f(x)\overline{g(x)}w(x)\,dx$ is an inner product for any well-behaved positive w. □

The following very important result is the basis for the statement that inner product spaces contain all the elements of Euclidean geometry, while normed spaces have length, but nothing corresponding to angle. It is an abstract version of the Cauchy-Schwarz inequality of Chapter 4.

Theorem 7.7 (Schwarz Inequality) (a) $|(x,y)|^2 \leqslant (x,x)(y,y)$ for any x,y in an inner product space.

(b) Equality holds in (a) if and only if x and y are linearly dependent.

Proof (a) For any scalar c (real or complex according to whether the space is real or complex), we have

$$(x - cy, x - cy) = (x,x) - 2\mathcal{Re}\{\bar{c}(x,y)\} + (y,y)|c|^2 \geqslant 0 \quad .$$

If $(y,y) \neq 0$, take $c = (x,y)/(y,y)$, and the result follows. If $(y,y) = 0$, Definition 7.1(a) implies that $y = 0$ and the result is obviously true in this case.

(b) If x and y are linearly dependent, then either $y = 0$, in which case the result is obvious, or $x = cy$ for some scalar c, in which case $|(x,y)|^2 = |c|^2|(y,y)|^2 = (cy,cy)(y,y) = (x,x)(y,y)$ as required.

Now suppose that $|(x,y)|^2 = (x,x)\,(y,y)$; we must show that x and y are linearly dependent. Consider the following linear combination of x and y:

$$u = (x,x)y - (y,x)x.$$

We have $(u,x) = (x,x)\,(y,x) - (y,x)\,(x,x) = 0$. Also $(u,y) = 0$ under the assumption that $|(x,y)|^2 = (x,x)\,(y,y)$. Hence $(u,u) = (x,x)\,(u,y) - (x,y)\,(u,x) = 0$, and therefore $u = 0$. Now, either $x = 0$, in which case $\{x,y\}$is linearly dependent, or $(x,x) \neq 0$, in which case u is a linear combination of x and y which vanishes but has a nonzero coefficient. In both cases it follows that $\{x,y\}$ is linearly dependent. □

Corollary 7.8 $\sqrt{(x+y,x+y)} \leqslant \sqrt{(x,x)} + \sqrt{(y,y)}$ for any x,y in an inner product space.

Proof
$$(x+y,x+y) = (x,x) + 2\mathcal{Re}(x,y) + (y,y)$$
$$\leqslant (x,x) + 2|(x,y)| + (y,y)$$

since $\mathcal{Re}(z) \leqslant |z|$ for any z,

$$\leqslant (x,x) + 2\sqrt{(x,x)(y,y)} + (y,y)$$
$$= [\sqrt{(x,x)} + \sqrt{(y,y)}]^2 \quad .$$
□

Remark 7.9 In any inner product space we can now define a norm by

$$\|x\| = \sqrt{(x,x)} \quad . \tag{7.2}$$

Corollary 7.8 ensures that the triangle inequality is satisfied, and the other properties required of a norm follow from the definition of inner product. Any inner product space can thus be made a normed space in a natural way, and the norms obtained in this way for Examples 7.3–7.5 are the same as the norms we have previously used in these spaces. In future we shall always adopt the norm (7.2) in an inner product space without further comment; we can then use the theory of convergence given in Chapter 4. □

The Schwarz inequality can now be written in the form

$$|(x,y)| \leqslant \|x\| \|y\| \quad . \tag{7.3}$$

This means that in a real inner product space, $(x,y)/\|x\| \|y\|$ is between -1 and 1. This may be compared with the formula $\cos\theta = x.y/|x| |y|$ from three-dimensional geometry, where θ is the angle between the vectors x and y. It allows us to define an angle between any two elements of a real inner product space by

$$\theta = \cos^{-1} [(x,y)/\|x\| \|y\|] \quad .$$

In this way one could build up essentially all Euclidean geometry in the context of real inner product spaces. The most useful notion for our purposes is that of a right angle, where $\cos\theta = 0$ and so $(x,y) = 0$.

Definition 7.10 Two vectors x,y in an inner product space are said to be **orthogonal** if $(x,y) = 0$. We write $x \perp y$ to mean x is orthogonal to y. □

This definition applies to both real and complex vector spaces. The geometrical interpretation is not as clear in complex spaces, though 7.12 below shows that the main theorem about right-angled triangles applies to complex inner product spaces as well as to $\mathbb{R}^2$.

Examples 7.11 (a) In $\mathbb{C}^2$ with the inner product of Example 7.3, the vectors $(1,i)$ and $(1,-i)$ are orthogonal, since $((1,i), (1,-i)) = 1.\overline{1} + i\overline{(-i)} = 1 + i^2 = 0$.

(b) In $C[0,\pi]$ with the inner product $(f,g) = \int_0^\pi f(x)\overline{g(x)}\,dx$, the functions $\sin(mx)$ and $\sin(nx)$ are orthogonal for any positive integers m,n with $m \neq n$.

(c) In any space, the zero vector is orthogonal to every vector. □

The following simple observation is an abstract version of Pythagoras' theorem.

Proposition 7.12 (Pythagoras) If $x \perp y$, then $\|x + y\|^2 = \|x\|^2 + \|y\|^2$.

Proof

$$\begin{aligned}(x + y, x + y) &= (x,x) + (x,y) + (y,x) + (y,y) \\ &= (x,x) + (y,y)\end{aligned}$$

if $(x,y) = 0$. □

The last result in this section is a simple lemma which will be useful in various proofs later in the chapter.

Lemma 7.13 (Continuity of the Inner Product) If $x_n \to x$, then $(x_n,y) \to (x,y)$ for any y. If $\Sigma_1^\infty u_n = U$, then $\Sigma_1^\infty (u_n,y) = (U,y)$ for any y.

Proof $|(x_n,y) - (x,y)| = |(x_n - x,y)|$

$$\leqslant \|x_n - x\| \, \|y\|$$

$$\to 0 \quad \text{as} \quad n \to \infty \quad .$$

The second statement of the theorem follows immediately from the first, setting $x_n = \Sigma_1^n u_r$ and $x = U$. □

This result can be interpreted as saying that the inner product (x,y) is a continuous function of x for any fixed y; it is also a continuous function of y for any fixed x, as follows immediately from Lemma 7.13 and Definition 7.1(b).

7.2 Orthogonal Bases

When we discussed normed vector spaces, much of the theory of finite-dimensional spaces was taken over to the infinite-dimensional case; addition of finitely many elements was extended to the definition of convergence of infinite series. You may have noticed, however, that we did not extend the idea of a basis to the infinite-dimensional case, even though we had the machinery for interpreting expressions like $x = \Sigma_1^\infty a_i e_i$ representing a vector as an infinite linear combination of basis vectors e_i. The reason is that there is more than one way to extend the ideas of linear independence and basis to infinite-dimensional spaces, and they differ in quite subtle ways. It is an intricate subject; a brief and clear account will be found in Korevaar. However, in an inner product space we can use orthogonality to replace the more subtle idea of linear independence, and the theory is then quite straightforward.

Definition 7.14 A set of vectors $\{x_i\}$ in an inner product space is called an **orthogonal set** if (i) $(x_i,x_j) = 0$ whenever $i \neq j$, and (ii) for each $i, x_i \neq 0$. □

We exclude the zero vector from an orthogonal set by clause (ii) of this definition, so that finite orthogonal sets are linearly independent (any set containing the zero vector would be linearly dependent).

Remark 7.15 We shall often refer to an orthogonal 'set' of vectors but in fact we usually mean a sequence, that is, a countable set arranged in a definite order.

Proposition 7.16 A finite orthogonal set is linearly independent.

Proof If $\{x_1, \ldots, x_n\}$ is an orthogonal set, we must show that $\Sigma_1^n c_i x_i = 0$ implies that $c_j = 0$ for $j = 1, \ldots, n$. If $\Sigma c_i x_i = 0$, then for any j,

$$(\Sigma_i c_i x_i, x_j) = 0 \quad ,$$

$$\therefore \Sigma_i c_i (x_i, x_j) = 0 \quad ,$$

$$\therefore c_j \| x_j \|^2 = 0$$

by orthogonality, therefore $c_j = 0$ using 7.14(ii). □

Example 7.17 Consider the space $C[0,\pi]$ with inner product $(f,g) = \int_0^\pi f(x)\overline{g(x)}\,\mathrm{d}x$. Easy integration shows that the set of functions $\sin(nx)$ for $n = 1,2,\ldots$ is orthogonal. □

The importance of Example 7.17 lies in the fact that every continuous function can be represented in the form $\Sigma A_n \sin(nx)$ on the interval $(0,\pi)$; we assume the reader is acquainted with Fourier series, of which this is an example. In the language of normed spaces we can say that every element of $C[0,\pi]$ can be represented as an infinite linear combination of the sine functions (in the sense of mean convergence), and we shall express this fact by saying that the sine functions in Example 7.17 form a basis for the space.

Definition 7.18 An **orthogonal basis** for an inner product space S is an orthogonal set (e_n) such that for any $x \in S$ there are scalars c_n such that $x = \Sigma c_n e_n$. We shall often refer to an orthogonal basis simply as a **basis.** □

Not every space has a basis; a space may be in some sense so large that no countable set can generate the whole space by its infinite linear combinations. See Problem 7.17 for an example. But the spaces commonly occurring in applications usually have bases, just as they are usually complete. Complete inner product spaces with a basis are thus particularly important.

Definition 7.19 A complete inner product space with a basis is called a **Hilbert space.**

Remark 7.20 Some authors use the term 'Hilbert space' to mean any complete inner product space. Furthermore, some authors use the term 'basis' differently, including uncountable sets which are not allowed under our definition. Extreme care is therefore needed when reading the literature.

Example 7.21 $\mathbb{R}^n$ and $\mathbb{C}^n$ are inner product spaces, as in Example 7.3; we saw that they are complete in Chapter 4, and the set of n unit vectors given in Example 4.19 is an orthogonal basis for both $\mathbb{R}^n$ and $\mathbb{C}^n$, so they are Hilbert spaces.

Example 7.22 The space $C[0,\pi]$ of Example 7.17 has an orthogonal basis, as we stated above without proof. It is not complete, as we found in section 4.5, and is therefore not a Hilbert space.

Example 7.23 $L_2[a,b]$ is complete in the norm generated by the inner product of Example 7.5 (see Example 4.60). The functions $\sin[n\pi(x-a)/(b-a)]$ (obvious modifications of those in Example 7.17) form a basis, as we shall prove in 7.34 below, therefore $L_2[a,b]$ is a Hilbert space.

Definition 7.24 **A subspace** of a Hilbert space is a subset which is itself a Hilbert space (with the same inner product as the first space). We shall sometimes call it a Hilbert subspace, to emphasise the distinction from the following definition.

Definition 7.25 **A vector subspace** of a Hilbert space is a subset which is a vector space (but may or may not be complete).

Remark 7.26 Some authors use the term 'subspace' to mean vector subspace, and refer to a Hilbert subspace as a closed subspace (cf. Problem 7.4).

Example 7.27 In $L_2[a,b]$ the set of all polynomials is a vector subspace (by Lemma 4.8), but not a Hilbert subspace. $f_n(x) = \Sigma_0^n x^r/r!$ is a sequence of polynomials converging to e^x in $L_2[a,b]$, and is therefore a Cauchy sequence by Theorem 4.55. It does not converge to a polynomial, therefore the set of polynomials is not complete. It is a vector subspace but not a Hilbert subspace.

Example 7.28 The set E of all even functions in $L_2[-a,a]$ is a vector subspace, by Lemma 4.8; we shall show that it is a Hilbert subspace. If (f_n) is a Cauchy sequence in E, it converges to a member of $L_2[-a,a]$ because $L_2[-a,a]$ is complete; but $f_n(x) - f_n(-x) = 0$ for all x and all n if f_n is even, so if $f_n \to f$, we have $f(x) - f(-x) = 0$, hence $f \in E$. This shows that E is complete and a subspace of the Hilbert space $L_2[-a,a]$.

This argument assumes that the elements of $L_2[a,b]$ are functions. But as we saw in 4.60, they should properly be regarded as equivalence classes of almost-everywhere-equal functions. We then define an element f of $L_2[a,b]$ to be 'even' if $f(x) = f(-x)$, where f or $f(x)$ denotes an equivalence class, and $f(-x)$ is the class of functions equivalent to $\phi(-x)$, where $\phi(x)$ is any member of the equivalence class f. With this understanding, and with the = symbol interpreted as equality of equivalence classes, or almost-everywhere equality, the argument above becomes valid.

Finally, we should show that E has a basis before concluding that it is a Hilbert space. It is intuitively clear that a vector subspace of a Hilbert space has a basis, which is a subset of a basis for the Hilbert space. It is remarkably hard to prove this; see Theorem 7.36 below. It follows from 7.36 that E, being a vector subspace of $L_2[-a,a]$, has a basis. □

In finite-dimensional spaces a basis is useful because instead of manipulating vectors we can manipulate their components, which are real or complex numbers and more accessible to computation. The same is true in infinite-dimensional spaces. It is therefore important to ask how orthogonal bases can be constructed. There is a systematic procedure by which, given an infinite set of vectors, one can replace it by an orthogonal set.

Theorem 7.29 (Gram-Schmidt Orthogonalisation) Given any sequence (f_n) of elements of an inner product space, there is an orthogonal sequence (g_n) such that every finite linear combination of the f_n is a finite linear combination of the g_n and vice versa.

Proof First we construct a new sequence (F_n) from (f_n) by removing the zero vector each time it occurs in (f_n), and also removing any f_n which is a linear combination of the preceding members of the sequence. Then (F_n) is a subsequence of (f_n), and any finite subset of (F_n) is linearly independent.

Now we construct (g_n) from (F_n) step by step. Take $g_1 = F_1$. Take $g_2 = F_2 + cg_1$, where c is a scalar to be determined so as to make g_2 orthogonal to g_1. Setting $(g_2,g_1) = 0$ gives

$$(F_2,g_1) + c(g_1,g_1) = 0 \quad ,$$

which determines c. Now take

$$g_3 = F_3 + dg_2 + eg_1 \quad ,$$

where d and e are determined so as to make g_3 orthogonal to g_2 and g_1:

$$(F_3,g_2) + d(g_2,g_2) = 0 \quad ,$$

$$(F_3,g_1) + e(g_1,g_1) = 0 \quad .$$

This determines d and e. This procedure can obviously be extended to give the sequence (g_n). g_n is a linear combination of $F_1, \ldots, F_n$, and each F_n is a member of the original sequence (f_n), hence any finite linear combination of the g_n is a finite linear combination of the f_n.

The transformation from $F_1, \ldots, F_n$ to $g_1, \ldots, g_n$ is invertible, because the transformation matrix is triangular with ones on the principal diagonal, and therefore has determinant 1. Therefore each F_n is a linear combination of $g_1, \ldots, g_n$. Any finite linear combination of the f_i is reducible to a finite linear combination of the F_i (by replacing any f_i which is a linear combination of the preceding f_j by that combination), and is therefore a finite linear combination of the g_n. This completes the proof. The procedure given here for computing the g_n from the f_n is called the **Gram-Schmidt** process. □

7.3 Orthogonal Expansions

We now consider the process of representing a vector in a Hilbert space as an infinite series of basis vectors. We shall see that it is very similar to expanding a function in Fourier series: the coefficients can be obtained in a way very similar to Fourier coefficients.

Theorem 7.30 Let (e_n) be an orthogonal basis for an inner product space. For any x in the space, $x = \Sigma c_n e_n$, where the c_n are uniquely determined by

$$c_n = (x, e_n)/\| e_n \|^2 \quad . \tag{7.4}$$

Proof By definition of a basis, there are numbers c_n such that $x = \Sigma c_r e_r$. Using Lemma 7.13, we have, for any n,

$$(x, e_n) = \Sigma_r c_r (e_r, e_n) \quad ,$$

and all the terms in this series vanish by orthogonality except the term with $r = n$, giving (7.4). □

We see that for inner-product spaces, orthogonality ensures the uniqueness of expansion coefficients, as linear independence does in finite-dimensional vector spaces. The formula (7.4) is a generalisation of the formula for calculating Fourier coefficients; the series $x = \Sigma c_r e_r$ is sometimes called a generalised Fourier series.

Definition 7.31 Let (e_n) be an orthogonal set in an inner product space V. For any $x \in V$ the numbers $c_n = (x, e_n)/\| e_n \|^2$ are called the **generalised Fourier coefficients** or **expansion coefficients** of x with respect to the set (e_n). □

The expansion coefficients (7.4) have the property that $\Sigma_1^N c_r e_r$ can be made as close as we please to x by taking N large enough, and in this sense they give the best approximation to x. But in any numerical calculation, we are normally restricted to some finite number of terms, and the question of what happens as $N \to \infty$ is of strictly academic interest. This raises the following question: given a finite set of elements $e_1, \ldots, e_N$ of a Hilbert space, what linear combination of them is the best approximation to a given x in the space?

Note carefully the difference between this question and that answered by Theorem 7.30 and paraphrased at the beginning of the last paragraph. Here we ask for the best approximation for a given N, with no consideration of the limit as $N \to \infty$. The answer will be of the form

$$\Sigma_{r=1}^N c_{rN} e_r \quad , \tag{7.5}$$

in which the coefficient of e_r may depend on N; hence the two subscripts. If one chooses the c_{rN} to give the best fit to x using $e_1, \ldots, e_N$, one expects to have to readjust all the coefficients to get the best fit when another vector e_{N+1} is included. However, in the special case where the e_r are orthogonal, it turns out that the coefficients in (7.5) are independent of N and are just the expansion coefficients (7.4). This is why orthogonal expansions are so much simpler than nonorthogonal expansions. Given the problem of approximating a vector by a combination of nonorthogonal vectors, one can orthogonalise them by the Gram-Schmidt process, and then use the following result.

Theorem 7.32 (Best Approximation) Let $\{e_1, \ldots, e_N\}$ be an orthogonal set in an inner product space. For any x, the numbers c_n which minimise $\| x - \Sigma_1^N c_n e_n \|$ are given by (7.4).

Proof Write $c_n = (x,e_n)\| e_n \|^{-2} + d_n$; we shall show that the best approximation is obtained by taking $d_n = 0$. We have

$$\| x - \sum_1^N c_n e_n \|^2 = \left(x - \Sigma \frac{(x,e_n)}{\| e_n \|^2} e_n - \Sigma d_n e_n , x - \Sigma \frac{(x,e_n)}{\| e_n \|^2} e_n - \Sigma d_n e_n \right)$$

$$= \| x \|^2 - \sum_1^N \frac{|(x,e_n)|^2}{\| e_n \|^2} + \sum_1^N |d_n|^2 \| e_n \|^2$$

after some calculation. This obviously takes its smallest value when all the d_n are zero. □

This result is useful in practical approximation problems; see Problem 7.18. But it is also essential in the theory of Hilbert space, being the main ingredient in the proof of the following useful theorem.

Theorem 7.33 Let S be a dense subset of an inner product space V. If (e_n) is a basis for S, then it is a basis for V.

Proof For any $v \in V$ and any $\epsilon > 0$ there is an $s \in S$ with $\| v - s \| < \epsilon/2$, that is,

$$\| v - \sum_1^\infty c_r e_r \| < \epsilon/2$$

where $c_r = (s,e_r)\| e_r \|^{-2}$. Since the series is convergent, there is an N such that $\| \Sigma_1^\infty c_r e_r - \Sigma_1^n c_r e_r \| < \epsilon/2$ for $n > N$. The triangle inequality then gives

$$\| v - \sum_1^n c_r e_r \| < \epsilon$$

for $n > N$. Theorem 7.32 now shows that

$$\| v - \sum_1^n (v,e_r)\| e_r \|^{-2} e_r \| < \epsilon$$

for $n > N$. Hence $v = \Sigma_1^\infty (v,e_r)\| e_r \|^{-2} e_r$, showing that (e_n) is a basis for V. □

Example 7.34 As a first application of this theorem we shall prove a version of Fourier's theorem for square-integrable functions, showing that $L_2[a,b]$ is a Hilbert space. The idea is first to prove that sufficiently well-behaved functions can be represented by a Fourier series, and then use 7.33 to extend the result to L_2.

In 4.51 we showed that the set $C_0^\infty[a,b]$ of smooth functions vanishing at a and b is dense in $C[a,b]$ with the L_2 norm. Since $C[a,b]$ is dense in $L_2[a,b]$, by 4.60, it follows from Problem 4.15 that $C_0^\infty[a,b]$ is dense in $L_2[a,b]$. Now, it is easy to prove that any smooth function vanishing at a and b can be expanded as a uniformly convergent series in the functions $\sin[n\pi(x-a)/(b-a)]$ (see Appendix H). Uniform convergence implies convergence in the mean, so the sine functions form a basis for $C_0^\infty[a,b]$. Theorem 7.33 now shows that they form a basis for $L_2[a,b]$. This completes the proof that $L_2[a,b]$ is a Hilbert space, and shows that any square-integrable function has a Fourier series which converges to it in the mean. □

We shall now use Theorem 7.33 to prove that if a space has a basis, so does any vector subspace. This is intuitively obvious. Spaces without a (countable) basis are in some sense bigger than spaces with a basis, just as in ordinary linear algebra spaces without a finite basis are bigger than finite-dimensional spaces. Since no subspace can be bigger than the whole space, the result is obvious. But it is not easy to prove; the reader willing to take 7.36 on trust may omit the rest of this section.

The first step in the proof is the following result, which has a more prominent place in purely theoretical accounts of functional analysis, though from our point of view it is only a stepping stone to Theorem 7.36.

Proposition 7.35 An inner product space has a basis if and only if it contains a countable† dense set.

Proof (a) Suppose that the inner product space V contains a countable dense set $\{f_n\}$; we shall show that V has a basis. Let S be the vector subspace of V spanned by $\{f_n\}$ (Definition 4.15), that is, the set of all finite linear combinations of the f_n. Then S is dense in V because it contains the dense set $\{f_n\}$. By the Gram-Schmidt process we can construct an orthogonal sequence (g_n) such that each f_n is a linear combination of finitely many g's. Each element of S is a linear combination of finitely many f's, therefore of finitely many g's. Hence (g_n) is a basis for S, and it follows from 7.33 that (g_n) is a basis for V, as required.

(b) Suppose now that V has a basis (e_n). Consider the set of all finite linear combinations of the e_n with rational coefficients; in the case of a complex space, take the coefficients to be rational complex numbers, that is, complex numbers whose real and imaginary parts are rational. We shall show that this is a countable dense set.

In Appendix D it is shown that the rational numbers are countable, and that

† For the meaning of 'countable', see Appendix D.

for any fixed N, a set whose members can be labelled by a set of N integers is countable. The complex rationals can be regarded as pairs of rationals, hence can be labelled by a pair of integers, and are therefore countable. Now take a fixed N and consider all linear combinations of $e_1, \ldots, e_N$ with (real or complex) rational coefficients. Since the rationals are countable, these linear combinations can be labelled by N integers, one for each coefficient, and are therefore countable. Finally, any rational linear combination of finitely many (e_n) can be indexed by two integers: the largest number N for which e_N appears in the combination, and the integer assigned to this particular combination of $e_1, \ldots, e_N$ by the scheme of counting whose existence we have just proved. This establishes the countability of the set of rational linear combinations of the (e_n).

Now we must show that for any $x \in V$ and any $\epsilon > 0$ there is a rational combination of the e_n within a distance ϵ of x. Because the e_n are basis vectors, for any $x \in V$ and any $\epsilon > 0$ there is an integer n such that

$$\| x - \sum_1^n c_r e_r \| < \epsilon/2 \quad ,$$

where the c_r are the expansion coefficients of x. But for any real or complex c_r, there is a real or complex rational number as close to it as we please, and in particular there is a rational b_r within a distance $\epsilon/2n \| e_r \|$ of c_r. Then

$$\begin{aligned} \| x - \sum_1^n b_r e_r \| &= \| x - \Sigma c_r e_r - \Sigma (b_r - c_r) e_r \| \\ &\leqslant \| x - \Sigma c_r e_r \| + \Sigma | b_r - c_r | \, \| e_r \| \\ &< (\epsilon/2) + n(\epsilon/2n) = \epsilon \quad . \end{aligned}$$

This completes the proof. □

Theorem 7.36 If an inner product space V has a basis, so does any vector subspace of V.

Proof By Proposition 7.35 there is a sequence (x_n) dense in V. We shall show that the vector subspace S also contains a dense sequence.

For each n, set

$$a_n = \inf \{ \| y - x_n \| : y \in S \} \quad .$$

Then S contains vectors whose distance from x_n is less than any number greater than a_n. In particular, for any integer p there is an element y_{np} of S such that

$$\| y_{np} - x_n \| < a_n + 1/p \quad .$$

The vectors y_{np} are indexed by two integers and are therefore countable. We now show that for any $y \in S$ and any $\epsilon > 0$ there is a member of the set $\{ y_{np} \}$ within a distance ϵ of y.

Because (x_n) is dense in V, there is an n such that

$$\| y - x_n \| < \epsilon/3 \quad . \tag{7.6}$$

Choose an integer p such that $1/p < \epsilon/3$. Then

$$\| y - y_{np} \| \leqslant \| y - x_n \| + \| x_n - y_{np} \|$$
$$< \epsilon/3 + a_n + 1/p \quad .$$

But (7.6) shows that $|a_n| \leqslant \epsilon/3$, and $1/p < \epsilon/3$, hence $\| y - y_{np} \| < \epsilon$ and we have shown that the countable set $\{y_{np}\}$ is dense in S. The theorem now follows from Proposition 7.35. □

7.4 The Bessel, Parseval, and Riesz-Fischer Theorems

In this section we consider more carefully the representation of members of a space as sums of basis vectors. It is convenient, just as in finite-dimensional spaces, to work with basis vectors of unit magnitude.

Definition 7.37 A **normalised** vector is a vector whose norm is 1. An **orthonormal** set is an orthogonal set of normalised vectors.

Remark 7.38 The expansion coefficients of a vector x with respect to an orthonormal basis (e_n) are $c_n = (x,e_n)$.

Remark 7.39 An orthogonal set can always be turned into an orthonormal set by the very simple operation of **normalisation**, that is, dividing each vector by its norm. We shall therefore assume from now on that our bases are orthonormal. □

A natural question to ask about the sequence of expansion coefficients of a given vector is how it behaves as $n \to \infty$. The answer is that it always tends to zero, as the following result shows.

Theorem 7.40 (Bessel's Inequality) If (e_n) is an orthonormal set in an inner product space, then for any x in the space

$$\sum_n |c_n|^2 \leqslant \| x \|^2 \tag{7.7}$$

where $c_n = (x,e_n)$.

Proof Note that if (e_n) is an infinite set, this theorem says two things: the series $\Sigma |c_n|^2$ converges, which is not at all obvious a priori (and not true if the set is not normalised); and its sum satisfies (7.7). If the orthonormal set is finite, then of course there is no question of convergence, and the inequality is easily proved by algebraic manipulation:

$$(x - \sum_1^n (x,e_r)e_r, x - \sum_1^n (x,e_s)e_s)$$

$$= (x,x) - \Sigma(x,e_r)(e_r,x) - \Sigma(x,e_s)(x,e_s)$$

$$+ \Sigma(x,e_r)(\overline{x,e_s})(e_r,e_s)$$

using (7.1),

$$= (x,x) - \sum_1^n |(x,e_r)|^2$$

using orthonormality. But the left hand side of this equation is of the form (X,X), which is non-negative, hence

$$(x,x) - \sum_1^n |(x,e_r)|^2 \geqslant 0 \quad ,$$

which is Bessel's inequality for a finite set.

If (e_n) is an infinite orthonormal set, consider any partial sum $s_n = \Sigma_1^n |c_r|^2$ of the series in (7.7). The argument given above for the finite case shows that

$$s_n = \sum_1^n |c_r|^2 \leqslant \|x\|^2 \quad ,$$

hence the sequence (s_n) is bounded above, and it is monotonic increasing, because the terms of the series are positive, hence (s_n) converges by a basic theorem on monotonic bounded sequences (see Appendix B), and its limit is $\leqslant \|x\|^2$. □

This result means that the expansion coefficients die away faster than $n^{-\frac{1}{2}}$, roughly speaking, as $n \to \infty$ (but see Problem 7.15). It is a very general result, which applies whether or not the space is a Hilbert space, and whether or not the set is a basis. If it is a basis, then we have equality in (7.7), giving a criterion for deciding whether a given orthonormal set is a basis.

Theorem 7.41 (Parseval's Relation) Let (e_n) be an orthonormal set in an inner product space. It is a basis if and only if for each x in the space

$$\sum_1^\infty |c_n|^2 = \|x\|^2 \quad , \tag{7.8}$$

where (c_n) are the expansion coefficients of x with respect to (e_n), $c_n = (x,e_n)$.

Proof In the proof of Bessel's inequality we showed that

$$\|x - \sum_1^n c_r e_r\|^2 = \|x\|^2 - \sum_1^n |c_r|^2 \quad . \tag{7.9}$$

If (e_r) is a basis, then the series $\Sigma_1^\infty c_r e_r$ converges to x, hence $\|x - \Sigma_1^n c_r e_r\|^2 \to 0$ as $n \to \infty$, and (7.9) gives $\|x\|^2 - \Sigma_1^n |c_r|^2 \to 0$ as $n \to \infty$ which proves (7.8).

Conversely, if (7.8) holds for all x in the space, then $\|x\|^2 - \Sigma_1^n |c_r|^2 \to 0$ as $n \to \infty$; (7.9) gives $\|x - \Sigma_1^n c_r e_r\|^2 \to 0$. This means that $x = \Sigma_1^\infty c_r e_r$, and shows that any x can be represented as an infinite linear combination of the (e_r); thus (e_r) is a basis. □

In $\mathbb{R}^2$, (7.8) reads $\|x\|^2 = x_1{}^2 + x_2{}^2$, where x_1 and x_2 are the components of the vector x; this is just Pythagoras' theorem. In three dimensions, (7.8) is the familiar result that the magnitude of a vector is the root of the sum of the squares of its components; (7.8) is the obvious generalisation to infinite-dimensional inner product spaces.

Given an orthonormal set, one way of finding out whether it is a basis is to calculate the expansion coefficients of some member of the space, sum their squares, and test whether (7.8) is satisfied. If not, then more vectors must be added to the orthonormal set to make it a basis, so that more terms will be added to the left hand side of (7.8) to make it up to the value of the right hand side. If (7.8) is satisfied for some particular x, on the other hand, it may increase one's confidence in the set's being a basis, but it does not prove anything, unless (7.8) can be verified for all x in some dense subspace (Theorem 7.33).

Example 7.42 In the real space $L_2[0,\pi]$, the set of functions $\cos(t)$, $\cos(2t)$, . . . is an orthogonal set; it can be turned into an orthonormal set by multiplying each cosine by a normalisation factor $(2/\pi)^{\frac{1}{2}}$. For the function $g(t) = t$ it is easy to evaluate the expansion coefficients $c_n = (2/\pi)^{\frac{1}{2}} \int_0^\pi t \cos(nt)\, dt$. Summing the first few terms gives $\Sigma |c_n|^2 \cong 2.58$ while $\|g\|^2 \cong 10.33$; clearly Parseval's relation is not satisfied, so the set of cosines is not a basis. The reader familiar with Fourier series will have seen this already. The functions $\cos(nt)$ are appropriate for a half-range expansion on the interval $(0,\pi)$; but the half-range cosine series has a constant term too, so a constant function should be added to the orthogonal set in order to form a basis. If this is done then Parseval's relation is satisfied. □

It should be added that this reasoning is not often used in practice to distinguish bases from other orthogonal sets. Bases are usually obtained as sets of eigenvectors of operators, in which case there are general theorems saying that the set of eigenvectors is a basis. The simplest of these theorems is proved in Chapter 9. For other ways of proving that a set of functions is a basis, see Higgins.

The position so far can be summed up as follows. In an inner product space with a basis, every vector has associated with it a set of expansion coefficients c_n, its components with respect to the basis, and the coefficients satisfy (7.8), or a corresponding relation if the basis is not normalised. Now, in a finite-dimensional space there is a converse result: given any numbers $c_1, \ldots, c_n$ there is a vector whose components are c_i, namely, $x = \Sigma c_i e_i$, where e_i is the basis. Is the same true in infinite-dimensional spaces? Obviously not, since the expansion coefficients of any vector with respect to an orthonormal basis must tend to zero fast enough for the sum of their squares to converge and satisfy (7.8), hence the sequence of numbers 1,1,1, . . . cannot be the expansion coefficients of any vector. Consider then a sequence (c_n) such that $\Sigma |c_n|^2$ converges, so that the above objection does not apply; is there always a vector whose expansion coefficients are (c_n)? The answer is yes, provided that the space is complete, as the following theorem shows. It is an important result, because it gives a criterion for convergence of a series involving an orthonormal basis. It is the first link in a chain of theorems leading up to the extremely useful spectral theorem of Chapter 9.

Theorem 7.43 (The Riesz-Fischer Theorem) Let (e_n) be an orthonormal basis for a (real or complex) infinite-dimensional Hilbert space H. If (c_n) is a sequence of (real or complex) numbers such that $\Sigma |c_n|^2$ converges, then there is an $x \in H$ such that $x = \Sigma_1^\infty c_n e_n$, and $c_n = (x,e_n)$.

Proof We shall show that the sequence of partial sums is a Cauchy sequence. We have

$$\left\| \sum_n^{n+p} c_i e_i \right\|^2 = \sum_n^{n+p} |c_i|^2$$

using orthonormality. For any $\epsilon > 0$ there is an N such that the right hand side is less than ϵ^2 for $n > N$ because $\Sigma |c_i|^2$ is given to be convergent. Hence $\| \Sigma_n^{n+p} c_i e_i \| < \epsilon$ for $n > N$, showing that the sequence of partial sums of $\Sigma c_i e_i$ is a Cauchy sequence, hence convergent to some $x \in H$ since H is complete. $c_n = (x,e_n)$ now follows from Theorem 7.30. □

It follows from the Riesz-Fischer theorem that all infinite-dimensional Hilbert spaces are the same, in the way that all n-dimensional spaces can be considered as the same (but all Banach spaces cannot). The reasoning is as follows.

Consider the space l_2 defined in Example 7.4. For any Hilbert space H with basis (e_n), Theorem 7.40 says that for each $x \in H$ there is a corresponding element $\boldsymbol{c} = (c_n)$ of l_2, and Theorem 7.43 says that for each $\boldsymbol{c} = (c_n)$ in l_2 there is a corresponding element of H; Theorem 7.30 shows that this element is unique. We thus have a one-to-one correspondence between l_2 and any Hilbert space H. It is easy to show that if $x,y \in H$ and $\boldsymbol{c},\boldsymbol{d}$ are the corresponding elements of l_2, then $ax + by$ corresponds to the sequence $a\boldsymbol{c} + b\boldsymbol{d}$, and that $\|x\| = \|\boldsymbol{c}\|$ (Parseval's relation) and $(x,y) = (\boldsymbol{c},\boldsymbol{d})$ (Problem 7.11). Thus all the algebraic properties of H are mirrored exactly in l_2; the two spaces are said to be **isomorphic**, and can be regarded as the same space in two different guises. It is easy to see that isomorphism is an equivalence relation. Since any Hilbert space is isomorphic to l_2, it follows that all Hilbert spaces are isomorphic to each other: there is really only one Hilbert space. More precisely, we should say that all real Hilbert spaces are isomorphic to the real space l_2, and all complex Hilbert spaces are isomorphic to the complex space l_2, so there is one real and one complex Hilbert space. For an extremely wide-ranging discussion of the idea of isomorphism, see Hofstadter.

7.5 Orthogonal Decomposition

In this section we develop the geometrical aspect of Hilbert space theory.

Definition 7.44 Let X be any subset of a Hilbert space H. We say $x \perp X$ if $x \perp \xi$ for all $\xi \in X$. The **orthogonal complement** of X is the set $X^\perp = \{x \in H: x \perp X\}$. □

Examples 7.45 If X consists of a single vector, $X = \{\xi\}$, then $x \perp X$ if and only if $x \perp \xi$. In $\mathbb{R}^3$ the orthogonal complement of a set consisting of a single vector $\boldsymbol{a} \neq \mathbf{0}$ is the plane through the origin perpendicular to that vector. The orthogonal complement of $\{\mathbf{0}\}$ is the whole space (and vice versa). If X consists of two nonzero nonparallel vectors, $X = \{\boldsymbol{a}, \boldsymbol{b}\}$, then $X^\perp$ is the set of vectors perpendicular to both of them, that is, the intersection of the orthogonal complements of the individual vectors, therefore a line through the origin; see Fig. 17. □

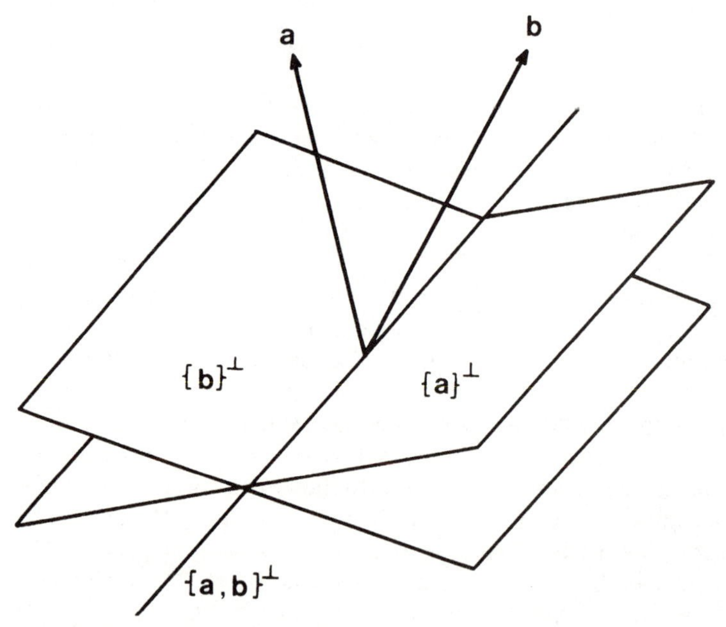

Fig. 17. Orthogonal complements of the sets $\{\boldsymbol{a}\}$, $\{\boldsymbol{b}\}$, and $\{\boldsymbol{a},\boldsymbol{b}\}$ in $\mathbb{R}^3$.

All the above examples are subspaces. This is true in general.

Theorem 7.46 (Orthogonal Complement) $X^\perp$ is a subspace of the Hilbert space H, for any subset X of H.

Proof $X^{\perp}$ is a vector subspace because for any numbers a,b and any $x,y \in X^{\perp}$, $(ax + by,u) = 0$ for all $u \in X$, hence $ax + by \in X^{\perp}$. We must prove that $X^{\perp}$ is complete. Let (x_n) be a Cauchy sequence in $X^{\perp}$; it converges to some $x \in H$ because H is complete. We must show that $x \in X^{\perp}$.

By Lemma 7.13, for any $u \in X$, $(x,u) = \lim_{n\to\infty}(x_n,u) = \lim_{n\to\infty}(0) = 0$, therefore $x \in X^{\perp}$. Thus every Cauchy sequence in $X^{\perp}$ converges to an element of $X^{\perp}$, so it is complete. Theorem 7.36 shows that it has a basis, so it is a Hilbert space and a subspace of H. □

Now, in three-dimensional space, given any plane, any vector can be projected on to the plane; that is, given a plane P and a vector x, there is a vector y in the plane and a vector z normal to the plane such that $x = y + z$; y is the projection of x on to P (see Fig. 18). The same is true in any Hilbert space.

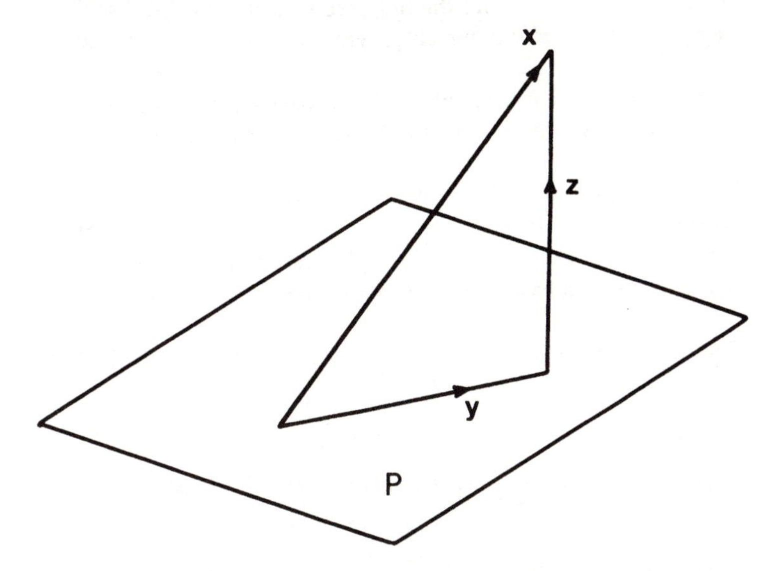

Fig. 18. The projection of a vector x on to a plane P.

Theorem 7.47 (Orthogonal Projection) If E is a subspace of a Hilbert space H, then every $x \in H$ can be uniquely written as $x = y + z$ with $y \in E$ and $z \in E^{\perp}$. y is called the projection of x on to E.

Proof Let (e_n) be an orthonormal basis for E. It is an orthonormal set in H, therefore $\Sigma|(x,e_n)|^2$ converges by Theorem 7.40, so $\Sigma(x,e_n)e_n$ converges by the Riesz-Fischer theorem. Let

$$\Sigma(x,e_n)e_n = y$$

with $y \in E$. We have now constructed the projection y of x on to E. We must show that $x - y \in E^{\perp}$.

Set $z = x - y$. Then

$$(z,e_n) = (x,e_n) - (y,e_n)$$
$$= (x,e_n) - (x,e_n)$$
$$\therefore (z,e_n) = 0 \quad \text{for all} \quad n \quad . \tag{7.10}$$

But every $u \in E$ can be written in the form $\Sigma(u,e_n)e_n$, hence

$$(z,u) = \Sigma(e_n,u)(z,e_n) = 0$$

for all $u \in E$ by (7.10), so $z \in E^{\perp}$.

Finally we must show that this decomposition is unique. Suppose $x = y + z = y' + z'$, with $y,y' \in E$ and $z,z' \in E^{\perp}$. Then $y - y' = z' - z$, but $y - y' \in E$ and is therefore orthogonal to $z' - z \in E^{\perp}$, and the only vector orthogonal to itself is the zero vector, hence $y - y' = z - z' = 0$, which proves the uniqueness of y and z. □

Corollary 7.48 If E is a subspace of a Hilbert space H, and (e_n) and (f_n) are bases for E and $E^{\perp}$ respectively, then (e_n) and (f_n) together form a basis for H.

Remark 7.49 The meaning of the statement above should be clear, though it is worded informally. More formally one might say that the sequence obtained by arranging the elements of the set $\{e_n\} \cup \{f_n\}$ in any order is a basis for H. The order of the elements in a basis does not matter; see Problem 7.14.

Proof of 7.48 For any $x \in H$, we have $x = y + z$ with $y \in E$ and $z \in E^{\perp}$, by 7.47. Hence

$$x = \Sigma(y,e_n)\| e_n \|^{-2} e_n + \Sigma(z,f_n)\| f_n \|^{-2} f_n$$

which gives the expansion of x in terms of $\{e_n\} \cup \{f_n\}$, and shows that this set is a basis. □

Notation 7.50 We symbolise the statement of Theorem 7.47, that every element of H can be written uniquely as the sum of an element of E and an element of $E^{\perp}$, by writing

$$H = E \oplus E^{\perp} \quad ,$$

and say that H is the **direct sum** of E and $E^{\perp}$. This can also be interpreted as saying that a basis for E plus a basis for $E^{\perp}$ is a basis for H. □

In three-dimensional space we are used to regarding the motion of a particle, say, as a combination of three rectilinear motions along the coordinate axes, and in a sense these motions are independent: the variation of the x-coordinate is governed solely by the x-component of the force. In this way, a three-dimensional problem can be thought of as a combination of three one-dimensional problems. The idea of the direct sum of subspaces is related to this. H can be regarded as built up from the two subspaces E and $E^\perp$ in the sense of Theorem 7.47. The procedure can be repeated, E and $E^\perp$ being decomposed into sums of smaller subspaces, so that H is a direct sum of many subspaces. One aim of the theory of Chapters 8 and 9 is to show how Hilbert spaces appropriate for the solution of certain kinds of differential and integral equations can be expressed as direct sums of finite-dimensional subspaces in such a way that the solution of the equation reduces to the solution of a number of finite-dimensional problems in those subspaces.

There is another way of looking at the orthogonal projection theorem: it can be regarded as an extension of the approximation theorem 7.32 to infinite orthogonal sets. To see this, let us rewrite the statement of 7.32 as follows. Let E be the subspace spanned by $\{e_1, \ldots, e_n\}$. Then for any x in the space, the element $y = \Sigma_r(x,e_r)e_r$ of E minimises $\|y - x\|$. It is intuitively clear that this minimising y is the vector y shown in Fig. 18, and that $y - x$ is perpendicular to E. The proof is easy: $(y - x,e_r) = (y,e_r) - (x,e_r) = 0$ for $1 \leqslant r \leqslant n$, hence $y - x \in E^\perp$. Thus Theorem 7.32 is the finite-dimensional version of 7.47, with the added statement that y is the best approximation in E to x. This statement extends at once to the general case.

Corollary 7.51 (Best Approximation) Let E be a subspace of a Hilbert space H. Then for any given $x \in H$ there is exactly one vector in E closer to x than any other vector, namely, the projection of x on to E.

Proof By 7.47, we have for any x a unique $y \in E$ and $z \in E^\perp$ such that $x = y + z$. Hence for any $u \in E$, $x - u = y - u + z$.

Now, $y - u \perp z$, hence by 7.12

$$\|x - u\|^2 = \|y - u\|^2 + \|z\|^2 = \|y - u\|^2 + \|y - x\|^2 \quad ,$$

$$\therefore \|x - u\| > \|x - y\|$$

if $u \neq y$. This shows that y is the element of E closest to x. □

This result may look very obvious. But it is not true in general in normed spaces: see Problem 7.28. Hilbert space has nicer geometry than other normed spaces.

7.6 Functionals on Normed Spaces

The next two sections are about functionals, which are the simplest operators on Hilbert space, and form the foundation of the theory of self-adjoint operators in Chapter 8. In this section we discuss the basic ideas of the theory of functionals, leading up to the important Theorem 7.57; these ideas do not involve the inner product, and can therefore be discussed in the general context of normed spaces. In the next section we shall return to the special case of Hilbert space.

Definition 7.52 **A functional** on a (real or complex) normed space is a map from the space to the (real or complex) numbers. A linear functional f is one such that $f(ax + by) = af(x) + bf(y)$ for any (real or complex) numbers a,b and any x,y in the space.

Example 7.53 For any fixed $\boldsymbol{v} \in \mathbb{R}^3$ we can define a functional $f_{\boldsymbol{v}}: \boldsymbol{x} \to \boldsymbol{v}.\boldsymbol{x}$ for all $\boldsymbol{x} \in \mathbb{R}^3$, where $\boldsymbol{v}.\boldsymbol{x}$ is the ordinary scalar product. $f_{\boldsymbol{v}}$ is clearly linear. Now, in section 6.1 we defined continuity of an operator. Since a functional on a space N is just an operator $N \to \mathbb{C}$, we can apply the ideas of continuity to functionals. $f_{\boldsymbol{v}}$ is uniformly continuous on $\mathbb{R}^3$: for any $\epsilon > 0$ the definition 6.3 is satisfied by taking $\delta = \epsilon/|\boldsymbol{v}|$ (unless $\boldsymbol{v} = \boldsymbol{0}$, in which case any $\delta > 0$ will do).

Theorem 7.54 (Continuity of Linear Functionals) If a linear functional is continuous at any one point, then it is uniformly continuous.

Proof Suppose f is continuous at a. Then for any $\epsilon > 0$ there is a $\delta > 0$ such that $|f(a + h) - f(a)| < \epsilon$ whenever $\|h\| < \delta$. Using the linearity of f, for any x we have

$$\begin{aligned} |f(x+h) - f(x)| &= |f(h)| \\ &= |f(a+h) - f(a)| \\ &< \epsilon \end{aligned}$$

whenever $\|h\| < \delta$. Since δ is independent of x, this shows that f is uniformly continuous. □

It follows that a linear functional is either discontinuous everywhere or uniformly continuous everywhere. We can therefore call a uniformly continuous linear functional simply 'continuous' without ambiguity.

Examples 7.55 (a) In $C[-1,1]$ with the sup norm, the functional $\Delta: \phi \mapsto \phi(0)$ is continuous. *Proof*: for any $\phi, \psi \in C[-1,1]$,

$$\begin{aligned} |\Delta(\phi) - \Delta(\psi)| &= |\phi(0) - \psi(0)| \\ &\leqslant \|\phi - \psi\| \quad , \end{aligned}$$

so the '$\epsilon - \delta$' definition is satisfied with $\delta = \epsilon$. This functional is clearly a close relative of the delta distribution of Chapter 1, but is not the same thing – it is a functional defined on a different space, and the definition of continuity is different.

(b) In $C[-1,1]$, with the integral norm of Example 4.28, we define a functional Δ_1 by the same rule as in (a); it is a different functional because it is defined over a different space. It is discontinuous, even though the very similar functional Δ above is continuous. *Proof*: the sequence of functions $f_n(x) = n^{\frac{1}{8}}e^{-nx^2}$ is convergent: $\|f_n\| \to 0$ as $n \to \infty$. But Δ_1 maps f_n into $n^{\frac{1}{8}}$, which is a divergent sequence of numbers. If Δ_1 were continuous. then Theorem 6.10 would be violated. □

Example 7.55(b) displays a common feature of discontinuous functionals: vectors of small norm are mapped into large numbers. Functionals which always map small vectors into small numbers are called bounded.

Definition 7.56 A functional f is **bounded** if there is a number A such that $|f(x)| \leqslant A\|x\|$ for all x in the space. □

This use of the word 'bounded' is different from its use in the phrase 'bounded function'. The function $x \mapsto 2x$ is not a bounded function on $\mathbb{R}$, it tends to infinity as $x \to \infty$. But regarded as a functional on $\mathbb{R}$, it satisfies Definition 7.56 with $A = 2$, and is therefore a bounded functional. It is the ratio $f(x)/\|x\|$ that must be bounded if f satisfies 7.56.

It is easy to see that the discontinuous functional of Example 7.55(b) is unbounded, while the continuous functional of Example 7.55(a) is bounded. It is true in general that continuity and boundedness go hand in hand for linear functionals (though not for functionals in general).

Theorem 7.57 (Bounded Linear Functionals) A linear functional is continuous if and only if it is bounded.

Proof (a) Suppose f is linear and bounded, $|f(x)| \leqslant A\|x\|$ for all x. For any x,y in the space,

$$\begin{aligned} |f(x) - f(y)| &= |f(x - y)| \\ &\leqslant A\|x - y\| \quad . \end{aligned}$$

Hence for any $\epsilon > 0$, $|f(x) - f(y)| < \epsilon$ whenever $\|x - y\| < \epsilon/A$, which proves continuity (unless $A = 0$, in which case we have $f(x) = 0$ for all x, which is trivially continuous).

(b) Now suppose that f is linear and continuous. Taking $\epsilon = 1$ in the definition of continuity at 0, it follows that there is a $\delta > 0$ such that

$$|f(x)| < 1 \quad \text{for} \quad \|x\| < \delta \quad . \tag{7.11}$$

Hence for any y in the space,

$$\begin{aligned} |f(y)| &= (2\|y\|/\delta)\,|f(y[\delta/2\|y\|])| \\ &< (2/\delta)\|y\| \end{aligned}$$

by (7.11), since $\| y(\delta/2\|y\|)\| = \delta/2 < \delta$. Hence Definition 7.56 is satisfied with $A = 2/\delta$. □

7.7 Functionals in Hilbert Space

In Hilbert space there is a particularly important class of functionals, analogous to Example 7.53: for any u in the space there is a corresponding functional $x \mapsto (x,u)$.

Proposition 7.58 If u is any fixed member of a Hilbert space H, the functional f: $x \mapsto (x,u)$ is linear and continuous.

Proof Linearity follows immediately from the properties of the inner product. Boundedness follows from the Schwarz inequality: $|f(x)| = |(x,u)| \leqslant \|u\|\|x\|$. Continuity follows from Theorem 7.57. □

Remark 7.59 In a real space the functional $x \mapsto (u,x)$ is also linear and continuous, since the inner product is symmetric. But in a complex space this functional is not linear, since $(u,ax) = \bar{a}(u,x)$. Functionals with this property are sometimes called **antilinear**. □

The aim of this section is to prove a kind of converse to the above proposition: every bounded linear functional in Hilbert space is of the form (x,u) for some u in the space. This result is an essential ingredient of the theory of operators in the next chapter. It is also related to the ideas of distribution theory. A regular distribution is a continuous functional on the space $\mathscr{D}$ (Definition 1.3), of the type $\phi \mapsto \int f(x)\phi(x)\,dx$. This integral has the same form as an inner product in the Hilbert space of real square-integrable functions. Regular distributions thus resemble functionals of the type $x \mapsto (x,u)$. Now, the whole point of distribution theory is the existence of singular functionals on $\mathscr{D}$ which are not of this form. We shall see that in Hilbert space, all continuous linear functionals are of the form $x \mapsto (x,u)$. One therefore needs a space more exotic than Hilbert space (such as $\mathscr{D}$) in order to obtain $\delta(x)$ as a continuous linear functional.

The following geometrical argument shows why every functional in Hilbert space is of the type discussed in Theorem 7.58. Consider the space $\mathbb{R}^3$; let f be a continuous linear functional $\mathbb{R}^3 \to \mathbb{R}$. In general there will be a region G_+ of $\mathbb{R}^3$ in which f takes positive values, and a region G_- where f is negative. The set N of points where f vanishes (the 'null space' of f) is a subspace of $\mathbb{R}^3$, because f is linear. Hence N is either a line or a plane through the origin. But N must divide $\mathbb{R}^3$ into two regions G_+ and G_-; any curve joining a point in G_+ to a point in G_- must pass through N, because a continuous function cannot pass from a positive to a negative value without vanishing. Thus N is a plane, not a line; a line cannot be the boundary between two regions of space. Now, any vector can be resolved into a vector in the plane N plus a vector perpendicular to N; for any x, $x = y + x_\perp$, where y lies in N and $x_\perp$ is perpendicular to N. Hence $f(x) = f(x_\perp) = f(\|x_\perp\|e) = \|x_\perp\| f(e)$ where e is a unit vector perpendicular to N and parallel to $x_\perp$. Then $\|x_\perp\| = (x,e)$ (ignoring the question of possible minus signs), and finally we have $f(x) = (x,u)$ where $u = f(e)e$. Thus every continuous linear functional can be represented as an inner product. The argument applies to any Hilbert space.

Theorem 7.60 (The Riesz Representation Theorem) For every continuous linear functional f on a Hilbert space H, there is a unique $u \in H$ such that $f(x) = (x,u)$ for all $x \in H$.

Proof Set $N = \{x \in H: f(x) = 0\}$. If $N = H$, then f is identically zero, and $u = 0$ gives $f(x) = (x,u)$ for all x. If $N \neq H$, we shall show that the orthogonal complement $N^\perp$ is one-dimensional.

We must show that every pair of vectors in $N^\perp$ is linearly dependent. Take any $x,y \in N^\perp$; $ax + by \in N^\perp$ for any numbers a,b by Theorem 7.46; but the vector $z = f(y)x - f(x)y \in N$ since $f(z) = 0$. Thus z lies in both N and $N^\perp$, and is therefore perpendicular to itself and therefore zero. Thus given any $x,y \in N$ we have constructed a linear combination of x and y which is zero, so $N^\perp$ is one-dimensional.

Let e be an element of $N^\perp$ with $\|e\| = 1$; $\{e\}$ is an orthonormal basis for $N^\perp$, and any $u \in N^\perp$ can be written $u = (u,e)e$. Now, by Theorem 7.47, any $x \in H$ can be written $x = y + z$, with $y \in N^\perp$ and $z \in N$. Then $f(x) = f(y) + f(z) = (y,e)f(e) + 0 = (x - z,e)f(e) = (x,e)f(e)$. Thus $f(x) = (x,u)$ as required, where $u = \overline{f(e)}e$.

To prove uniqueness, suppose $f(x) = (x,u) = (x,v)$ for all $x \in H$. Taking $x = u - v$ gives $(u - v,u) = (u - v,v)$, hence $(u - v, u - v) = 0$, so $\|u - v\| = 0$, and $u = v$. □

7.8 Weak Convergence

We have met several different types of convergence so far. The space $\mathscr{D}$ of test functions was given a very strong definition of convergence, uniform convergence of all derivatives. Normed spaces were given a definition of convergence very much like that of elementary analysis in principle, but taking various forms according to the particular norm used. We took this idea of convergence over to Hilbert space, using the inner product to define a norm. However, there is another kind of convergence in Hilbert space, whose counterpart in normed spaces is beyond the scope of this book; see Problem 7.25. In this section we define it and study its relation to ordinary convergence. We begin by showing how this new idea of convergence arises.

Consider a sequence (x_n) with $x_n \to 0$ in the usual sense, that is, $\|x_n\| \to 0$. By Lemma 7.13,

$$(x_n,y) \to 0 \quad \text{as} \quad n \to \infty \quad \text{for all} \quad y \in H \quad . \tag{7.12}$$

(x_n,y) can be interpreted as the component of x_n in the direction of y, times a factor $\|y\|$. Equation (7.12) then says that if $x_n \to 0$ then any component of x_n tends to zero. In finite-dimensional spaces the converse is true: if a sequence of vectors is such that every component tends to zero, that is, (7.12) holds, then the sequence itself tends to zero in the sense that $\|x_n\| \to 0$. Is it true in general that (7.12) implies $x_n \to 0$?

Example 7.61 Let (e_n) be an infinite orthonormal set in an inner product space. By Theorem 7.40, $\Sigma|(y,e_n)|^2$ converges, and hence $(y,e_n) \to 0$ for any y. But $\|e_n\| = 1$, so $e_n \nrightarrow 0$. □

This example shows that the answer to the question at the end of the last paragraph is a definite no. And yet, if every component of a sequence tends to zero, it seems reasonable to regard that sequence as converging to zero in some sense, even if it does not converge according to Definition 4.32. We shall therefore call such sequences 'weakly convergent'. We shall see that they have many properties in common with convergent sequences, and they play an important role in Hilbert space. Proving that a sequence converges weakly can be easier than proving that it converges in the usual sense. It can also be a stepping-stone to proving ordinary convergence, as we shall see. Remember that the ultimate aim of functional analysis in applied mathematics is to give solutions to equations, often by means of convergent sequences of approximations, so the theory of convergence is of direct practical relevance.

Definition 7.62 A sequence (x_n) in a Hilbert space H **converges weakly** if for every $y \in H$ the sequence of numbers (x_n,y) converges. □

This definition is like the definition 1.30 of convergence of distributions; it does not involve the idea of a limit element to which the sequence converges. However, such a limit element may exist.

Definition 7.63 If $X, x_1, x_2, \ldots$ belong to a Hilbert space H, we say $x_n \to X$ weakly as $n \to \infty$ if $(x_n,y) \to (X,y)$ for all $y \in H$. □

Ordinary (norm) convergence is often called **strong** convergence, to distinguish it from weak convergence. We can now say that the sequence in Example 7.61 converges weakly but not strongly to zero.

It is conceivable that a sequence might be weakly convergent (according to Definition 7.62), yet not converge weakly to any element of H (according to 7.63). But though conceivable, it is not true. We shall now prove that every weakly convergent sequence has a weak limit (cf. 1.31). This fact is useful because it is an existence theorem: it asserts that there exists an element of the space with certain properties.

The proof depends on the following proposition, which is the analogue of a result which is very easy to prove in the case of strong convergence. For weak convergence, however, the result is not at all obvious, and the proof is quite deep and subtle. In purely theoretical treatments of functional analysis, a version of this proposition occupies a central position, under the name of the Principle of Uniform Boundedness or the Banach-Steinhaus Theorem. From our point of view, however, it is just a stepping-stone to Theorem 7.65 and to the spectral theorem in Chapter 9. The reader who is uninterested in analytic subtleties and willing to take 7.65 for granted may omit the proof of 7.64 and 7.65.

Proposition 7.64 Every weakly convergent sequence in a Hilbert space is bounded.

Proof Suppose (x_n) converges weakly in a Hilbert space H. We wish to prove that the sequence of numbers $(\|x_n\|)$ is bounded. We know that for any $y \in H$ the sequence of numbers (x_n,y) is convergent and therefore bounded. We shall now show that the theorem follows easily if (x_n,y) is bounded uniformly for y in some sphere, that is, if there is a $y_0 \in H$ and numbers $M,r > 0$ such that

$$|(x_n,y)| < M \quad \text{for all} \quad y \in \Sigma(y_0;r) \quad \text{and all} \quad n \quad , \tag{7.13}$$

where $\Sigma(y_0;r)$ is the open sphere of radius r centred on y_0:

$$\Sigma(y_0;r) = \{y \in H: \|y - y_0\| < r\} \quad .$$

For suppose that (7.13) holds. Then for any z with $\|z\| < 1$ we have $|(x_n,y_0 + rz)| < M$, hence

$$|(x_n,y_0) + r(x_n,z)| < M \quad .$$

But $|(x_n,y_0)| < M$ since $y_0 \in \Sigma(y_0;r)$, hence by the triangle inequality

$$\begin{aligned} r|(x_n,z)| &\leqslant |r(x_n,z) + (x_n,y_0)| + |-(x_n,y_0)| \\ &< 2M \quad . \end{aligned}$$

Here z is an arbitrary vector of norm less than 1; taking $z = x_n/(2\|x_n\|)$ gives

$$\|x_n\| < 4M/r \quad ,$$

which is a bound on $\|x_n\|$ independent of n. Hence the theorem will be proved if we can establish that (7.13) holds for some y_0,r,M.

We prove this by contradiction: suppose that (x_n,y) is unbounded on every sphere. Then there exist n_1, y_1 such that

$$|(x_{n_1},y_1)| > 1 \quad .$$

Since the inner product is continuous, $|(x_{n_1},y)| > 1$ for all y in some sphere $\Sigma(y_1;r_1)$ centred on y_1 with radius $r_1 > 0$. Since we are supposing that (x_n,y) is unbounded on all spheres, there is an n_2 and a $y_2 \in \Sigma(y_1;r_1)$ such that

$$|(x_{n_2},y_2)| > 2 \quad .$$

Again, by continuity we have $|(x_{n_2},y)| > 2$ for $y \in \Sigma(y_2;r_2)$ for some r_2, which can be chosen small enough so that $\Sigma(y_2;r_2) \subset \Sigma(y_1;r_1)$ and $r_2 < \frac{1}{2}r_1$. Again, (x_n,y) unbounded on all spheres implies that there is an n_3 and a $y_3 \in \Sigma(y_2;r_2)$ with $|(x_{n_3},y_3)| > 3$. Proceeding in this way gives sequences (n_p), (y_p), (r_p) such that

$$|(x_{n_p},y)| > p \quad \text{for} \quad y \in \Sigma(y_p;r_p) \tag{7.14}$$

and $\quad \|y_p - y_{p+1}\| < r_p$

and $\quad r_p < \frac{1}{2}r_{p-1} < \ldots < (\frac{1}{2})^{p-1}r_1 \quad .$

Hence for any q

$$\begin{aligned} \|y_p - y_{p+q}\| &\leqslant \|y_p - y_{p+1}\| + \|y_{p+1} - y_{p+2}\| + \ldots + \|y_{p+q-1} - y_{p+q}\| \\ &< r_p(1 + \tfrac{1}{2} + \ldots + (\tfrac{1}{2})^{q-1}) \\ &< 2(\tfrac{1}{2})^{p-1}r_1 \to 0 \quad \text{as} \quad p \to 0 \quad \text{for any} \quad q \quad . \end{aligned}$$

This shows that (y_n) is a Cauchy sequence; and a Hilbert space is complete, hence $y_n \to \bar{y}$ for some $\bar{y} \in H$.

Since for any p the p-th sphere contains all members of the sequence (y_n) after the p-th, it must contain the limit of y_n, so

$$\bar{y} \in \Sigma(y_p;r_p) \quad \text{for all} \quad p \quad .$$

From (7.14) we have

$$|(x_{n_p},\bar{y})| > p \quad \text{for all} \quad p \quad . \tag{7.15}$$

But x_{n_p} converges weakly, therefore $(x_{n_p},\bar{y})$ is a convergent and hence bounded sequence of real numbers. This contradicts (7.15), and completes the proof. □

Theorem 7.65 If (x_n) is a weakly convergent sequence in a Hilbert space H, then there is an $X \in H$ such that $x_n \to X$ weakly.

Proof Proposition 7.64 shows that (x_n) is a bounded sequence, say $\|x_n\| \leqslant M$ for all n. Define a functional

$$f: y \mapsto \lim(y,x_n) \quad .$$

It is clearly linear, and is bounded because

$$|f(y)| \leqslant \max_n \{\|y\|\|x_n\|\} \leqslant M\|y\|$$

for all y. By the Riesz representation theorem there is an $X \in H$ such that $f(y) = (y,X)$, or $\lim(y,x_n) = (y,X)$, which shows that $x_n \to X$ weakly. □

We next prove the basic elementary properties of weak convergence.

Proposition 7.66 (a) If $x_n \to X$ weakly and $y_n \to Y$ weakly, then $ax_n + by_n \to aX + bY$ weakly for any scalars a,b.

(b) Weak limits are unique, that is, if $x_n \to X_1$ weakly and $x_n \to X_2$ weakly, then $X_1 = X_2$.

Proof (a) For any z in the space, $(ax_n + by_n,z) = a(x_n,z) + b(y_n,z) \to a(X,z) + b(Y,z) = (aX + bY,z)$.

(b) Using (a), $x_n - x_n \to X_1 - X_2$ weakly, therefore $(0,z) \to (X_1 - X_2,z)$ for all z. But $(0,z)$ is a sequence of numbers all of which are zero, so its limit is zero, and $(X_1 - X_2,z) = 0$ for all z in the space. Taking $z = X_1 - X_2$ gives $\|X_1 - X_2\|^2 = 0$, hence $X_1 = X_2$. □

We now justify the terms 'strong' and 'weak' convergence, by showing that weak convergence is indeed a weaker property than strong convergence, that is, strong implies weak convergence.

Proposition 7.67 If $x_n \to X$ strongly, then $x_n \to X$ weakly.

Proof Follows immediately from Lemma 7.13. □

Example 7.61 shows that the converse of this result is not true. However, if a sequence is contained in a compact set, then weak convergence implies strong convergence. In order to prove this, we need the following rather esoteric-sounding result. Readers willing to take Theorems 7.69 and 7.70 on trust may omit the rest of this section except for the statements of those theorems.

Lemma 7.68 Suppose (x_n) is a sequence for which every subsequence (y_n) has a sub-subsequence (z_n) which converges strongly to X. Then $x_n \to X$ strongly.

Proof We shall suppose that $x_n \nrightarrow X$ and obtain a contradiction. Then there is an $\epsilon > 0$ such that for any n_0 there is an $n_1 > n_0$ with $\|x_{n_1} - X\| \geqslant \epsilon$. Similarly there is an $n_2 > n_1$ such that $\|x_{n_2} - X\| \geqslant \epsilon$, and so on. Thus we construct a sequence $(y_r) = (x_{n_r})$ with $\|y_r - X\| \geqslant \epsilon$ for all r; (y_r) clearly cannot converge to X, and neither can any subsequence of (y_r), contradicting the given condition. □

Theorem 7.69 If (x_n) is a sequence contained in a compact set, and $x_n \to X$ weakly, then $x_n \to X$ strongly.

Proof By compactness, every subsequence (y_n) of (x_n) has a sub-subsequence (z_n) which converges strongly, say $z_n \to Z$. By 7.67, $z_n \to Z$ weakly; but $z_n \to X$

weakly because (z_n) is a subsequence of (x_n); hence $Z = X$ by 7.66(b). Therefore $z_n \to X$ strongly, and 7.68 completes the proof. □

The last theorem in this section is a kind of converse to Theorem 7.64. Of course, the strict converse of 7.64 is false: $(a,-a,a,-a, \ldots)$ is a bounded sequence which does not converge weakly. But we shall show that even if a bounded sequence does not converge weakly, at least some subsequence does.

Theorem 7.70 Every bounded sequence in a Hilbert space has a weakly convergent subsequence.

Proof Let (x_n) be a bounded sequence in a Hilbert space H, with $\| x_n \| \leqslant M$ for all n. We shall construct a weakly convergent subsequence by the 'diagonal method' (as in the proof of 6.33).

Take an orthonormal basis (e_n) for H. The sequence of complex numbers (e_1,x_n) lies in the circle $|z| \leqslant M$ in $\mathbb{C}$, which is compact. Hence there is a subsequence of these numbers which converges, say

$$(e_1,x_n^1) \quad \text{converges as} \quad n \to \infty \quad .$$

Again, the sequence of numbers (e_2,x_n^1) contains a convergent subsequence, (e_2,x_n^2), say, and in this way we obtain for any r a sequence $(x_1^r,x_2^r, \ldots)$ such that

$$(e_r,x_n^r) \quad \text{converges as} \quad n \to \infty \quad , \tag{7.16}$$

and $(x_1{}^r,x_2{}^r, \ldots)$ is a subsequence of $(x_1{}^{r-1},x_2{}^{r-1}, \ldots)$. Write

$$y_n = x_n^n \quad ;$$

(y_n) is the diagonal sequence of the array $(x_n{}^r)$. Since each member of (y_n) after the r-th belongs to the sequence $(x_1{}^r,x_2{}^r, \ldots)$, it follows from (7.16) that for each r,

$$(e_r,y_n) \quad \text{converges as} \quad n \to \infty \quad . \tag{7.17}$$

Now, (7.17) implies that if z is a finite linear combination of the basis vectors (e_r), then (z,y_n) converges. We shall extend this to all $z \in H$. For any $\epsilon > 0$ and any $z \in H$ we can write

$$z = v + w \quad ,$$

where $\quad v = \sum_1^m (z,e_r)e_r \quad ,$

$$w = \sum_{m+1}^{\infty} (z,e_r)e_r \quad ,$$

and m can be chosen large enough so that $\| w \| < \epsilon/M$, where M is the bound on (x_n) and therefore on (y_n). For any n, we have

$$(z,y_n) = (v,y_n) + (w,y_n) \quad ,$$

therefore

$$|(z,y_n) - (v,y_n)| \leqslant \| w \| \| y_n \| \leqslant M\epsilon/M = \epsilon \quad .$$

But the sequence (v,y_n) converges by (7.17), because v is a finite combination of (e_n). Hence for any $\epsilon > 0$ there is a convergent sequence of numbers whose difference from (z,y_n) is $\leqslant \epsilon$. By Problem 4.27 it follows that (z,y_n) converges; thus (y_n) is weakly convergent. □

This theorem can be thought of as a weak-convergence analogue of a compactness statement; it says that bounded sets in Hilbert space are 'weakly relatively compact'. We shall use it in the proof of convergence of the Rayleigh-Ritz approximation method in Chapter 10.

7.9 Summary and References

In section 7.1 we defined the notion of inner product, and showed how it induces a natural norm on the space. In section 7.2 we introduced orthogonal basis sets, and showed how a nonorthogonal set can be orthogonalised. In section 7.3 we derived a very simple formula for the components or expansion coefficients of a vector in terms of an orthonormal basis, and considered a related problem in approximation theory. We also proved the surprisingly difficult theorem that if a space has a basis, then so does any subspace. In section 7.4 we showed that the expansion coefficients of a vector x with respect to an orthonormal basis (e_n) are square-summable to $\|x\|^2$, and that conversely, if $\Sigma|c_n|^2$ converges, then $\Sigma c_n e_n$ converges to a vector in the space (the Riesz-Fischer theorem).

In section 7.5 we defined the orthogonal complement of a set of vectors, and showed that it is a subspace, and that a Hilbert space can be split into a direct sum of any subspace and its orthogonal complement. We also considered the relation of orthogonal projection to approximation theory. In section 7.6 we considered bounded and continuous functionals on a normed space, and showed that boundedness and continuity are equivalent for linear functionals. In section 7.7 we showed, using the orthogonal decomposition theorem, that every linear continuous functional on a Hilbert space is generated by some element u of the space as an inner product $x \mapsto (x,u)$ (the Riesz Representation Theorem). Finally, in section 7.8 we discussed weak convergence in a Hilbert space.

All this basic Hilbert space theory can be found in various forms in almost any book on functional analysis. There are many different ways of setting out and logically organising the material; in particular, the theory of weak convergence is often discussed in the context of Banach spaces, in a form that may not be easily recognised as corresponding to our section 7.8. This is the case in the otherwise very useful book of Kreyszig. Helmberg gives a lucid and self-contained account of Hilbert space theory with no reference to the more general normed or metric spaces. Riesz & Nagy is a classic, and the chapters on Hilbert space are highly recommended. Higgins has a particularly detailed account of the theory of orthogonal basis sets. The following may also be useful: Balakrishnan, Hutson & Pym, Kolmogorov & Fomin (1970), Luenberger, Lusternik & Sobolev, Milne, Naylor & Sell, Pryce, and Vulikh.

PROBLEMS 7

Sections 7.1, 7.2

7.1 Prove the following results for an inner product space, and illustrate them with diagrams.

(a) $\|x+y\| = \|x\| + \|y\|$ if and only if $ax = by$ for some non-negative numbers a,b, not both zero.

(b) $\|x-y\| + \|y-z\| = \|x-z\|$ if and only if $y = ax + (1-a)z$ for some real a in $[0,1]$.

(c) $\|x-y\| = |\,\|x\| - \|y\|\,|$ under the same conditions as in (a).

7.2 (a) Show that in an inner product space, $\|x+y\|^2 + \|x-y\|^2 = 2(\|x\|^2 + \|y\|^2)$. This is called the 'parallelogram law'; draw a diagram to explain the name. Give an example showing that it need not hold in a normed space.

(b) Show that for all x,y in a real inner product space, $4(x,y) = \|x+y\|^2 - \|x-y\|^2$. For a complex space, find a similar expression for (x,y) in terms of norms of combinations of x and y. (This shows that if the norms of all elements are given, then the inner products can be deduced.)

7.3 Let V be the vector space of those infinite sequences $x = (x_1, x_2, \ldots)$ for which only a finite number of terms x_n are non-zero. Show that $(x,y) = \Sigma x_n \bar{y}_n$ defines an inner product on V. By considering $x^{(k)} = (1, 1/2, 1/3, \ldots 1/k, 0, 0, \ldots)$ or otherwise, show that V is not a Hilbert space. Write down an orthonormal basis for V.

7.4 Show that if a vector subspace of a Hilbert space is closed, then it is a Hilbert subspace.

7.5 If (y_n) is a bounded sequence in an inner product space, and (x_n) is a sequence converging to zero, prove that $(x_n, y_n) \to 0$.

7.6 Show that the set of continuous functions on $[a,b]$ is not a subspace of $L_2[a,b]$, though it is a vector subspace.

7.7 Use the Gram-Schmidt process to orthogonalise the first few powers $1, x, x^2, \ldots$ in the space $C[-1,1]$. (The resulting functions are proportional to the Legendre polynomials, which arise in the solution of Laplace's equation in spherical polar coordinates.)

7.8 Show that the set of functions f such that $\int_0^\infty |f(t)|^2 e^{-t} dt$ converges, with inner product $(f,g) = \int_0^\infty f(t)\overline{g(t)} e^{-t} dt$, is an inner product space. Show that the functions $L_n(t) = (e^t/n!) D^n(t^n e^{-t})$, $n = 1,2, \ldots$, where D means d/dt, are polynomials, and are orthogonal in this space. (They are called the Laguerre polynomials; they can be shown to form a basis.)

Sections 7.3, 7.4

7.9 Show that the following sequences are orthonormal.

(a) $\sqrt{(2/a)} \sin(n\pi x/a)$ in the space $L_2[0,a]$.

(b)
$$\phi_n(x) = \begin{cases} 2\sqrt{3}(x - 2^{-n-1}) 2^{3(n+1)/2} & \text{if } 2^{-n-1} \leqslant x \leqslant \frac{1}{2}(2^{-n-1} + 2^{-n}) \\ -2\sqrt{3}(x - 2^{-n}) 2^{3(n+1)/2} & \text{if } \frac{1}{2}(2^{-n-1} + 2^{-n}) \leqslant x \leqslant 2^{-n} \\ 0 \text{ otherwise,} \end{cases}$$

in the space $L_2[0,1]$.

(c) (for readers acquainted with the theory of functions of a complex variable) $(2\pi)^{-\frac{1}{2}} z^n$ in the space of functions analytic for $|z| \leqslant 1$, with inner product $(f,g) = \oint_C f\bar{g}\, dz/iz$ where C is the unit circle.

7.10 Sketch the functions of Problem 7.9(b). Now write down an orthonormal sequence of functions in $L_2[0,1]$, each of which is constant in a certain interval and zero outside that interval.

7.11 If (c_n) and (d_n) are the expansion coefficients of vectors x and y with respect to an orthonormal basis, show that $(x,y) = \Sigma c_n \bar{d}_n$.

7.12 Show that if (ϕ_n) is an orthogonal sequence such that the only vector orthogonal to all the ϕ_n is the zero vector, then (ϕ_n) is a basis.

7.13 If (e_n) is a basis for the vector subspace S of a Hilbert space H, then any $x \in S$ can be expressed in the form $x = \Sigma c_n e_n$. Conversely, if $y = \Sigma c_n e_n$, does it follow that $y \in S$? Give a proof or a counterexample.

What if S is a Hilbert subspace of H?

7.14 Given a convergent infinite series, one cannot in general rearrange the terms; if the sequence (v_n) is a rearrangement of a sequence (u_n), and $\Sigma u_n = U$, then Σv_n need not equal U, unless Σu_n converges absolutely. But consider the special case of expansions in terms of an orthogonal basis. Prove that if (e_n) is an orthogonal basis, and (f_n) is a sequence obtained by arranging the (e_n) in a different order, then (f_n) is a basis, and therefore the series $x = \Sigma(x,e_n)\| e_n \|^{-2} e_n$ can be rearranged.

7.15 "Expansion coefficients with respect to an orthonormal basis must satisfy $c_n/n^{-\frac{1}{2}} \to 0$ as $n \to \infty$ in order that $\Sigma |c_n|^2$ may converge." True or false? Give a proof or counter-example.

7.16 Show that if (e_n) is an orthonormal set in a Hilbert space H, the set of all vectors of the form $x = \Sigma c_n e_n$ is a subspace of H.

7.17 (A space with no basis). The set of all periodic functions $\mathbb{R} \to \mathbb{C}$ is clearly not a vector space. But if we consider the set M of functions which are sums or products of finitely many periodic functions, we obtain a vector space. The elements of M are called 'almost periodic functions' (this is a simplified version of the idea of almost periodicity; for a fuller account, see Riesz & Nagy). It can be proved that for any $f,g \in M$,

$$\lim_{T\to\infty}([1/2T] \int_{-T}^{T} f(t)\overline{g(t)}\,\mathrm{d}t)$$

exists and defines an inner product on M.

Verify that any two members of the family of functions e^{iat}, where a is real, are orthogonal in the inner product space M. Deduce that M has no countable basis.

Section 7.5

7.18 Use Theorem 7.32 to calculate the quadratic polynomial Q which gives the best fit to e^x over the interval $[0,1]$, in the sense that $\int_0^1 [e^x - Q(x)]^2 \mathrm{d}x$ is minimised. How does your result compare for accuracy with the Taylor polynomial about $x = 0$? Is it better than the Taylor polynomial about the mid-point of the interval? What is meant by 'better' here?

7.19 What is the orthogonal complement of the set of all even functions in $L_2[-1,1]$?

7.20 M, N are subspaces of a Hilbert space. If $M \subset N$, show that $N^\perp \subset M^\perp$. Show also that $(M^\perp)^\perp = M$.

7.21 M, N are vector subspaces of a Hilbert space. If $M \subset N$, show that $N^\perp \subset M^\perp$, $(M^\perp)^\perp \supset M$, but $(M^\perp)^\perp$ need not equal M (cf. Problem 7.20).

Sections 7.6–7.8

7.22 In the space $C[-1,1]$ with inner product $\int_{-1}^{1} u(x)\overline{v(x)}\,dx$, define a functional f by $f(u) = \int_0^1 u(x)\,dx$. Show that f is linear and continuous. Is there a $\phi \in C[-1,1]$ such that $f(u) = (u,\phi)$ for all u? Does your answer agree with, or contradict, the Riesz representation theorem?

7.23 Is the functional f on $C[a,b]$ defined by $f(u) = \max\{u(x): a \leqslant x \leqslant b\}$ linear? Is it bounded?

7.24 Let B denote the space of all bounded sequences, that is, all sequences (x_n) of complex numbers such that $|x_n| \leqslant M$ for all n, where M is a number independent of n, but different for different sequences. Show that $\|x\| = \sup_n |x_n|$ defines a norm on B. Now, define a functional f on B by $f(x) = x_N$, where N is a fixed integer. Is it linear? Is it continuous? What about the functional $g(x) = \sup_n \{x_n\}$?

7.25 Given a Banach space B, define B^* to be the set of all bounded linear functionals on B. We can define the notion of weak convergence for Banach spaces as follows: if $x, x_1, x_2, \ldots$ belong to B, we say $x_n \to x$ weakly if $f(x_n) \to f(x)$ for all $f \in B^*$. Prove the Banach-space analogues of the basic results 7.66(a), 7.67 (the analogue of 7.66(b) is true, but much more difficult to prove). Prove that if B is a Hilbert space, then the above definition of weak convergence is equivalent to Definition 7.63.

7.26 Show that in a Hilbert space $x_n \to x$ strongly if and only if $x_n \to x$ weakly and $\|x_n\| \to \|x\|$.

7.27 Try to give an example of a linear functional on a Hilbert space which is not continuous. (This is really difficult; see Balakrishnan, Example 1.7.2.)

More problems will be found on page 377.

Chapter 8

The Theory of Operators

The first two sections of this chapter discuss the theory of operators on Banach spaces, including a discussion of operator power series, with an application to the approximate solution of equations of the type $Ax = b$. The rest of the chapter deals with the eigenvalue problem for operators on Hilbert space, and introduces the apparatus needed in order to state and prove the important spectral theorem of section 9.1.

8.1 Bounded Operators on Normed Spaces

In section 7.6 we discussed those properties of linear functionals which do not involve the inner product, and can therefore be discussed in the setting of normed spaces; we then went on to the special properties of functionals on Hilbert space. In the same way, this section and the next discuss the general theory of linear operators on normed spaces; in section 8.3 we shall return to Hilbert space.

Operators on vector spaces were introduced in section 5.1. In this chapter we concentrate on linear operators.

Definition 8.1 Let N, M be vector spaces. An operator $A: N \to M$ is **linear** if $A(ax + by) = aAx + bAy$ for all scalars a, b and all $x, y \in N$. □

The operators in Examples 5.4, 5.5, 5.7 and 5.10 are linear; those in 5.3 and 5.6 are not. Linear functionals, defined in section 7.6, are further examples of linear operators.

Proposition 8.2 (Range of a Linear Operator) If N, M are vector spaces and $A: N \to M$ is a linear operator, then the range of A is a vector subspace of M.

Proof Set $R(A) = \{Au: u \in N\}$ (cf. Definition 5.2). If $x, y \in R(A)$, then $x = Au$, $y = Av$ for some $u, v \in N$. Hence for any scalars α, β,

$$\begin{aligned} \alpha x + \beta y &= \alpha Au + \beta Av \\ &= A(\alpha u + \beta v) \\ &\in R(A) \quad , \end{aligned}$$

hence $R(A)$ is a subspace by 4.8. □

Note that this result applies to linear operators only. Consider, for example, the nonlinear $\mathbb{R} \to \mathbb{R}$ operator $x \mapsto x^2$; its range is the set of non-negative numbers, which is not a subspace of $\mathbb{R}$.

Definition 8.3 The **rank** of a linear operator is the dimension of its range.

Examples 8.4 If A is a real nonsingular $n \times n$ matrix, it gives a rank n linear transformation $\mathbb{R}^n \to \mathbb{R}^n$. A singular $n \times n$ matrix may have any rank less than n. For example, the matrix

$$B = \begin{pmatrix} 1 & 2 & 1 \\ 1 & 3 & 1 \\ 0 & 1 & 0 \end{pmatrix}$$

has rank 2: for any $u \in \mathbb{R}^3$, it can be verified that Bu is a linear combination of $(1,0,-1)$ and $(0,1,1)$, so $\{Bu: u \in \mathbb{R}^3\}$ is two-dimensional.

Most operators on infinite-dimensional spaces have infinite rank, just as most $n \times n$ matrices are nonsingular and have rank n. Projections on to finite-dimensional subspaces have finite rank: if E is an n-dimensional subspace of a Hilbert space H, then the operator P which maps any vector in H into its projection on E (defined in 7.47) is linear and has rank n. See Example 8.40 for another finite-rank operator on an infinite-dimensional space. □

In 6.1 we defined what it means to say that an operator is continuous, and in section 7.6 we considered functionals, which are operators from a Banach space to $\mathbb{R}$ or $\mathbb{C}$, and showed that continuity is equivalent to boundedness for linear functionals. We shall now extend this result to operators in general. We begin by showing that, just as for functionals, a linear operator continuous at one point is uniformly continuous.

Theorem 8.5 (Continuity of Linear Operators) If a linear operator $A: N \to M$ is continuous at a point $x \in N$, then it is uniformly continuous on N.

Proof Identical to the proof of 7.54. □

Now we extend the idea of boundedness (Definition 7.56) from functionals to operators in general.

Definition 8.6 Let A be an operator $N \to M$, where N, M are normed spaces. A is **bounded** if there is a number m such that

$$\|Ax\| \leqslant m \|x\|$$

for all $x \in N$. If A is bounded, we define its **norm** to be the least such m; more precisely, we define

$$\|A\| = \sup\{\|Ax\|/\|x\| : x \in N,\ x \neq 0\} \quad . \qquad \square \quad (8.1)$$

It follows from this definition that

$$\|Ax\| \leqslant \|A\|\|x\| \quad . \qquad (8.2)$$

The norm symbol is used here in three different ways: in $\|x\|$ it denotes the norm in the space N, in $\|Ax\|$ it denotes the norm in the space M, and in $\|A\|$ it denotes the number defined in (8.1), which is a different kind of thing altogether. No confusion will arise between the different kinds of norm, provided that the reader keeps the meaning of all the symbols A,x etc. clearly in mind.

Example 8.7 Define $A : C[a,b] \to C[a,b]$ by

$$(Af)(x) = \int_a^b K(x,y,f(y))\,dy \quad ,$$

where K is a continuous function of three variables. Using the sup norm in $C[a,b]$, we have

$$\|Af\| \leqslant (b-a)\sup\{|K(x,y,f(y))| : a \leqslant x,y \leqslant b\} \quad .$$

The operator A is not linear, and in general is not bounded. However, if $|K(x,y,z)| \leqslant c|z|$ for some c, then

$$\|Af\| \leqslant c(b-a)\sup\{|f(y)| : a \leqslant y \leqslant b\}$$
$$\leqslant c(b-a)\|f\|$$

for all $f \in C[a,b]$. Hence A is bounded, and

$$\|A\| \leqslant c(b-a) \quad .$$

Note that this argument does not give the value of $\|A\|$, only an inequality for it.

An important special case arises when the dependence of K on its third argument is linear,

$$K(x,y,z) = K_1(x,y)z \quad .$$

Then it is easy to verify that A is a linear operator, and we have

$$\|Af\| \leqslant (b-a)\sup|K_1(x,y)f(y)|$$
$$\leqslant (b-a)\sup|K_1(x,y)|\|f\| \quad ,$$

so that $\|A\| \leqslant (b-a)\sup|K_1(x,y)| \quad . \qquad (8.3)$

Example 8.8 Consider the integral operator of Example 8.7 acting on $L_2[a,b]$ instead of $C[a,b]$. The calculation near the end of the proof of Theorem 5.17, with u_2 set equal to zero, shows that the operator is bounded provided that K satisfies a Lipschitz condition with respect to its third argument. We can say more in the case of the linear operator

$$T: f \mapsto \int_a^b K(x,y) f(y)\, \mathrm{d}y \quad .$$

Equation (5.3) of Example 5.5 shows that

$$\| T \| \leqslant \left(\int_a^b \int_a^b |K(x,y)|^2 \, \mathrm{d}x \, \mathrm{d}y \right)^{\frac{1}{2}} \qquad \square \quad (8.4)$$

For linear operators, just as for linear functionals, continuity and boundedness are equivalent.

Theorem 8.9 (Bounded Linear Operators) A linear operator is bounded if and only if it is continuous.

Proof Identical to the proof of Theorem 7.57. □

Example 8.10 In Example 6.11 we showed that the operator of differentiation is not continuous on $C^1[a,b]$. It follows from Theorem 8.9 that it is not bounded either. A direct proof that D is unbounded can easily be constructed using the functions f_n of 6.11. In Problem 6.3, however, it is shown that if a different norm is used, then D becomes continuous and hence bounded. □

There is another way of looking at bounded operators. Instead of considering their effect on one element of the space at a time, we can consider their effect on a whole subset of the space. They can then be characterised as those operators which transform bounded sets into bounded sets.

Theorem 8.11 (Bounded Linear Operators) A linear operator is bounded if and only if it maps every bounded set into a bounded set.

Proof (a) If $A: N \to M$ is bounded, then $\|x\| \leqslant C$ implies that $\|Ax\| \leqslant \|A\| C$, hence a bounded set is mapped into a bounded set.

(b) If A is linear and maps bounded sets into bounded sets, then in particular the unit sphere $\{x: \|x\| \leqslant 1\}$ is mapped into a bounded set, so there is an m such that

$$\|Ax\| \leqslant m \quad \text{whenever} \quad \|x\| \leqslant 1 \quad .$$

Now, for any nonzero $x \in N$,

$$\begin{aligned} \|Ax\|/\|x\| &= \|A(x/\|x\|)\| \\ &\leqslant m \quad , \end{aligned}$$

which proves that A is bounded. Notice the essential role of linearity in this proof; the theorem is not true for nonlinear operators. □

8.2 The Algebra of Bounded Operators

In this section we shall see how operators can be added, multiplied, and generally manipulated algebraically. We begin by considering linear combinations of operators, and then we shall justify the notation $\|A\|$ introduced in Definition 8.6 by showing that the set of bounded linear operators is a normed space.

The obvious way to define sums and scalar multiples of operators is as follows.

Definition 8.12 If A, B are operators $N \to M$, where N,M are vector spaces, we define the operator $A + B$ by

$$(A + B)x = Ax + Bx \quad \text{for all} \quad x \in N \quad ,$$

and for any scalar c we define the operator cA by

$$(cA)x = c(Ax) \quad \text{for all} \quad x \in N \quad .$$

□

It is very easy to verify that the set of all operators $N \to M$, with this definition of addition and scalar multiplication, is a vector space; the zero element of the space is the operator $\tilde{0}$ which maps every element of N into the zero element of M. We are particularly interested in bounded linear operators; they form a subspace of the space of all operators $N \to M$. We shall show that this subspace is a Banach space with the norm of Definition 8.6. First we show that it is a normed space.

Proposition 8.13 If N,M are normed spaces, the set of all bounded linear operators $N \to M$ is a normed space with the operations of Definition 8.12 and the norm of Definition 8.6. This space is denoted by $B(N,M)$.

Proof We first show that it is a vector space. We must show that a linear combination of bounded linear operators is bounded and linear. If $A, B \in B(N,M)$, then for any scalars c,d,e,f, and $x,y \in N$,

$$\begin{aligned}(cA + dB)(ex + fy) &= cA(ex + fy) + dB(ex + fy) \\ &= ceAx + cfAy + deBx + dfBy \\ &= e(cA + dB)x + f(cA + dB)y \quad ,\end{aligned}$$

using 8.12 and the linearity of A and B. Hence the operator $cA + dB$ is linear. Now,

$$\begin{aligned}\|(cA + dB)x\| &= \|cAx + dBx\| \\ &\leqslant |c|\|Ax\| + |d|\|Bx\| \\ &\leqslant (|c|\|A\| + |d|\|B\|)\|x\|\end{aligned}$$

for any $x \in N$, by (8.2). Hence $cA + dB$ is bounded, and

$$\|cA + dB\| \leqslant |c|\|A\| + |d|\|B\| \quad . \tag{8.5}$$

We have now shown that $B(N,M)$ is a vector space (apart from the trivial verification of the algebraic rules (a)–(h) of Definition 4.2).

We must show that Definition 8.6 gives a norm for this space. $\|A\| \geqslant 0$ by definition for any A; the zero element of $B(N,M)$ is the operator $\widetilde{0}: x \mapsto 0$ for all $x \in N$, and clearly $\|\widetilde{0}\| = 0$; if $A \neq \widetilde{0}$ then there is an $x_1 \in N$ with $Ax_1 \neq 0$, therefore $\|Ax_1\|/\|x_1\| > 0$, so $\|A\| > 0$ if $A \neq \widetilde{0}$. $\|cA\| = |c|\,\|A\|$ follows immediately from the definitions, and the triangle inequality follows from (8.5) with $c = d = 1$. This completes the proof. □

We can now construct infinite sequences and series of operators; their convergence is interpreted in terms of the operator norm: $A_n \to A$ means $\|A_n - A\| \to 0$. The idea of convergence raises one of the main questions about any normed space: is it complete? The answer here is that $B(N,M)$ is complete if M is.

Theorem 8.14 (Banach Space of Operators) The space $B(N,M)$ of bounded linear operators $N \to M$ is a Banach space if M is.

Proof Let (A_n) be a Cauchy sequence in $B(N,M)$; we shall show that $(A_n x)$ is a convergent sequence in M for any x. For any integers n,r and any $x \in N$ we have

$$\begin{aligned} \|A_n x - A_{n+r} x\| &\leqslant \|A_n - A_{n+r}\|\|x\| \\ &\to 0 \quad \text{as} \quad n \to \infty \end{aligned} \tag{8.6}$$

because (A_n) is a Cauchy sequence. Hence $(A_n x)$ is a Cauchy sequence in the Banach space M, hence it converges. We can therefore define an operator A: $x \mapsto \lim(A_n x)$. We must show that $A \in B(N,M)$ and $A_n \to A$.

It is very easy to show that A is linear; we now show that A is bounded. Because (A_n) is a Cauchy sequence, it follows from (8.6) that for any $\epsilon > 0$ there is an n_0 such that

$$\|A_n x - A_{n+r} x\| < \|x\|\epsilon \quad \text{for all} \quad n > n_0, x \in N \quad .$$

Taking the limit as $r \to \infty$ gives

$$\|A_n x - Ax\| \leqslant \|x\|\epsilon \quad \text{for all } n > n_0, x \in N \quad . \tag{8.7}$$

This shows that $A_n - A$ is a bounded operator for $n > n_0$, and since $B(N,M)$ is a vector space it follows that

$$A = A_n + (A - A_n) \in B(N,M) \quad .$$

It remains to be shown that $A_n \to A$. From (8.7) it follows that for any $\epsilon > 0$,

$$\|A_n - A\| = \sup\{\|A_n x - Ax\|/\|x\|\} \leqslant \epsilon$$

for $n > n_0$, which shows that $\|A_n - A\| \to 0$, that is, $A_n \to A$. □

We shall use Theorem 8.14 to construct a theory of convergent power series of operators. This requires us to multiply operators as well as add them. In Definition 5.8 we defined the product of two operators A and B to be the operator corresponding to performing the operations B and A in turn, so that $ABx = A(Bx)$. Here we shall only consider operators mapping a space N into itself, because for operators $A: N \to M$, we cannot define $A^2, A^3, \ldots$ unless $M \subset N$. We now set out the main properties of operator multiplication.

Proposition 8.15 (Product of Operators) If $A, B \in B(N,N)$, then the operator AB defined by $(AB)x = A(Bx)$ for all $x \in N$ is bounded and linear, and

$$\|AB\| \leqslant \|A\| \|B\| \quad . \tag{8.8}$$

For any operators $A, B, C \in B(N,N)$ we have

$$(AB)C = A(BC) \quad ,$$

$$(A + B)C = AC + BC \quad ,$$

$$A(B + C) = AB + AC \quad .$$

Proof Linearity of AB is trivial to prove. Boundedness is also easy:

$$\|ABx\| = \|A(Bx)\| \leqslant \|A\| \|Bx\| \leqslant \|A\| \|B\| \|x\| \quad ,$$

using (8.2) twice, which shows that AB is bounded and proves (8.8). The algebraic rules set out in the proposition follow easily from the definitions. For example, for any $x \in N$,

$$\begin{aligned}
[(A + B)C]x &= (A + B)(Cx) && \text{by definition of multiplication,}\\
&= A(Cx) + B(Cx) && \text{by definition of addition,}\\
&= (AC)x + (BC)x && \text{by definition of multiplication,}\\
&= [AC + BC]x && \text{by definition of addition,}
\end{aligned}$$

showing that $(A + B)C = AC + BC$. The others are just as easily proved. □

The usual rules of algebraic manipulation apply to $B(N,N)$ except that multiplication is noncommutative, that is, AB and BA are not necessarily equal.

Definition 8.16 The **identity** $I_N: N \to N$ is the operator which maps every vector in N into itself. When there is no danger of confusion, we simply call it I. □

I clearly satisfies $IA = AI = A$ for any operator $A: N \to N$. We can now define powers of an operator by the obvious rule $A^2 = AA$, $A^3 = A^2A, \ldots$, and note that $A^rA^s = A^sA^r = A^{r+s}$ for any positive integers r, s; we define $A^0 = I$ for any operator A. We can now consider power series of operators. We begin with the theory of absolute convergence in Banach spaces.

Definition 8.17 If $x_1, x_2, \ldots$ are members of any normed space, we say the series Σx_n **converges absolutely** if the series of numbers $\Sigma \| x_n \|$ converges. □

Note that the definition of absolute convergence says nothing about whether the series of vectors Σx_n is convergent in the usual sense in the normed space. In other words, an absolutely convergent series need not be convergent. However, in a Banach space absolute convergence does imply convergence; this is not true in an incomplete normed space (cf. Problem 4.26).

Theorem 8.18 An absolutely convergent series in a Banach space is convergent.

Proof We must show that if $\Sigma \| x_n \|$ converges, then there is an x such that $\Sigma x_n = x$, that is, $s_n \to x$ where $s_n = \Sigma_{r=1}^n x_r$. We use the Cauchy convergence criterion. We have

$$\begin{aligned} \| s_{n+p} - s_n \| &= \| \sum_{r=n+1}^{n+p} x_r \| \\ &\leqslant \sum_{n+1}^{n+p} \| x_r \| \\ &\to 0 \quad \text{as} \quad n \to \infty \end{aligned}$$

because $\Sigma \| x_n \|$ converges. Hence (s_n) is a Cauchy sequence in a Banach space, and therefore converges to a member of the space. □

In ordinary analysis, absolutely convergent series can be rearranged and multiplied together – that is, if $x = \Sigma x_n$ and $y = \Sigma y_n$, then $x = \Sigma \xi_n$ where the ξ_n are the same numbers as the x_n arranged in a different order, and

$$xy = x_0y_0 + x_0y_1 + x_1y_0 + x_0y_2 + x_1y_1 + x_2y_0 + \ldots \quad .$$

These results do not necessarily hold for non-absolutely convergent series. Series of operators behave in the same way.

Proposition 8.19 Absolutely convergent series of bounded linear operators on a Banach space can be rearranged, and multiplied together term by term. □

We do not give the proof of this; with norm signs replaced by modulus signs, it is the same as the proof for series of numbers, which can be found in any thorough textbook of analysis.

The last two results have concerned series in any Banach space. We shall now concentrate on power series, which cannot be defined in all Banach spaces, but only in those, such as $B(N,N)$, for which multiplication is defined. The following theorem gives a criterion for convergence of a power series of operators, in terms of the convergence of an associated series of numbers.

Theorem 8.20 (Operator Power Series) Let $A \in B(N,N)$, and let (c_n) be a sequence of numbers. If the series $\Sigma_0^\infty c_n \| A \|^n$ converges absolutely, then the series of operators $\Sigma_0^\infty c_n A^n$ converges absolutely to a member of $B(N,N)$.

Proof From (8.8) we have $\| A^2 \| \leqslant \| A \|^2$, hence $\| A^3 \| \leqslant \| A^2 \| \| A \| \leqslant \| A \|^3$, and generally

$$\| A^n \| \leqslant \| A \|^n \quad . \tag{8.9}$$

But $\Sigma |c_n| \| A \|^n$ is given to be convergent, hence $\Sigma \| c_n A^n \|$ converges, so $\Sigma c_n A^n$ converges absolutely, and by Theorem 8.18, $\Sigma c_n A^n$ converges in the Banach space $B(N,N)$. □

We shall now apply these results to one of the standard problems in applied mathematics: solving the equation

$$Ax = b \tag{8.10}$$

where x is an unknown member of a normed space, b a given member, and A a linear operator. If there is a unique solution x of (8.10) for any b, we write $x = A^{-1}b$, and call A^{-1} the inverse operator.

Definition 8.21 An operator $A : N \to M$ is **invertible** if for each $x \in M$ there is one and only one $y \in N$ such that $Ay = x$. The mapping $x \mapsto y$ is called the **inverse** of A, and we write $y = A^{-1}x$. □

Note that if A maps N into M, then A^{-1} maps M into N. The following elementary facts follow immediately from the definition.

Proposition 8.22 If $A : N \to M$ is invertible, then

(a) A^{-1} is invertible, and $(A^{-1})^{-1} = A$;

(b) $AA^{-1} = I_M$ and $A^{-1}A = I_N$, using the notation of 8.16;

(c) If A is linear, then A^{-1} is linear;

(d) If $B: P \to N$ is invertible, then AB is an invertible operator $P \to M$ and

$$(AB)^{-1} = B^{-1}A^{-1} \quad . \tag{8.11}$$

Proof (a) is obvious: if A^{-1} exists, there is a one-to-one correspondence between N and M. The first part of (b) follows by applying A to the equation $y = A^{-1}x$, where $Ay = x$, and the second part by taking $x = Ay$ in the equation $y = A^{-1}x$. To prove (c), let $x = A\xi$, $y = A\eta$, then for any scalars c,d,

$$\begin{aligned} A^{-1}(cx + dy) &= A^{-1}(cA\xi + dA\eta) \\ &= A^{-1}(Ac\xi + Ad\eta) \\ &= A^{-1}A(c\xi + d\eta) \\ &= c\xi + d\eta \\ &= cA^{-1}x + dA^{-1}y \quad . \end{aligned}$$

Finally, the invertibility of AB is clear: for each $x \in M$ there is a unique $y \in N$ such that $x = Ay$, and there is a unique $z \in P$ such that $y = Bz$, hence $x = ABz$. We have $z = B^{-1}y$ and $y = A^{-1}x$, hence $z = B^{-1}A^{-1}x$ when $x = ABz$, which proves (8.11). □

Remark 8.23 The sum of two invertible operators is not in general invertible; for example, consider $I + (-I)$. □

The following is a converse to 8.22(b).

Proposition 8.24 If $A: N \to M$ and $B: M \to N$, and $AB = I_M$ and $BA = I_N$, then A and B are invertible, and $A^{-1} = B$, $B^{-1} = A$.

Proof For each $x \in M$ there is a $y \in N$ such that $Ay = x$, namely $y = Bx$. And if y is such that $Ay = x$, then $BAy = y = Bx$, thus y is uniquely determined by x. □

The identity is the most obvious example of an invertible operator. The following result shows that operators which are not too far from the identity, as measured by the operator norm, are also invertible.

Theorem 8.25 (Inverse Operator) Let A be a bounded linear operator $N \to N$ where N is a Banach space. If $\|A\| < 1$ then $I - A$ is invertible, $(I - A)^{-1}$ is bounded, and $\|(I - A)^{-1}\| \leq 1/(1 - \|A\|)$.

Proof We construct the inverse using the binomial series. Since $\|A\| < 1$, $\Sigma_0^\infty \|A\|^n$ converges absolutely, hence by Theorem 8.20, $\Sigma_0^\infty A^n$ converges to a bounded linear operator. Multiplying out $(I - A)\Sigma A^n$ and $(\Sigma A^n)(I - A)$ (justified by 8.19) shows that they both equal I, hence

$$(I - A)^{-1} = \sum_0^\infty A^n \quad .$$

Hence $\|(I - A)^{-1}\| = \|\Sigma A^n\| \leq \Sigma \|A^n\| \leq \Sigma \|A\|^n$ using (8.9), and $\Sigma_0^\infty \|A\|^n = (1 - \|A\|)^{-1}$. □

Corollary 8.26 Let N be a Banach space and let A, B be bounded linear operators $N \to N$, with A invertible. Then $A - B$ is invertible if $\| B \| \leqslant \| A^{-1} \|^{-1}$, and $\| (A - B)^{-1} \| \leqslant (\| A^{-1} \|^{-1} - \| B \|)^{-1}$.

Proof Since A is invertible, we can write

$$A - B = A(I - A^{-1}B) \quad .$$

Now, $\| A^{-1}B \| \leqslant \| A^{-1} \| \| B \| < 1$ if $\| B \| < \| A^{-1} \|^{-1}$. Thus $I - A^{-1}B$ is invertible by Theorem 8.25, hence $A - B$ is invertible by 8.22(d) and $(A - B)^{-1} = (I - A^{-1}B)^{-1}A^{-1}$. Hence

$$\| (A - B)^{-1} \| \leqslant \| (I - A^{-1}B)^{-1} \| \, \| A^{-1} \|$$

$$\leqslant \| A^{-1} \| / (1 - \| A^{-1}B \|)$$

by Theorem 8.25,

$$\leqslant \| A^{-1} \| / (1 - \| A^{-1} \| \| B \|) \quad .$$

□

It follows from this corollary that the set of invertible operators is an open set in $B(N, N)$, that is, for any invertible A, all the operators sufficiently close to it (that is, in a sphere of radius $\| A^{-1} \|^{-1}$ in the operator norm) are also invertible. From the point of view of the applied mathematician or numerical analyst this is useful because an approximate or numerical procedure for solving an equation $Ax = b$ means replacing A by some approximate operator A_a. Corollary 8.26 says that if A is invertible, then the approximate operator A_a is invertible, provided that A_a is not too far from A, and 8.26 gives the following estimate for the error produced in the solution x by replacing A by A_a.

Corollary 8.27 (Error Estimate) Let A and $A_a = A + \Delta$ be invertible bounded linear operators, with $\| A_a^{-1} \| \leqslant \alpha$ and $\| \Delta \| \leqslant \delta$, where $\alpha\delta < 1$. If x and x_a satisfy $Ax = b$ and $A_a x_a = b$, then

$$\| x - x_a \| \leqslant (\alpha\delta / 1 - \alpha\delta) \| x_a \| \quad . \tag{8.12}$$

Proof

$$x - x_a = [(A_a - \Delta)^{-1} - A_a^{-1}]b$$

$$= [(I - A_a^{-1}\Delta)^{-1} - I]A_a^{-1}b$$

$$= [I - (I - A_a^{-1}\Delta)](I - A_a^{-1}\Delta)^{-1}A_a^{-1}b$$

$$= A_a^{-1}\Delta(I - A_a^{-1}\Delta)^{-1}x_a \quad .$$

Hence

$$\| x - x_a \| \leqslant \| A_a^{-1} \| \| \Delta \| \| (I - A_a^{-1}\Delta)^{-1} \| \| x_a \|$$

$$\leqslant \{\alpha\delta / (1 - \| A^{-1} \| \| \Delta \|)\} \| x_a \|$$

using (8.8) and Theorem 8.25,

$$\leqslant (\alpha\delta / 1 - \alpha\delta) \| x_a \| \quad .$$

□

The statement of Corollary 8.27 has been carefully framed to make it useful. The introduction of α and δ is not logically necessary; we have

$$\| x - x_a \| \leqslant (\| A_a^{-1} \| \| \Delta \| / [1 - \| A_a^{-1} \| \| \Delta \|]) \| x_a \| \tag{8.13}$$

as an expression for the error in the solution. But in general it is difficult to find the norm of an operator exactly; in the examples in section 8.1 we generally obtained inequalities of the form $\| A \| \leqslant \alpha$. The expression (8.12), unlike (8.13), can be evaluated when we are given such partial information about the norms. Again, the result (8.12) only uses information about the approximate operator A_a and the approximate solution x_a; thus it gives a usable and rigorous bound on the error even when the exact solution x is unknown, and even though the inverse of the exact operator A cannot be calculated. A value for α is presumably calculable because the process of solving $A_a x_a = b$ involves calculating A_a^{-1}. Finally, note that if $\alpha\delta$ is close to 1, then (8.12) permits a large error even when δ, the norm of the difference of the operators A and A_a, is small. This is the phenomenon of 'ill-conditioning'; see Sawyer, or books on numerical analysis.

8.3 Self-Adjoint Operators

We now consider operators in Hilbert space, which has more structure than a normed space, and a correspondingly richer operator theory. Because we are interested in the eigenvalue problem, we shall consider only operators which map a Hilbert space into itself. The main idea of this section is that of the adjoint of an operator; this is a generalisation of a well-known idea in matrix theory.

Symmetric real matrices occur very often in applications of matrix theory, and they have particularly nice properties: an $n \times n$ symmetric real matrix has a set of n orthogonal eigenvectors, which form a basis for $\mathbb{R}^n$. Using this basis, the matrix can be transformed into a diagonal matrix Λ, which is much easier to deal with. The rest of this chapter, and the next, are devoted to generalising this to Hilbert space.

A symmetric matrix is defined as a matrix A such that $\widetilde{A} = A$, where $\widetilde{A}$ is the 'transposed matrix' defined by $\widetilde{A}_{ij} = A_{ji}$. It follows that for any $x,y \in \mathbb{R}^n$,

$$(x, Ay) = (\widetilde{A}x, y) \quad . \tag{8.14}$$

One can invert the logic, and given a matrix A, *define* a new matrix $\widetilde{A}$ by the requirement that it satisfy (8.14); it is then easy to deduce that $\widetilde{A}_{ij} = A_{ji}$. This way of defining the transpose is harder, because one has to prove that a matrix $\widetilde{A}$ satisfying (8.14) can always be found, which is not obvious. But this method can be more easily generalised to operators on Hilbert space, since (8.14) makes sense in a Hilbert space, while the operation of interchanging the rows and columns of a matrix has no immediately obvious meaning for general Hilbert space operators. The following proposition carries out the programme of defining a new operator by means of (8.14). A different notation is used, A^* rather than $\widetilde{A}$, because we wish to include the case of complex spaces, where the appropriate operation corresponds to the combination of transposing and complex-conjugating a matrix.

Proposition 8.28 (Adjoint Operator) If $A: H \to H$ is a bounded linear operator on a Hilbert space H, then there is a unique operator $A^*: H \to H$ such that

$$(x,A^*y) = (Ax,y) \quad \text{for all} \quad x,y \in H \quad . \tag{8.15}$$

A^* is linear and bounded, $\|A^*\| = \|A\|$, and $(A^*)^* = A$. A^* is called the **adjoint** of A.

Proof Consider a fixed $y \in H$. The functional f defined by $f(x) = (Ax,y)$ is linear (proof is trivial), and bounded because $|f(x)| = |(Ax,y)| \leq \|Ax\| \|y\| \leq \|A\| \|x\| \|y\|$. By the Riesz representation theorem 7.60 there is a unique $u \in H$ such that

$$(Ax,y) = (x,u) \quad \text{for all} \quad x \in H \quad . \tag{8.16}$$

We thus have a rule which, given any $y \in H$, defines an element $u \in H$. This is an operator $H \to H$; call it A^*, so that $A^*y = u$ where u satisfies (8.16). We shall now show that A^* has the required properties.

A^* satisfies (8.15) by definition. To prove that it is linear, take any $x,y,z \in H$ and any scalars a,b. Then

$$\begin{aligned}(x,A^*[ay + bz]) &= (Ax, ay + bz) \\ &= \bar{a}(Ax,y) + \bar{b}(Ax,z) \\ &= \bar{a}(x,A^*y) + \bar{b}(x,A^*z) \\ &= (x,aA^*y) + (x,bA^*z) \\ &= (x,aA^*y + bA^*z) \quad .\end{aligned}$$

But if $(x,u) = (x,v)$ for all $x \in H$, then $(x,u - v) = 0$ for all x, which implies $u - v = 0$, or $u = v$. Hence we can deduce from the above that

$$A^*(ay + bz) = aA^*y + bA^*z \quad ,$$

which shows that A^* is a linear operator.

The uniqueness of A^* is easy to prove: if A_1^* and A_2^* are both adjoints of the same operator A, then from (8.15) we have $(x, [A_1^* - A_2^*]y) = 0$ for all x,y, hence $[A_1^* - A_2^*]y = 0$ for all y, so $A_1^* = A_2^*$. Now $(A^*)^* = A$ follows from (8.15) by the symmetry (or conjugate symmetry in a complex space) of the inner product: reversing the inner products in (8.15) gives

$$(y,Ax) = (A^*y,x)$$

for all x,y, which is just the condition for A to be the adjoint of A^*.

Finally we must show that A^* is bounded, and find its norm. For any $y \in H$ we have

$$\begin{aligned}\|A^*y\|^2 &= (A^*y, A^*y)\\ &= (AA^*y, y)\\ &\leqslant \|AA^*y\|\|y\|\\ &\leqslant \|A\|\|A^*y\|\|y\| \quad .\end{aligned}$$

If $\|A^*y\| \neq 0$, we can cancel it from this inequality, getting

$$\|A^*y\| \leqslant \|A\|\|y\| \quad ,$$

which is trivially true also when $\|A^*y\| = 0$. Hence A^* is bounded, and

$$\|A^*\| \leqslant \|A\| \quad . \tag{8.17}$$

This applies to any operator, hence in particular to A^*: $\|A^{**}\| \leqslant \|A^*\|$. But $A^{**} = A$, so we have $\|A^*\| \geqslant \|A\|$. Combining this with (8.17) gives $\|A^*\| = \|A\|$. □

Example 8.29 Let $A: \mathbb{R}^n \to \mathbb{R}^n$ be the operator which multiplies each vector by a real $n \times n$ matrix M, so that $A(x) = Mx$ for $x \in \mathbb{R}^n$. Let M^* be the matrix corresponding to the adjoint operator A^*. Taking $x = e_r$ and $y = e_s$ in (8.15), where e_i is the unit vector with 1 in the i-th place and 0 elsewhere, gives $M^*_{rs} = M_{sr}$, so M^* is the transpose of M. In the complex space $\mathbb{C}^n$, the same calculation shows that M^* is the complex conjugate of the transpose of M. □

Example 8.30 Define $A: L_2[a,b] \to L_2[a,b]$ by

$$(Af)(x) = \int_a^b K(x,y) f(y)\,\mathrm{d}y \quad ,$$

where K is a continuous function. The defining relation (8.15) for A^* here reads

$$\begin{aligned}\int_a^b (A^*f)(x)\overline{g(x)}\,\mathrm{d}x &= \int_a^b f(x)\{\int_a^b \mathrm{d}y\ \overline{K(x,y)g(y)}\}\,\mathrm{d}x\\ &= \int_a^b \{\int_a^b \mathrm{d}y\ f(y)\ \overline{K(y,x)}\}\overline{g(x)}\,\mathrm{d}x\end{aligned}$$

on inverting the order of integration and relabelling the variables. But this equation is clearly satisfied if we take

$$(A^*f)(x) = \int_a^b \overline{K(y,x)} f(y)\,\mathrm{d}y \quad , \tag{8.18}$$

and in view of the uniqueness clause of Proposition 8.28, we deduce that (8.18) defines the adjoint of A. It is an integral operator of the same kind as A, but with a kernel obtained from the kernel of A by interchanging the variables and taking the complex conjugate (compare this with Example 8.29). In the particular case where K is real and symmetric, or more generally when

$$\overline{K(x,y)} = K(y,x) \tag{8.19}$$

A^* coincides with A. This is a particularly important case.

Definition 8.31 A bounded linear operator A is **self-adjoint** if $A^* = A$.

Example 8.32 We saw in Chapter 2 that Green's function for a Sturm-Liouville operator is symmetric, so it is the kernel of a self-adjoint integral operator. Green's functions for Laplace's equation are also symmetric and give rise to self-adjoint operators on the space $L_2[V]$ (see section 5.5). □

We shall see that self-adjoint operators not only are very common in applications, but also have particularly simple mathematical properties.

Lemma 8.33 If A is self-adjoint, then (u,Au) is real for any u.

Proof $(u,Au) = \overline{(Au,u)} = \overline{(u,Au)}$. □

The next result gives, for self-adjoint operators, an expression for the norm similar to that in Definition 8.6 but simpler. In particular, it is linear in A, whereas that in 8.6 involves $\|Au\| = \sqrt{(Au,Au)}$, the square root of a quadratic, and is therefore more difficult to use. Although the statement of Proposition 8.34 is straightforward, the proof is not; the remarks in the proof of Lemma 4.24 apply here.

Proposition 8.34 If A is a bounded self-adjoint operator, then

$$\|A\| = \sup\{|(x,Ax)| : \|x\| = 1\} \quad .$$

Proof Set $M = \sup\{|(x,Ax)| : \|x\| = 1\}$. Then for any x with $\|x\| = 1$ we have

$$(x,Ax) \leqslant \|x\|\,\|Ax\| \leqslant \|x\|^2\,\|A\| = \|A\| \quad .$$

Therefore

$$M \leqslant \|A\| \quad .$$

To prove that $M \geqslant \|A\|$, consider the following identity:

$$\begin{aligned} 4\|Au\|^2 &= (A[\beta u + \beta^{-1}Au], \beta u + \beta^{-1}Au) \\ &\quad - (A[\beta u - \beta^{-1}Au], \beta u - \beta^{-1}Au) \quad . \end{aligned} \tag{8.20}$$

This is true for any u and any real $\beta \neq 0$, as can be verified by multiplying out the right hand side. Now, $\beta u \pm \beta^{-1}Au$ can be expressed as $\|\beta u \pm \beta^{-1}Au\|$ times a vector of unit norm; from the definition of M, (8.20) therefore gives

$$4\|Au\|^2 \leqslant M\|\beta u + \beta^{-1}Au\|^2 + M\|\beta u - \beta^{-1}Au\|^2$$

$$\therefore 4\|Au\|^2 \leqslant 2M(\beta^2\|u\|^2 + \beta^{-2}\|Au\|^2) \tag{8.21}$$

by the parallelogram law (Problem 7.2). Now take $\beta^2 = \|Au\|/\|u\|$. Then (8.21) reduces to

$$\|Au\| \leqslant M\|u\| \quad . \tag{8.22}$$

All this holds for any nonzero u in the space, so (8.22) implies that $\|A\| \leqslant M$; and we showed above that $\|A\| \geqslant M$, so $\|A\| = M$. □

The last result in this section is a generalisation of the following fact of linear algebra. If A is a linear operator on $\mathbb{R}^n$, then by Proposition 8.2 the range of A is either (i) a proper subspace of $\mathbb{R}^n$ or (ii) $\mathbb{R}^n$ itself; in case (i) A is not invertible, and in case (ii) A is invertible (see Problem 8.21). The position in infinite-dimensional spaces is more complicated. It is not in general true that an operator on H whose range is the whole of H must be invertible; see Problem 8.22. But the following result shows that it is true in the special case of self-adjoint operators, and the condition $R(A) = H$ can be weakened to $R(A)$ dense in H.

Proposition 8.35 (Invertibility) Let A be a self-adjoint operator on H, and let V be the range of A. If V is dense in H, then $A: H \to V$ is invertible.

Proof We begin by showing that A maps all nonzero vectors into nonzero vectors. Suppose that $Ax = 0$; we shall show that $x = 0$. Let ϵ be any positive number. Since $R(A)$ is dense in H there is a $u \in H$ with $\|x - Au\| < \epsilon$. Write $x - Au = z$; then we have

$$(x, x - z) = (x, Au)$$

$$= (Ax, u)$$

since A is self-adjoint,

$$= 0$$

since $Ax = 0$. Hence

$$(x,x) = (x,z) \quad .$$

Therefore

$$\|x\|^2 \leqslant \|x\|\,\|z\| \quad ,$$

or $$\|x\| \leqslant \|z\| = \epsilon \quad .$$

But ϵ is an arbitrary positive number; thus $\|x\|$ is less than all positive numbers, and so $\|x\| = 0$. This shows that $Ax = 0$ implies $x = 0$.

We can now show that A is invertible. For any $y \in V$ there is at least one $x \in H$ with $Ax = y$. Suppose that $Ax_1 = y$ and $Ax_2 = y$. Then $A(x_1 - x_2) = y - y = 0$, and from the result of the previous paragraph it follows that $x_1 - x_2 = 0$, so

$x_1 = x_2$, and we have proved that there is one and only one $x \in H$ such that $Ax = y$. Hence A is invertible. □

We shall use this result in our treatment of differential equations in section 9.2.

8.4 Eigenvalue Problems for Self-Adjoint Operators

Definition 8.36 If $A: V \to V$ is a linear operator on a vector space V, the number λ is called an **eigenvalue** of A if there is a nonzero $u \in V$ such that

$$Au = \lambda u \quad . \tag{8.23}$$

Any such nonzero u is called an **eigenvector** of A, and is said to **belong** or **correspond** to the eigenvalue λ. If the elements of V are functions, u is called an **eigenfunction.** □

The **eigenvalue problem** means the problem of finding the eigenvalues and eigenvectors of a given operator; we have met such problems already, in Chapter 5. Authors concerned for the purity of the English language prefer the terms 'characteristic value' or 'proper value' instead of the German-English hybrid 'eigenvalue'.

One of the main sources of eigenvalue problems in mechanics is the theory of vibrating systems, which can be summarised very roughly as follows. The state of a system at any given time t may be represented by an element $u(t)$ of a space H, and the equation of motion of classical mechanics is of the form

$$\mathrm{d}^2 u/\mathrm{d}t^2 = Au \quad ,$$

where A is an operator on H. When the system vibrates, the time-dependence of u will be sinusoidal: $u(t) = (\sin\omega t)v$ where v is a fixed element of H. When A is linear, the equation of motion becomes

$$-\omega^2 v = Av \quad ,$$

which says that $-\omega^2$ is an eigenvalue of A. Thus the eigenvalues of the operator A correspond to the possible frequencies of vibration. In the case of an atomic system, the vibration frequencies are visible as the bright lines in the spectrum of the light it emits. For this reason, the theory of the eigenvalue problem is called **spectral theory**, and the set of eigenvalues of an operator is called its **point spectrum.** We shall refer to the point spectrum simply as the **spectrum**; for some operators there is a distinction between the point spectrum and other types of spectrum, but for the operators considered in this book the point spectrum is the only type, and there is no danger of confusion.

In this section we discuss some simple properties and examples of the eigenvalue problem. Note first that eigenvectors are never unique, since any multiple of an eigenvector is an eigenvector belonging to the same eigenvalue.

Proposition 8.37 The set of all eigenvectors belonging to one particular eigenvalue of an operator is a vector space.

Proof If $Au = \lambda u$ and $Av = \lambda v$, then for any scalars a and b, $A(au + bv) = \lambda(au + bv)$ because A is linear, so $au + bv$ is also an eigenvector belonging to eigenvalue λ, so the set of eigenvectors is a vector space by 4.8. □

Remark 8.38 The observant reader will have noticed that the above proposition is not strictly correct; the set of eigenvectors does not include the zero vector. The true statement is 'the set of eigenvectors, together with the zero vector, is a vector space'. But the statement 8.37 is simpler and more memorable. It is an example of the useful technique, sometimes called 'abuse of language', whereby the meaning of terms is stretched beyond their strict definition, in a way which can be understood by applying common sense. Another common example is to speak of 'the function $f(x)$' instead of the correct but clumsy 'the function $f: x \mapsto f(x)$'. Remark 7.15 refers to another example.

Definition 8.39 The set of all eigenvectors belonging to one particular eigenvalue of an operator is called the **eigenspace** of that eigenvalue. Its dimension is called the **multiplicity** of the eigenvalue. An eigenvalue of multiplicity one is called **simple** or **non-degenerate**, and an eigenvalue of multiplicity greater than one is called **multiple** or **degenerate**.

Example 8.40 Consider the integral operator of Example 8.30 with $K(x,y) = \cos(x - y)$, and $a = 0$, $b = 2\pi$. The eigenvalue equation (8.23) is

$$\int_0^{2\pi} \cos(x - y) u(y)\, dy = \lambda u(x)$$

or

$$\cos x \int_0^{2\pi} \cos y\, u(y)\, dy + \sin x \int_0^{2\pi} \sin y\, u(y)\, dy = \lambda u(x) \quad . \tag{8.24}$$

If $\lambda \neq 0$, this says that $u(x)$ is a linear combination of $\sin x$ and $\cos x$ with constant coefficients. Thus every eigenfunction for $\lambda \neq 0$ has the form

$$u(x) = a \cos x + b \sin x \tag{8.25}$$

for some constants a and b. Substituting this in (8.24) gives

$$\pi a = \lambda a, \; \pi b = \lambda b \quad . \tag{8.26}$$

Thus λ must be π, and a and b can take any values. In other words, this operator has exactly one nonzero eigenvalue, namely π, and its eigenfunctions are all functions of the form $a \cos x + b \sin x$. This is a two-dimensional eigenspace, so the eigenvalue has multiplicity 2. Equations (8.26) are also satisfied for any λ by $a = b = 0$, but this is irrelevant to the eigenvalue problem because it means that u is the zero vector, which is not allowed as an eigenvector. Equation (8.24) shows that zero is also an eigenvalue, its eigenfunctions being all functions orthogonal to $\cos y$ and $\sin y$; it has infinite multiplicity.

Theorem 8.41 (Eigenvalues of a Self-Adjoint Operator) If A is a self-adjoint operator on a Hilbert space, then all its eigenvalues are real, and eigenvectors belonging to different eigenvalues are orthogonal.

Proof If $Au = \lambda u$, then $\lambda \| u \|^2 = \lambda(u,u) = (Au,u) = (u,Au) = (u,\lambda u) = \bar{\lambda} \| u \|^2$. Since $\| u \| \neq 0$, we deduce that $\lambda = \bar{\lambda}$, so λ is real. If also $Av = \mu v$, then

$$(v,Au) = (Av,u) \quad ,$$

$$\therefore \lambda(v,u) = \mu(v,u) \quad .$$

Hence if $\lambda \neq \mu$ then $(v,u) = 0$ □

Example 8.42 Consider again the operator of Example 8.30, with $a = 0$, $b = 2\pi$, but now $K(x,y) = \cos(x+y)$. This kernel, like that in Example 8.40, is symmetric and real, and the integral operator is therefore self-adjoint. The analysis is very similar to that of Example 8.40; in the same way we deduce that if $\lambda \neq 0$ then u must have the form given in (8.25), and substitution in the eigenvalue equation gives

$$\pi a = \lambda a, \; \pi b = -\lambda b \quad . \tag{8.27}$$

There are two ways of satisfying (8.27): either $\lambda = \pi$ and $b = 0$, or $\lambda = -\pi$ and $a = 0$. Thus there are two nonzero eigenvalues; the eigenvalue π has eigenfunctions $a\cos x$, and the eigenvalue $-\pi$ has eigenfunctions $b \sin x$, for any a,b. These eigenvalues are non-degenerate, and it is easy to verify that their eigenfunctions are orthogonal, as predicted by Theorem 8.41. Zero is also an eigenvalue; as in Example 8.40, its eigenspace is the orthogonal complement of $\{\sin x, \cos x\}$. □

The examples above have been chosen for their simplicity: they have only a small number of eigenvalues. In the theory of matrices, an $n \times n$ matrix is generally expected to have n eigenvalues; similarly an operator on an infinite-dimensional Hilbert space may be expected to have an infinite number of eigenvalues. This immediately raises questions about the behaviour of the n-th eigenvalue as $n \to \infty$: can it tend to infinity, or is it bounded? The following result answers this question for bounded operators.

Theorem 8.43 (Eigenvalue Bound) If A is a bounded operator and λ an eigenvalue of A, then $|\lambda| \leqslant \| A \|$.

Proof If $Au = \lambda u$, then $\| \lambda u \| = \| Au \|$, hence $|\lambda| \, \| u \| \leqslant \| A \| \, \| u \|$ by (8.2); the theorem now follows. □

If the eigenvalues are regarded as points in the complex plane, this result means that they all lie inside a circle of radius $\| A \|$ (or an interval of length $2\| A \|$ in the case of real spaces, where $\lambda \in \mathbb{R}$). For a stronger result of this type see Problem 8.14. In the case of self-adjoint operators we can go further, by using Proposition 8.34. This immediately gives the following result, which is the basis of Rayleigh's approximation method for finding eigenvalues, as well as the starting point for the proof of the spectral theorem in section 9.1.

Corollary 8.44 (Eigenvalue Bound for Self-Adjoint Operators) All eigenvalues λ of a bounded self-adjoint operator A satisfy

$$|\lambda| \leqslant \sup\{|(x,Ax)| : \|x\| \leqslant 1\} \quad . \qquad \square \tag{8.28}$$

It is natural to ask if any eigenvalue actually equals the upper bound given by (8.28). In general the answer is no, but for a particular class of operators, called compact operators, the answer is yes. Another natural question is whether the eigenvectors of a self-adjoint operator form an orthogonal basis for Hilbert space in the same way as for finite-dimensional spaces. Again, the answer in the special case of compact operators is yes. In the next section we shall define compact operators, but first we shall pause to work out the application of these ideas to Sturm-Liouville systems.

Example 8.45 (Sturm-Liouville Systems) A Sturm-Liouville system is a differential equation of the form

$$\mathrm{D}[p(x)\mathrm{D}u] + q(x)u = \lambda r(x)u \quad , \tag{8.29}$$

with boundary conditions at $x = a$ and b, which we shall take to be separated end-point conditions, though the theory applies also to other boundary conditions. In (8.29) λ is a parameter, and we are interested in finding values of λ for which the system has non-trivial solutions. Equation (8.29) can be interpreted as an eigenvalue equation for a differential operator, but the theory of this chapter does not apply to that operator because differential operators are unbounded. We therefore replace the differential equation by an equivalent integral equation.

In Theorem 2.13 we proved the existence of Green's function for the operator $\mathrm{D}p\mathrm{D} + q$ on the left of (8.29), under the assumption that zero is not an eigenvalue. If this is not true, we can make it true by adding $cr(x)u$ to both sides of (8.29) for a suitable constant c; we suppose this done. Then there is a Green's function g for (8.29), and the equation is equivalent to the integral equation $u = \lambda\Gamma u$, where the operator Γ is defined by

$$(\Gamma u)(x) = \int_a^b g(x;y)r(y)u(y)\,\mathrm{d}y. \tag{8.30}$$

This operator is of the type discussed in Example 8.30, with $K(x,y) = g(x,y)r(y)$. Now, Green's function is symmetric and real, so unless r is constant the kernel K does not satisfy the condition (8.19) for Γ to be self-adjoint.

We remove this difficulty by using a different inner product. We define a Hilbert space $L_2[a,b;r]$ whose elements are functions square-integrable on $[a,b]$, with inner product

$$(u,v) = \int_a^b u(x)\overline{v(x)}r(x)\,dx \quad . \tag{8.31}$$

We assume that r is positive and piecewise continuous, say, and is bounded away from zero, that is, $r(x) \geqslant m$ for $a \leqslant x \leqslant b$, for some $m > 0$. Then (8.31) satisfies all the rules for an inner product, and the corresponding norm is easily shown to be equivalent to the usual norm on $L_2[a,b]$, in the sense of section 4.6. It follows from 4.67 that $L_2[a,b;r]$ is complete. The functions

$$\phi_n(x) = [r(x)]^{-\frac{1}{2}} \sin[n\pi(x-a)/(b-a)]$$

are orthogonal with respect to the inner product (8.31). They form a basis because the sine functions form a basis (Example 7.23): for any function f, $f\sqrt{r}$ can be expanded in terms of the sines, hence f can be expanded in terms of the ϕ_n. We have now shown that $L_2[a,b;r]$ is a Hilbert space.

In this space the operator Γ is self-adjoint, since for any functions u,v,

$$\begin{aligned}(u,\Gamma v) &= \int_a^b u(x)\int_a^b g(x;y)r(y)\overline{v(y)}\,dy\; r(x)\,dx \\ &= \int_a^b\int_a^b g(y;x)r(x)u(x)\,dx\; \overline{v(y)}r(y)\,dy \\ &= (\Gamma u,v) \quad .\end{aligned}$$

The Sturm-Liouville problem is thus reduced to a self-adjoint eigenvalue problem for a bounded operator.

8.5 Compact Operators

Definition 8.46 An operator A on a normed space is **compact** if it maps every bounded set into a relatively compact set. □

This definition should be compared with Theorem 8.11. For linear operators, compactness is a stronger condition than boundedness; every compact operator is bounded, but not vice versa.

Examples 8.47 Every bounded linear operator in a finite-dimensional space is compact, by Theorem 8.11, because bounded sets in finite-dimensional spaces are relatively compact. In an infinite-dimensional space, any bounded linear operator whose range is finite-dimensional (that is, which maps every element of the space into a member of a certain finite-dimensional subspace) is compact for the same reason. In particular, the operators in Examples 8.40 and 8.42 map every function into a linear combination of $\sin x$ and $\cos x$, therefore into a two-dimensional subspace of $L_2[0,2\pi]$; they are therefore compact operators.

Example 8.48 If $K(x,y)$ is continuous for $a \leqslant x,y \leqslant b$, and $f(y,z)$ is continuous for $a \leqslant y \leqslant b$ and all z, then the operator $T: C[a,b] \to C[a,b]$ defined by

$$(Tu)(x) = \int_a^b K(x,y)f(y,u(y))\,dy$$

is compact. The proof of this fact is tedious, but fortunately we have done it already. The operator T appears in Theorem 6.45 on Hammerstein equations; the proof of that theorem uses Schauder's theorem, and therefore involves relative compactness of the image of a bounded set. If you read that proof carefully, you should be able to extract from it a proof of the compactness of T.

Example 8.49 The simplest of all operators, the identity, is bounded but not compact in an infinite-dimensional space. The unit sphere, for example, is bounded but not compact (see the second half of the proof of 6.22), and the identity operator therefore maps a bounded set (the unit sphere) into a non-compact set (the unit sphere). □

In Example 8.48 we considered an integral operator on the space of continuous functions. When applying the fixed-point theorems of Chapters 5 and 6 to integral equations, we had a choice between working in $C[a,b]$ or $L_2[a,b]$. But $C[a,b]$ is not a Hilbert space, so if we wish to use Hilbert space theory we are forced to treat integral operators in $L_2[a,b]$. We therefore need a result telling us when an integral operator is compact on $L_2[a,b]$.

A useful way of proving that an operator A is compact is to approximate it by a sequence of simpler operators which are known to be compact. Theorem 8.51 below shows that if this can be done, then A itself is compact. By 'simpler operators' we mean operators of finite rank, in the sense of Definition 8.3.

Proposition 8.50 (Finite-Rank Operators) A bounded linear operator of finite rank is compact.

Proof The operator maps bounded sets into bounded subsets of its range, which are relatively compact by 6.34. □

Our plan for proving an integral operator A compact is to approximate it by a sequence of finite-rank operators A_n, where A_n has rank n, and then use the following theorem. Its proof may be omitted by readers uninterested in the technicalities of analysis.

Theorem 8.51 (Limit of Compact Operators) Let $A, A_1, A_2, \ldots$ be operators $N \to M$, where N is a normed space and M a Banach space. If A_i is compact for each i, and $A_i \to A$ as $i \to \infty$, then A is compact.

Proof We must show that if B is any bounded subset of N, then $\{Ax: x \in B\}$ is relatively compact, that is, for any sequence (x_n) in B there is a subsequence (y_n) such that (Ay_n) is convergent. We use the diagonal method, as in the proof of 7.70, constructing an infinite number of subsequences, and then taking the diagonal sequence.

Since A_1 is compact, (x_n) has a subsequence (x_n^1) such that $(A_1 x_n^1)$ is convergent. Since A_2 is compact, the sequence (x_n^1) has a subsequence (x_n^2) such that $(A_2 x_n^2)$ is convergent. Proceeding in this way gives, for all r, a sequence $(x_n^r) = (x_1^r, x_2^r, \ldots)$ such that $A_r x_n^r$ converges as $n \to \infty$ with r fixed, and the sequence (x_n^r) is a subsequence of (x_n^{r-1}). They are all subsequences of the original bounded sequence (x_n).

Now consider the diagonal sequence $(y_n) = (x_n^n)$. We shall show that it is a Cauchy sequence. For any i,j,r we have

$$\|Ay_i - Ay_j\| \leqslant \|Ay_i - A_r y_i\| + \|A_r y_i - A_r y_j\| + \|A_r y_j - Ay_j\|$$

$$\leqslant \|A - A_r\| \|y_i\| + \|A_r y_i - A_r y_j\| + \|A_r - A\| \|y_j\| \quad . (8.32)$$

Now, each y_i is a member of the sequence (x_n) in the bounded set B. Hence there is a constant c such that $\|y_i\| \leqslant c$ for all i. Roughly speaking, this means that each term on the right of (8.32) tends to zero. We shall now spell this out precisely.

Because $A_n \to A$, it follows that given any $\epsilon > 0$ there is an r such that $\|A - A_r\| < \epsilon/3c$. Furthermore, for $i,j > r$, the vectors $y_i = x_i^i$ and $y_j = x_j^j$ are members of the sequence $(x_1^r, x_2^r, \ldots)$, hence $(A_r y_1, A_r y_2, \ldots)$ is a subsequence of $(A_r x_1^r, A_r x_2^r, \ldots)$, which is convergent by construction. It follows that $(A_r y_1, A_r y_2, \ldots)$ is a Cauchy sequence, so there is a $k > 0$ such that

$$\|A_r y_i - A_r y_j\| < \epsilon/3 \quad \text{for} \quad i,j > k \quad .$$

Thus for any $\epsilon > 0$ there exist r,k such that

$$\|Ay_i - Ay_j\| < 2\epsilon/3 + \epsilon/3$$

for $i,j > \max\{k,r\}$. This shows that (Ay_i) is a Cauchy sequence, and it belongs to the Banach space M, so it is convergent. This proves the theorem. □

We now apply this theorem to prove the compactness of linear integral operators on $L_2[a,b]$. We shall approximate the operator by a sequence of finite-rank operators of the type defined below.

Definition 8.52 The operator $A: L_2[a,b] \to L_2[a,b]$ defined by

$$(Af)(x) = \int_a^b K(x,y) f(y)\, dy$$

is said to have a **separable kernel** if

$$K(x,y) = \sum_{i=1}^{n} p_i(x)q_i(y) \quad ,$$

where $p_i, q_i \in L_2[a,b]$ for $i = 1, \ldots, n$. □

Example 8.40 is an operator with a separable kernel.

Lemma 8.53 An operator with a separable kernel is compact.

Proof It is easy to show that a separable kernel is square-integrable, and hence gives a continuous and bounded operator by 6.8. It has finite rank because its range is spanned by $\{p_1, \ldots, p_n\}$. Hence it is compact by 8.50. □

Theorem 8.54 (Compact Integral Operators) The linear integral operator on $L_2[a,b]$ with kernel K is compact if

$$\int_a^b \int_a^b |K(x,y)|^2 \, dx \, dy \quad \text{converges.} \tag{8.33}$$

Proof $L_2[a,b]$ is a Hilbert space, and therefore has an orthonormal basis (p_n). It is easy to show that the set of functions $p_i(x)p_j(y)$ is a basis for the space of functions of two variables satisfying (8.33) (take $q_j = p_j$ in Problem 8.28). We can therefore expand K:

$$K(x,y) = \sum_{i,j=1}^{\infty} a_{ij} p_i(x)p_j(y) \quad , \tag{8.34}$$

where $a_{ij} = \int_a^b \int_a^b K(x,y)\overline{p_i(x)}\,\overline{p_j(y)} \, dx \, dy$.

The sum in (8.34) is to be interpreted as converging with respect to the norm in the space of square-integrable functions of two variables; that is, (8.34) implies

$$\int_a^b \int_a^b |K(x,y) - \sum_{i,j=1}^{n} a_{ij} p_i(x)p_j(y)|^2 \, dx \, dy \to 0 \tag{8.35}$$

as $n \to \infty$. Let A_n be the separable-kernel operator whose kernel is the function obtained by truncating the series in (8.34):

$$(A_n f)(x) = \int_a^b \sum_1^n a_{ij} p_i(x)p_j(y)f(y) \, dy \quad .$$

We shall show that $\|A - A_n\| \to 0$ as $n \to \infty$, and then use 8.51 and 8.53.

In Example 8.8 we showed that for any operator B with kernel $L(x,y)$, $\|B\|^2 \leqslant \int\int |L(x,y)|^2 \, dx \, dy$. Applying this to $A - A_n$ gives

$$\|A - A_n\|^2 \leqslant \int\int |K(x,y) - \sum_1^n a_{ij} p_i(x)p_j(y)|^2 \, dx \, dy$$

$$\to 0 \quad \text{as} \quad n \to \infty$$

by (8.35). Since A_n is compact by Lemma 8.53, A is compact by Theorem 8.51. □

There is another way of looking at the property of compactness. Bounded operators map bounded sets into bounded sets; compact operators map bounded sets into relatively compact sets, so compactness for operators is a stronger version of boundedness. Now, for linear operators, boundedness is equivalent to continuity (Theorem 8.9). Continuous operators are those which map convergent sequences into convergent sequences. Compact linear operators have a stronger property than this: they map weakly convergent sequences into convergent sequences. Compactness can thus be regarded as a stronger version of continuity.

Theorem 8.55 (Complete Continuity) If $x_n \to x$ weakly and A is a compact linear operator, then $Ax_n \to Ax$ strongly.

Proof The weakly convergent sequence (x_n) is bounded, by 7.64. Therefore the sequence (Ax_n) is contained in some compact set (Definition 8.46). And $Ax_n \to Ax$ weakly, since for any y in the space,

$$\begin{aligned}(Ax_n,y) &= (x_n,A^*y)\\ &\to (x,A^*y) = (Ax,y) \quad .\end{aligned}$$

Theorem 7.69 now shows that $Ax_n \to Ax$ strongly. □

Operators with this property of mapping weakly convergent sequences into convergent sequences are often called **completely continuous**; then Theorem 8.55 says that every compact linear operator is completely continuous. For linear operators the converse is also true, although we shall not need it. For nonlinear operators, however, complete continuity is not equivalent to compactness.

We have now assembled the main definitions and results of operator theory that we shall need. In the next chapter we shall use this theory to prove that the eigenvectors of compact self-adjoint operators on Hilbert space form a basis for the space.

8.6 Summary and References

In section 8.1 we defined bounded and linear operators, and showed that a linear operator is continuous if and only if it is bounded. In section 8.2 we showed that the set of bounded linear operators on a Banach space is itself a Banach space; we considered the theory of power series of operators, and applied it to the inversion of operator equations. In section 8.3 we defined the adjoint of an operator, and in section 8.4 we showed that the eigenvectors of a self-adjoint operator are orthogonal and the eigenvalues are real and bounded by $\|A\|$. In section 8.5 we defined compact operators, and proved that integral operators with square-integrable kernels are compact. We also showed that compact operators map weakly convergent sequences into strongly convergent sequences.

This theory is given in most books on functional analysis; see the references given in section 7.9. For applications in computing, see Moore (1985).

PROBLEMS 8

Section 8.1

8.1 Consider the normed space $\mathbb{R}^2$ with $\|x\| = \max\{|x_1|, |x_2|\}$. Given a real number a, define $f: \mathbb{R}^2 \to \mathbb{R}$ by $f(x) = ax_1$. Find the norm of the operator f. Now find the norm of $g: x \mapsto a_1x_1 + a_2x_2$. Generalise to $\mathbb{R}^n$.

8.2 Consider the normed space $\mathbb{R}^n$ with norm $\|x\| = \max_i |x_i|$. Let A be the operator corresponding to multiplication by a square matrix (a_{ij}) thus: $(Ax)_i = \Sigma_j a_{ij}x_j$. Show that $\|A\| = \max_k \Sigma_j |a_{kj}|$. (You may find it easier to begin with the case where all $a_{ij} > 0$, and then drop this assumption.)

8.3 Rework Problem 8.2 using the norm $\|x\| = \Sigma |x_i|$; show that the operator norm is then $\max_k \Sigma_j |a_{jk}|$. Compare these results with each other and with Problem 4.10.

8.4 (a) What is the greatest possible value of $\int_0^1 x^2 f(x)\,dx$ subject to the condition $|f(x)| \leqslant 1$ for $0 \leqslant x \leqslant 1$? Deduce the norm of the operator $C[0,1] \to \mathbb{R}$ defined by $f \mapsto \int_0^1 x^2 f(x)\,dx$.

(b) What is the greatest possible value of $\int_0^1 \sin(2\pi x)f(x)\,dx$ subject to the condition $|f(x)| \leqslant 1$ for $0 \leqslant x \leqslant 1$ (note: f is not required to be continuous)? Deduce the norm of the operator $C[0,1] \to \mathbb{R}$ defined by $f \mapsto \int_0^1 \sin(2\pi x)f(x)\,dx$. (Note: a discontinuous function can be approximated arbitrarily closely by continuous functions.)

(c) What is the norm of the operator $C[0,1] \to \mathbb{R}$ defined by $f \mapsto \int_0^1 \phi(x)f(x)\,dx$, where ϕ is a given continuous function?

8.5 In Example 8.10 we noted that the differentiation operator $\mathrm{D}: C^1[a,b] \to C[a,b]$ is not bounded. Problem 6.3 shows that D is bounded when regarded as an operator $C_1^1[a,b] \to C[a,b]$, where $C_1^1[a,b]$ is defined in Problem 4.23 and $C[a,b]$ has the L_2 norm. Show that for $\mathrm{D}: C_1^1[a,b] \to C[a,b]$ we have $\|\mathrm{D}\| \leqslant 1$ (easy), and in fact $\|\mathrm{D}\| = 1$ (fairly easy).

8.6 Consider the vector space P_n of polynomials of degree $\leqslant n$, where n is a given integer. If $f(t) = \Sigma_0^n a_i t^i$, show that $\|f\| = \max_i |a_i|$ defines a norm on P_n. Is P_n a Banach space?

Show that the differentiation operator D is a bounded operator $P_n \to P_n$, and find its norm.

8.7 An 'infinite matrix' is an infinite array of numbers (a_{ij}), $i,j = 1,2,\ldots$. You may prefer to think of it as a mapping from pairs of positive integers to the real or complex numbers, just as a sequence can be regarded as a mapping from positive integers to the real or complex numbers. An infinite matrix gives an operator on spaces of sequences, just as an $n \times n$ matrix gives an operator on $\mathbb{R}^n$ or $\mathbb{C}^n$.

Let l_1 be the space of sequences $x = (x_n)$ such that $\Sigma|x_i|$ converges, with norm $\|x\| = \Sigma|x_i|$. Let b be the space of bounded sequences $x = (x_n)$ with $\|x\| = \sup|x_i|$. Let (a_{ij}) be an infinite matrix whose entries are bounded, $|a_{ij}| \leqslant a$ for all i,j. Define an operator A on sequences by $(Ax)_i = \Sigma_j a_{ij}x_j$. Show that A is a bounded operator $l_1 \to b$, and $\|A\| \leqslant a$. Show also that A does not generally belong to $B(l_1,l_1)$.

If (a_{ij}) is not only bounded but banded, that is, all entries vanish except those in a finite band surrounding the principal diagonal ($a_{ij} = 0$ if $|i-j| > k$ for some fixed k), show that A is a bounded operator $l_1 \to l_1$. Find an upper bound for its norm.

Section 8.2

8.8 If $A_n \to A$ and $B_n \to B$, prove that $A_nB_n \to AB$, where A_n, A, B_n, B are bounded operators from a normed space N into itself.

8.9 (The exponential of an operator.) If A is a bounded linear operator $N \to N$, where N is a Banach space, show that $\Sigma_0^\infty A^n/n!$ converges. Call its sum e^A. Show that for any integer $p > 0$, $(e^A)^p = e^{pA}$. Show that $e^O = I$ (where O is the zero operator), and that e^A is always invertible (even if A is not), and its inverse is e^{-A}. Show that if $AB = BA$, then $e^A e^B = e^{A+B}$. Does $e^A e^B = e^{A+B}$ hold for all A,B?

8.10 (Differential equations) Let $u(t)$ be a vector, in a Banach space N, which depends on a real parameter t (think of t as time). Define $\dot{u}(t) = \lim_{h\to 0}\{[u(t+h)-u(t)]/h\}$, if the limit exists. Show that for any fixed $u_0 \in N$, the vector function $u(t) = e^{At}u_0$ satisfies the differential equation $\dot{u} = Au$, where e^{At} is defined in Problem 8.9. Can you solve the equation $\dot{u} = Au + b$, where b is a fixed element of N?

(Applications of this when $N = \mathbb{R}^n$ can be found in Bellman (1960) or books on differential equations or linear systems and control theory. For differential equations on infinite-dimensional spaces, see Showalter, or Curtain & Pritchard.)

8.11 In the notation of Corollary 8.27, show that $\|x_a - x\| \leqslant (\alpha^2\delta/1 - \alpha\delta)\|b\|$. This formula has the advantage that it can be evaluated before x_a has been evaluated, and is therefore called an 'a priori estimate' for the error. Explain why it will give a more pessimistic error estimate, in general, than equation (8.12).

8.12 Invent a 3×3 set of linear algebraic equations on which to test the results of Corollary 8.27. If you make the matrix almost diagonal, then A_a will be diagonal and easy to invert. Use the norm of Problem 8.2, so as to give estimates of the greatest possible error in the solution, and check against an exact computed solution.

8.13 A sequence of real numbers u_n such that $u_{n+m} \leqslant u_n + u_m$ for all n,m has the property that u_n/n either converges or tends to $-\infty$ as $n \to \infty$ (the proof of this is not easy; see Pólya & Szego (1972) p. 23). Use this fact to show that for any bounded operator A, the sequence of numbers $\| A^n \|^{1/n}$ is convergent. Call its limit $r(A)$, the **spectral radius** of A (the name is explained in Problem 8.14). Prove that $0 \leqslant r(A) \leqslant \| A \|$. Show that the Volterra integral operator defined in Problem 5.10 has spectral radius zero.

8.14 N is a Banach space and $A \in B(N,N)$. Show that if λ is an eigenvalue of A, then the operator $\lambda I - A$ is not invertible. (This is the starting-point of an extensive theory of noncompact operators, to be found in textbooks under the heading 'resolvent operator'.)

Use the power-series expansion to prove that for any complex number μ, $\mu I - A$ is invertible if $|\mu| > r(A)$ (see Problem 8.13). Deduce that all eigenvalues of A, regarded as points in the complex plane, lie inside a circle whose radius is the spectral radius. Deduce from Problem 8.13 that a Volterra operator has no nonzero eigenvalues.

Section 8.3

8.15 A, B are bounded linear operators on a Hilbert space. Show that $(AB)^* = B^*A^*$, that $(cA)^* = \bar{c}A^*$ for any scalar c, and that if A and B are self-adjoint, then AB is self-adjoint if and only if $AB = BA$.

8.16 On the space l_2 of Example 7.4 we define an operator T by $Tx = (x_1, x_2/2, x_3/3, \ldots)$. Show that T is bounded, and find its adjoint.

8.17 Let A be a bounded linear operator on a Hilbert space H. Define $R(A) = \{Ax : x \in H\}$ and $N(A) = \{x \in H : Ax = 0\}$. Prove that $N(A)$ is a subspace and $R(A)$ a vector subspace of H, and that $[R(A)]^\perp = N(A^*)$.

8.18 A self-adjoint bounded linear operator P on a Hilbert space H is called a **projection** if $P^2 = P$. In the notation of Problem 8.17, prove that if P is a projection then $R(P)$ is a subspace of H, and $H = R(P) \oplus N(P)$ in the notation of 7.50. Draw a diagram to explain the term 'projection'.

8.19 Let E be a subspace of a Hilbert space H. We use Theorem 7.47 to define an operator $P: x \mapsto$ the unique $y \in E$ such that $x = y + z$ for some $z \in E^{\perp}$. Show that P is a projection in the sense of Problem 8.18.

8.20 Show that every bounded linear operator T on a Hilbert space can be written in the form $T = A + iB$ where A, B are self-adjoint.

8.21 Show that a linear operator $A: \mathbb{R}^n \to \mathbb{R}^n$ is invertible if and only if the range of A is $\mathbb{R}^n$.

8.22 Give an example of a non-self-adjoint operator on a Hilbert space H whose range is H and which is not invertible. (Note: this shows that the self-adjointness condition in 8.35 is essential. Problem 8.21 shows that your example must be infinite-dimensional.)

Sections 8.4, 8.5

8.23 If you are not familiar with methods of finding eigenvalues and eigenvectors of real symmetric matrices (that is, self-adjoint operators on $\mathbb{R}^n$), read about them in linear algebra books and work some examples.

8.24 A bounded linear operator A on a Hilbert space H is called **unitary** if $A^*A = AA^* = I$. Show that if A is unitary then $\|Ax\| = \|x\|$ for all $x \in H$ (thus unitary operators are like rotations, which do not change lengths). Deduce that all eigenvalues of a unitary operator have modulus 1, and eigenvectors belonging to different eigenvalues are orthogonal. Show that all unitary operators are invertible.

The exponential of an operator is defined in Problem 8.9. If B is a self-adjoint operator, show that e^{iB} is unitary. (Unitary operators can be used to transform from one basis in Hilbert space to another, in the same way that rotations in $\mathbb{R}^3$ transform from one set of axes to another; see textbooks on quantum mechanics.)

8.25 If a sequence of self-adjoint linear operators is convergent, show that its limit is self-adjoint.

8.26 If λ is an eigenvalue of an operator A, show that for any polynomial p, $p(\lambda)$ is an eigenvalue of the operator $p(A)$.

8.27 Show that an operator of rank n can have at most n nonzero eigenvalues.

8.28 Let (p_i) and (q_i) be two orthonormal bases for $L_2[a,b]$. Let H be the space of square-integrable functions of two variables on the square $a \leqslant x,y \leqslant b$, with inner product $\int_a^b \int_a^b f(x,y)\overline{g(x,y)}\,dx\,dy$.

(a) Show that the set of functions $p_i(x)q_j(y)$ is orthonormal in H.

(b) Show that if $\phi \in H$ and $\int_a^b \int_a^b \phi(x,y)p_i(x)q_j(y)\,dx\,dy = 0$ for all i,j, then $\phi = 0$.

(c) The set of functions $p_i(x)q_j(y)$ is labelled by two integers and is therefore countable (see Appendix D), and can be arranged in a sequence. Prove that this sequence is a basis for H.

Remark: Since orthogonal expansions can be rearranged (Problem 7.14), we write expansions in terms of this basis in the form $\Sigma_{i,j=1}^{\infty} a_{ij}p_i(x)q_j(y)$, the value being independent of the order of the terms.

More problems will found on page 377.

Chapter 9

The Spectral Theorem

In the first section of this chapter we use the machinery assembled in Chapter 8 to prove that the eigenvectors of a compact self-adjoint operator form a basis for the Hilbert space; this theorem is the culmination of the work in Chapters 7 and 8. In section 9.2 we apply it to ordinary differential operators, and in section 9.3 to the Laplacian. In section 9.4 we discuss another consequence of the spectral theorem, the 'Fredholm Alternative', governing the existence and uniqueness of solutions of linear operator equations. In section 9.5 we introduce a new idea, that of a projection operator, and use it first to reformulate the spectral theorem, and then to give a general technique for calculating functions of an operator.

Readers who are interested in applications but not in proofs may skip sections 9.1, 9.2 and 9.3, except for the statements of 9.4, 9.5, 9.6, 9.7, 9.16 and 9.28. The definition and basic properties of projection operators given at the beginning of section 9.5 will be used in the discussion of approximation techniques in the next chapter, but the later parts of section 9.5 may be omitted if desired.

9.1 The Spectral Theorem

We build up to the basic Theorem 9.4 in a series of propositions, starting from Corollary 8.44, which gives an upper bound for the eigenvalues of a self-adjoint bounded operator. This corollary suggests a connection between the eigenvalue problem and the problem of maximising $|(x,Ax)|$ with $\|x\| = 1$. In general there need not be a vector maximising $|(x,Ax)|$, but we can certainly construct a sequence of vectors (x_n) such that $|(x_n,Ax_n)|$ approaches its supremum as $n \to \infty$. If this sequence converges, its limit, as we shall see, is an eigenvector. Even if it does not converge, the sequence gives an 'approximate solution' to the eigenvalue problem, in the following sense.

Proposition 9.1 If A is a self-adjoint bounded operator, then there is a number λ and a sequence (x_n) with $\|x_n\| = 1$ for all n, such that $Ax_n - \lambda x_n \to 0$ as $n \to \infty$. The number λ equals either $\|A\|$ or $-\|A\|$.

Proof Write $S = \{x : \|x\| = 1\}$, and

$$f(x) = (x,Ax) \quad .$$

Then Proposition 8.34 says that $\| A \| = \sup\{|f(x)| : x \in S\}$. Hence given any $z_1 \in S$ there is a $z_2 \in S$ such that $|f(z_2)|$ is at most half as far from $\| A \|$ as $|f(z_1)|$ is. Again, there is a $z_3 \in S$ such that $|f(z_3)|$ is at most half as far from $\| A \|$ as $|f(z_2)|$ is. In this way we define a sequence (z_n) in S such that

$$|f(z_n)| \to \| A \| \quad . \tag{9.1}$$

Now, $f(z_n)$ is real for all n (Lemma 8.33), therefore either infinitely many of the $f(z_n)$ are non-negative or infinitely many are non-positive or perhaps both. So either (i) (z_n) has a subsequence (y_n) with $f(y_n) \geqslant 0$ for all n, or (ii) (z_n) has a subsequence (y'_n) with $f(y'_n) \leqslant 0$, or (iii) it has both. In case (i), set $x_n = y_n$ and $\lambda = \| A \|$; in cases (ii) and (iii) set $x_n = y'_n$ and $\lambda = -\| A \|$. In all cases, (9.1) gives

$$f(x_n) = (x_n, Ax_n) \to \lambda \quad \text{as} \quad n \to \infty \quad . \tag{9.2}$$

Now we must show that $\| Ax_n - \lambda x_n \| \to 0$. We have

$$\begin{aligned} \| Ax_n - \lambda x_n \|^2 &= \| Ax_n \|^2 - 2\lambda(Ax_n, x_n) + \lambda^2 \| x_n \|^2 \\ &\leqslant \| A \|^2 - 2\lambda(Ax_n, x_n) + \lambda^2 \\ &\to 2\lambda^2 - 2\lambda^2 = 0 \quad , \end{aligned}$$

using (9.2) and the fact that $|\lambda| = \| A \|$ by definition. This shows that $Ax_n - \lambda x_n \to 0$ and completes the proof. □

In the case of compact operators we can say more: the sequence of Proposition 9.1 converges to a solution of the eigenvalue problem.

Theorem 9.2 If A is a compact self-adjoint linear operator, then it has an eigenvalue λ equal to either $\| A \|$ or $-\| A \|$; there is a corresponding normalised eigenvector which maximises $|(x,Ax)|$ under the condition $\| x \| = 1$, and the maximum value of $|(x,Ax)|$ is $|\lambda|$.

Proof Let (x_n) and λ be as in Proposition 9.1; (x_n) is normalised and therefore bounded. Because A is compact, the sequence (Ax_n) contains a convergent subsequence, say

$$Au_n \to v \tag{9.3}$$

where each u_n equals some x_m. We now show that (u_n) is convergent. We have

$$\begin{aligned} &\lambda u_n - v = (\lambda u_n - Au_n) + (Au_n - v) \\ &\therefore \lambda u_n - v \to 0 \quad \text{as} \quad n \to \infty \quad , \end{aligned} \tag{9.4}$$

using (9.3) and the fact that (u_n) is a subsequence of (x_n) which satisfies $Ax_n - \lambda x_n \to 0$.

If $\lambda = 0$, then $\|A\| = 0$, and A is the zero operator; the theorem is then trivially true, zero being the only eigenvalue and every nonzero vector being an eigenvector. If $\lambda \neq 0$, then (9.4) gives $u_n \to v/\lambda$, hence $Au_n \to Av/\lambda$ because A is continuous. Comparing this with (9.3), we see that $Av = \lambda v$, so v is an eigenvector belonging to the eigenvalue $\lambda = \pm\|A\|$.

If w is a normalised eigenvector, then

$$\begin{aligned}|(w,Aw)| &= |(w,\lambda w)| \\ &= |\lambda|\|w\|^2 \\ &= \|A\| \\ &= \sup\{|(x,Ax)|: \|x\| = 1\}\end{aligned}$$

by 8.34. Hence w maximises $|(x,Ax)|$ and the maximum value is $|\lambda| = \|A\|$. □

This theorem not only guarantees the existence of an eigenvalue but gives a useful technique for finding it, by maximising a certain quadratic expression. In the next chapter we shall look at the practical aspects of this procedure. Now we shall extend it step by step to give an infinite sequence of eigenvalues.

Proposition 9.3 If A is a compact self-adjoint operator on a Hilbert space H, then A has an orthonormal set of eigenvectors $e_1, e_2, \ldots$ with corresponding eigenvalues $\lambda_1, \lambda_2, \ldots$, and for any $x \in H$

$$x = \sum_n c_n e_n + y \tag{9.5}$$

for some scalars c_n, where y satisfies $Ay = 0$. If H is infinite-dimensional, then $\lambda_n \to 0$ as $n \to \infty$.

Proof Theorem 9.2 gives an eigenvalue λ_1 and a normalised eigenvector e_1. Set

$$Q_1 = \{x \in H: x \perp e_1\} \quad .$$

Q_1 is the orthogonal complement of the set $\{e_1\}$, and is therefore a subspace of H (Theorem 7.46). If $x \in Q_1$, then

$$(Ax,e_1) = (x,Ae_1) = \lambda_1(x,e_1) = 0 \quad ,$$

so that $Ax \in Q_1$. Thus A maps the Hilbert space Q_1 into itself, so another application of Theorem 9.2, with H replaced by Q_1, gives an eigenvalue λ_2 with

$$|\lambda_2| = \max\{|(x,Ax)|: x \in Q_1 \quad \text{and} \quad \|x\| = 1\} \quad ,$$

and a corresponding normalised eigenvector e_2. Clearly $e_1 \perp e_2$. We can repeat this procedure, giving an orthonormal sequence of eigenvectors (e_n), a nested sequence of subspaces Q_n with each Q_n a subspace of Q_{n-1}, and eigenvalues λ_n satisfying

$$|\lambda_n| = \max\{|(x,Ax)|: x \in Q_{n-1} \quad \text{and} \quad \|x\| = 1\} \quad . \tag{9.6}$$

If H is n-dimensional, the procedure terminates after n steps, and we have a set of n orthogonal eigenvectors which form a basis, so that (9.5) holds with $y = 0$, and the theorem is proved. If H is infinite-dimensional, the sequence (e_n) converges weakly to zero (see 7.61), and A is compact, hence by 8.55 $Ae_n \to 0$ strongly. Therefore $|\lambda_n| = \|Ae_n\| \to 0$ as $n \to \infty$.

Now, let M be the subspace of H consisting of all infinite linear combinations of the e_n. The Orthogonal Projection Theorem shows that any $x \in H$ can be written in the form (9.5) where $y \perp M$. Write $N = M^\perp$, then N is contained in each subspace Q_n, since every element of N is orthogonal to all the e_n. We must show that for all $y \in N$, $Ay = 0$.

For any $y \in N$, write $y_1 = y/\|y\|$; then $\|y_1\| = 1$ and

$$(y,Ay) = \|y\|^2 (y_1,Ay_1) \quad .$$

But $y_1 \in N \subset Q_n$ for each n, hence

$$(y,Ay) \leqslant \|y\|^2 \lambda_n$$

by (9.6). But $\lambda_n \to 0$ as $n \to \infty$, so for any $\epsilon > 0$ we have $|(y,Ay)| < \epsilon$, so $(y,Ay) = 0$ for all $y \in N$. Now Proposition 8.34 shows that if we consider A as an operator $N \to N$, then its norm as an operator on that space is zero, so it maps every element of N to the zero vector, which completes the proof. □

Theorem 9.4 (Spectral Theorem for Compact Self-Adjoint Operators) Let A be a compact self-adjoint operator on a Hilbert space H. Then H has a basis (e_n) consisting of orthonormal eigenvectors of A. If H is infinite-dimensional, the corresponding eigenvalues λ_n tend to zero as $n \to \infty$, and if

$$x = \Sigma c_n e_n$$

then

$$Ax = \Sigma c_n \lambda_n e_n \quad .$$

Proof Most of this theorem is contained in the preceding proposition, which gives a set of eigenvectors corresponding to nonzero eigenvalues, and a subspace N consisting of vectors mapped to zero by A. We can choose an orthonormal basis for the space N, each element of which is an eigenvector of A with eigenvalue zero. If we include these new eigenvectors in the sequence of eigenvectors constructed in Proposition 9.3, we obtain a basis for H. Since the new eigenvalues are zero, their inclusion does not destroy the convergence to zero of the sequence (λ_n) constructed in 9.3.

Finally, A is a continuous operator, therefore for any convergent sequence (u_n) we have $A \lim(u_n) = \lim(Au_n)$. In particular,

$$A\sum_{1}^{\infty}c_ne_n = A\lim_{r\to\infty}\sum_{1}^{r}c_ne_n$$

$$= \lim_{r\to\infty}\sum_{1}^{r}c_nAe_n = \lim_{r\to\infty}\sum_{1}^{r}c_n\lambda_ne_n = \sum_{1}^{\infty}\lambda_nc_ne_n$$

which completes the proof. □

The statement that H has an orthonormal basis consisting of eigenvectors of A can easily be shown to be equivalent to the following: if we take an orthonormal basis set for each eigenspace of A, then the union of all these sets is an orthonormal basis for H. In the language of 7.50, $H = E_1 \oplus E_2 \oplus \ldots$, where E_i are the eigenspaces.

Another common way of expressing the same fact is to say 'the eigenvectors of a compact self-adjoint operator are complete'. The word 'complete' is here being used in a sense quite different from that of Definition 4.56; completeness of a set of vectors must not be confused with completeness of a normed space.

Definition 9.5 **A complete set** of vectors in an inner product space is a set such that the only vector orthogonal to all members of the set is the zero vector. □

A complete orthogonal set is the same as a basis; it is obvious that a basis is complete, and the converse is easily proved:

Proposition 9.6 (Completeness) An orthogonal set of vectors in an inner product space is complete if and only if it is a basis.

Proof If (e_n) is a basis, then it is complete, for if $(x,e_n) = 0$ for all n, Theorem 7.30 gives $x = \Sigma\, 0e_n = 0$.

To prove the converse, let $\{e_n\}$ be a complete orthogonal set, and let x be any vector in the space. It is easy to verify that the vector $x - \Sigma(x,e_n)\|e_n\|^{-2}e_n$ is orthogonal to each e_r, hence it is zero, so $x = \Sigma(x,e_n)\|e_n\|^{-2}e_n$; this is an expression for an arbitrary vector as a combination of the e_n. □

Most of the rest of this chapter consists of applications of the theorem that compact self-adjoint operators have a complete set of eigenvectors. But first we shall give an extension of the theorem to pairs of operators. The following is a generalisation of the theorem that two matrices can be simultaneously diagonalised if they commute, that is, if $AB = BA$. It is important in quantum mechanics (it means that two observable quantities can be simultaneously measured with arbitrary precision if their operators commute), and will also be used later in this chapter.

Theorem 9.7 (Commuting Operators) If A and B are compact self-adjoint operators which commute (that is, $AB = BA$), then they have a complete orthogonal set of common eigenvectors (that is, vectors which are eigenvectors of both A and B).

Proof Let λ be an eigenvalue of A and S the corresponding eigenspace. For any $v \in S$ we have

$$ABv = BAv = \lambda Bv \quad ,$$

thus Bv is an eigenvector of A with eigenvalue λ, unless $Bv = 0$. In any case, $Bv \in S$, so B maps S into itself. Now $B: S \to S$ is a compact self-adjoint operator on S, and the spectral theorem shows that S has a basis consisting of eigenvectors of B; these vectors are also eigenvectors of A because they belong to S. If we take such a basis for each eigenspace of A and put them together, the spectral theorem for A shows that the resulting set is complete. □

9.2 Sturm-Liouville Systems

In Example 8.45 we showed that the Sturm-Liouville equation

$$\mathrm{D}(p\mathrm{D}u) + qu = \lambda ru \quad , \tag{9.7}$$

on the interval $[a,b]$ with separated end-point conditions, can be turned into a self-adjoint eigenvalue problem for an integral operator. The aim of this section is to show that the eigenfunctions of a Sturm-Liouville system are complete, by applying Theorem 9.4 to the compact integral operator, and then transferring the result to the related differential operator. We must therefore study in more detail the relation between the differential and integral eigenvalue problems. We shall assume that $p(x)$ and $r(x)$ are positive for $a \leqslant x \leqslant b$, and p, p', q, and r are continuous.

Equation (9.7) is the eigenvalue equation for the operator

$$L = \frac{1}{r(x)}\{\mathrm{D}p(x)\mathrm{D} + q\} \quad . \tag{9.8}$$

This operator can be applied only to twice-differentiable functions (in the present context of classical analysis), so the appropriate vector space is $C^2[a,b]$ (see Example 5.10). This can be made an inner product space, and we found in Example 8.45 that the appropriate inner product is

$$(u,v) = \int_a^b u(x)\overline{v(x)}r(x)\,\mathrm{d}x \quad . \tag{9.9}$$

We write $C^2[a,b;r]$ for the vector space $C^2[a,b]$ with this inner product. The norm in $C^2[a,b;r]$ is equivalent to the usual L_2 norm (Problem 4.34). We saw in section 4.5 that with the L_2 norm $C[a,b]$ is incomplete, therefore its subspace $C^2[a,b]$ is incomplete. It follows from 4.67 that $C^2[a,b;r]$ is incomplete. The theory of adjoint operators given in section 8.3 applies to bounded operators on a Hilbert space. The case of unbounded operators on incomplete subspaces is rather more complicated, but the central idea of self-adjointness can easily be extended to operators like L as follows.

Definition 9.8 Let $A: V \to H$ be an operator on a dense subspace V of a Hilbert space H. A is **symmetric** if

$$(x,Ay) = (Ax,y) \quad \text{for all} \quad x,y \in V \quad .$$ □

If A is a self-adjoint bounded operator on H, it is clearly symmetric (take $V = H$ in 9.8); symmetry is a natural generalisation of self-adjointness.

Theorem 9.9 The eigenvalues of a symmetric operator are real, and eigenvectors belonging to different eigenvalues are orthogonal.

Proof Identical to the proof of 8.41. □

Now consider the Sturm-Liouville operator $L: C^2[a,b;r] \to L_2[a,b;r]$ given by (9.8). Integrating twice by parts shows that

$$\begin{aligned}(u,Lv) &= \textstyle\int_a^b u[(p\bar{v}')' + q\bar{v}]\,\mathrm{d}x \\ &= [up\bar{v}' + u'p\bar{v}]_a^b + \textstyle\int_a^b [(u'p)'\bar{v} + qu\bar{v}]\,\mathrm{d}x \quad , \\ \therefore (u,Lv) &= [up\bar{v}' + u'p\bar{v}]_a^b + (Lu,v) \quad . \end{aligned} \tag{9.10}$$

Thus L is not symmetric. However, if u and v satisfy the boundary conditions associated with the Sturm-Liouville equation, then the first term on the right of (9.10) vanishes. We note that since the boundary conditions are linear homogeneous algebraic equations, the set of all functions satisfying them is a vector space. It is therefore reasonable to restrict L to the subspace of $C^2[a,b;r]$ consisting of functions satisfying the boundary conditions.

Definition 9.10 The domain $\mathscr{D}$ for a Sturm-Liouville system on $[a,b]$ with separated end-point conditions is the vector subspace of $C^2[a,b;r]$ consisting of functions satisfying the boundary conditions of the system. □

Proposition 9.11 The domain $\mathscr{D}$ defined in 9.10 is dense in $L_2[a,b;r]$.

Proof Note first that the norm in $L_2[a,b;r]$ is equivalent to the usual L_2 norm (Problem 4.34), and therefore $\mathscr{D}$ is dense in $L_2[a,b;r]$ if and only if it is dense with the L_2 norm (Problem 4.17). Now, in 4.51 we showed that the set of smooth functions vanishing at a and b is dense in $C[a,b]$. If the Sturm-Liouville boundary conditions are of the form $u(a) = u(b) = 0$, then all these functions belong to

$\mathscr{D}$, so $\mathscr{D}$ is dense in $C[a,b]$. The argument of 4.51 easily gives the stronger result that the set of all smooth functions u such that u and all its derivatives vanish at a and b is dense in $C[a,b]$. No matter what type of Sturm-Liouville boundary conditions we have, all such functions belong to $\mathscr{D}$, so $\mathscr{D}$ is dense in $C[a,b]$. And $C[a,b]$ is dense in L_2 (see 4.60), so $\mathscr{D}$ is dense in L_2 by Problem 4.15. □

We can now see that the Sturm-Liouville operator $L:\mathscr{D} \to C[a,b;r]$ is symmetric; 9.11 shows that its domain is dense, and equation (9.10) shows that the symmetry condition is satisfied when $u,v \in \mathscr{D}$. We shall now show that its inverse is the compact integral operator Γ whose kernel is Green's function; then we shall apply the spectral theorem to Γ and deduce that the eigenfunctions of L are complete.

Proposition 9.12 Let $L = \frac{1}{r}[DpD + q]$ be a Sturm-Liouville operator on the interval $[a,b]$, with domain space $\mathscr{D}$, and suppose that zero is not an eigenvalue of L. Define the operator Γ by

$$(\Gamma u)(x) = \int_a^b g(x;y)u(y)r(y)\,dy \quad ,$$

where g is Green's function for the operator $DpD + q$. Then Γ maps $C[a,b]$ on to $\mathscr{D}$, L maps $\mathscr{D}$ on to $C[a,b]$, and

$$L^{-1} = \Gamma: C[a,b] \to \mathscr{D} \quad .$$

Remarks 9.13 (a) In this section we shall frequently abbreviate $C[a,b;r]$ to C, etc. We also write it as $C[a,b]$ when the inner product is not important.

(b) $\Gamma: C \to \mathscr{D}$ denotes the integral operator regarded as acting on $C[a,b]$. It can be extended to the whole of L_2. As in 6.14, we shall write Γ' for the operator extended to L_2; thus $\Gamma'u = \Gamma u$ for all $u \in C$. Note that Γ' is not the inverse of L, because if $u \in L_2$ is singular, then $\Gamma'u$ is not in general twice-differentiable, and so $L(\Gamma'u)$ does not in general exist.

Proof of 9.12 We must show that L is invertible, that is, for any $v \in C[a,b]$ there is just one $w \in \mathscr{D}$ such that $v = Lw$. Take $w = \Gamma v$; we must show that $w \in \mathscr{D}$ and

$$Lw = L\Gamma v = v \quad . \tag{9.11}$$

There are two possible approaches. If the coefficients p,q,r are smooth, we can use distribution theory. It follows from 2.15 that w satisfies the boundary conditions and is a weak solution of (9.11); now 2.7 shows that w is a twice-differentiable classical solution. Distribution theory can be extended, along the lines suggested by Problem 2.11, to cover the case where p,q,r are not smooth but satisfy the conditions set out at the beginning of this section. Alternatively one can abandon the distributional approach, and proceed by straightforward analysis, as follows.

Given a continuous function v, we shall show that $\Gamma v \in \mathscr{D}$ and (9.11) is satisfied. We calculated Green's function in the proof of 2.13. In the notation of that proof we have

$$w(x) = (\Gamma v)(x) = Ku_2(x)\int_a^x u_1(y)v(y)r(y)\,dy + Ku_1(x)\int_x^b u_2(y)v(y)r(y)\,dy \quad . \tag{9.12}$$

Since the integrands in (9.12) are continuous, the integrals are differentiable, and $w \in C^1[a,b]$. Differentiating (9.12) gives

$$w'(x) = Ku_2'(x)\int_a^x u_1(y)v(y)r(y)\,dy + Ku_1'(x)\int_x^b u_2(y)v(y)r(y)\,dy$$

(the other two terms cancel out). Again it follows that w' is differentiable, and $w \in C^2[a,b]$. It follows from (9.12) that w satisfies the Sturm-Liouville boundary conditions, since u_1 satisfies the condition at a and u_2 satisfies the condition at b. Hence $w \in \mathscr{D}$. Finally, a simple calculation, using (2.31) and the fact that $Lu_1 = Lu_2 = 0$, shows that $Lw = v$.

To show uniqueness, suppose that $Lw_1 = Lw_2 = v$. Then $L(w_1 - w_2) = 0$, and since zero is not an eigenvalue of L we must have $w_1 = w_2$. This proves the uniqueness of w, and it follows that L is invertible. (9.11) now shows that $L^{-1} = \Gamma$. □

It follows that L has the same eigenfunctions as $\Gamma: C \to \mathscr{D}$. This operator is compact, because g is continuous and therefore square-integrable. Γ is symmetric because for any $u,v \in C$, $(u,\Gamma v) = (L\Gamma u, \Gamma v) = (\Gamma u, L\Gamma v) = (\Gamma u, v)$ (see 8.45 for another proof of the symmetry of Γ). However, Γ acts on the incomplete space $C[a,b]$, and we cannot apply Theorem 9.4. That theorem applies to the Hilbert-space operator $\Gamma': L_2 \to L_2$. Every eigenfunction of $\Gamma: C \to \mathscr{D}$ is obviously an eigenfunction of its extension Γ', but the converse is not obvious. It is conceivable that Γ' could have discontinuous eigenfunctions, in which case the eigenfunctions of L and Γ would be a proper subset of the eigenfunctions of Γ', and would therefore not be a complete set. We shall now show that this is not the case.

Lemma 9.14 If $u \in L_2$, then the function $\Gamma' u$ defined by

$$(\Gamma' u)(x) = \int_a^b g(x;y)u(y)r(y)\,dy$$

is continuous. In other words, Γ' maps L_2 into C.

Proof Set $v = \Gamma' u$. Then

$$|v(x) - v(\xi)| \leqslant \int |g(x;y) - g(\xi;y)|\sqrt{r(y)}\sqrt{r(y)}|u(y)|\,dy \leqslant \left[\int |g(x;y) - g(\xi;y)|^2 r(y)\,dy\right]^{\frac{1}{2}} \|u\| \quad , \tag{9.13}$$

using the Schwarz inequality. Now, g is continuous on the closed rectangle $a \leqslant x,y \leqslant b$, hence uniformly continuous by 6.37. So for any $\epsilon > 0$ there is a $\delta > 0$ such that

$$|g(x;y) - g(\xi;y)| < \epsilon / \|u\| \sqrt{(b-a)} \max\{r(x)\}$$

whenever $|x - \xi| < \delta$; (9.13) then gives $|v(x) - v(\xi)| < \epsilon$ whenever $|x - \xi| < \delta$, showing that v is continuous. □

It follows at once that eigenfunctions of Γ' are continuous, if the eigenvalue is nonzero, for $u = \lambda^{-1}\Gamma' u \in C$ by 9.14. The exceptional case $\lambda = 0$ is disposed of by the following result.

Lemma 9.15 The operator $\Gamma': L_2[a,b] \rightarrow C[a,b]$ defined in 9.14 is invertible.

Proof Proposition 8.35 says that Γ' is invertible if its range is dense in L_2. Now 9.12 shows that for any $u \in \mathscr{D}$, $\Gamma(Lu) = u$, therefore u belongs to the range of Γ. Thus $\mathscr{D}$ is contained in the range of Γ and therefore of its extension Γ'. But $\mathscr{D}$ is dense in L_2 by 9.11, hence so is the range of Γ', and the lemma follows from 8.35. □

Putting the above results together, we have the main theorem of this section.

Theorem 9.16 (Sturm-Liouville Expansions) The eigenfunctions of a regular Sturm-Liouville system form an orthogonal basis for $L_2[a,b;r]$, and the eigenvalues λ_n satisfy $|\lambda_n| \rightarrow \infty$.

Proof This theorem hinges on Proposition 9.12, which does not apply if zero is an eigenvalue of the given equation

$$(pu')' + [q - \lambda r]u = 0 \quad . \tag{9.14}$$

But for any constant c, the equation

$$(pu')' + [q + cr - \Lambda r]u = 0 \quad , \tag{9.15}$$

with $\Lambda = \lambda + c$, is completely equivalent to (9.14), and has eigenvalues Λ differing by c from those of (9.14). Choosing c so that $-c$ is not an eigenvalue of (9.14), we thus have a new Sturm-Liouville system (9.15), with the same eigenfunctions as (9.14), and not having zero as an eigenvalue.

Let L denote the operator $[\mathrm{D}p\mathrm{D} + (q + cr)]/r$. Then 9.12 shows that its inverse is the integral operator $\Gamma: \mathscr{D} \rightarrow C[a,b]$ whose kernel is the product of r and the Green's function for L. Let Γ' be the extension of this operator to L_2. In 8.45 we showed that Γ' is self-adjoint. Since its kernel is continuous, Γ' is compact. Hence its eigenfunctions are complete and orthogonal by 9.4.

If $\Gamma'\phi = \lambda\phi$ and $\phi \neq 0$, then $\lambda \neq 0$ by 9.15, hence $\phi = \lambda^{-1}\Gamma'\phi$ and $\phi \in C[a,b]$ by 9.14. Now, $\Gamma\phi = \Gamma'\phi$ for $\phi \in C[a,b]$, so $\phi = \lambda^{-1}\Gamma\phi \in \mathscr{D}$ by 9.12. Therefore $L\phi = \lambda^{-1}\phi$. Thus each eigenfunction of Γ', with eigenvalue λ, is an eigenfunction of L, with eigenvalue λ^{-1}, and the eigenfunctions of L are therefore complete and orthogonal. The eigenvalues of Γ' tend to zero by 9.4, so their inverses, the eigenvalues of L, satisfy $|\lambda_n| \to \infty$. The eigenvalues of the given system (9.14) differ by a constant from those of L, so they too satisfy $|\lambda_n| \to \infty$. □

Our proof of this theorem hinged on the existence of Green's function. We proved this in Chapter 2 for the case of separated end-point conditions. It is not difficult to extend the proof to the more general boundary conditions given in Appendix G, equations (5) and (6). Similar methods can be used for some singular equations for which $p(x)$ vanishes at one end of the interval; see Problem 9.8. But the general theory of singular Sturm-Liouville systems involves the spectral theorem for noncompact operators, which is beyond the scope of this book.

We note that the following version of Fourier's theorem is a direct consequence of Theorem 9.16.

Corollary 9.17 For any square-integrable function f, its Fourier series converges to f in the mean.

Proof Consider the sine series $\Sigma A_n \sin(n\pi x/a)$ on the interval $(0,a)$. The functions $\sin(n\pi x/a)$ are the eigenfunctions of the Sturm-Liouville system $-u'' = \lambda u$ with $u(0) = u(a) = 0$, so the result follows from 9.16. For the cosine series the result follows similarly from the system $-u'' = \lambda u$, $u'(0) = u'(a) = 0$, and the full-range series corresponds in the same way to the boundary conditions $u(0) = u(a)$, $u'(0) = u'(a)$. □

Notice that this result says nothing about pointwise convergence; the Fourier series need not converge pointwise to f in general. But of course it will if f is sufficiently well-behaved; see Problem 9.7.

Application 9.18 (Expansion of Green's Function) We can use Theorem 9.16 to derive an elegant representation of Green's function. Let (ϕ_n) be the orthogonal eigenfunctions of a Sturm-Liouville system, normalised with respect to the inner product (9.9). Since the eigenvalues are real, the eigenfunctions satisfy a real differential equation, with real boundary conditions, and therefore can be taken to be real.

For each fixed x, Green's function $g(x;y)$ is a square-integrable function of y, and can therefore be expanded as a series in $\phi_n(y)$ whose coefficients depend on x:

$$g(x;y) = \Sigma a_n(x)\phi_n(y) \quad .$$

The expansion coefficients a_n are determined by

$$a_n(x) = (g,\phi_n) = \int g(x;y)\phi_n(y)r(y)\,\mathrm{d}y$$
$$= \lambda_n^{-1}\phi_n(x)$$

since the ϕ_n are real eigenfunctions, with eigenvalues λ_n^{-1}, of the integral operator with kernel g. Hence

$$g(x;y) = \sum \frac{\phi_n(x)\phi_n(y)}{\lambda_n} \quad . \tag{9.16}$$

This elegant formula displays the symmetry of Green's function very clearly, and it also shows, in agreement with 2.13, that something goes wrong if zero is an eigenvalue of the differential equation. It gives a method of constructing Green's function without having to solve an inhomogeneous equation; the eigenfunctions ϕ_n and eigenvalues λ_n are found by solving the homogeneous equation.

If we use (9.16) to solve the equation $Lu = f$, where L is a Sturm-Liouville operator and f a given function, we have

$$u(x) = \int g(x;y)f(y)\,\mathrm{d}y = \Sigma\lambda_n^{-1}\phi_n(x)\int f(y)\phi_n(y)\,\mathrm{d}y \quad .$$

This result follows directly from the following general method for solving an inhomogeneous equation $Lu = f$, where L is an invertible operator with a complete orthogonal set of eigenvectors (e_n). Expand u as a series of the basis vectors (e_n), $u = \Sigma c_n e_n$; then

$$\Sigma\lambda_n c_n e_n = f = \Sigma(f,e_n)e_n \quad ,$$

$$\therefore \lambda_n c_n = (f,e_n) \quad \text{for all} \quad n \quad ,$$

$$\therefore u = \sum \frac{1}{\lambda_n}(f,e_n)e_n \quad .$$

This method must in general be justified, by considering the convergence of the series.

Finally, we remark that (9.16) leads to a representation of the delta function. Consider, for simplicity, the case $r(x) = 1$. Green's function satisfies the equation

$$(\mathrm{D}p\mathrm{D} + q)g = \delta \quad ,$$

and using the expression (9.16) for g gives

$$\delta(x - y) = \Sigma\phi_n(x)\phi_n(y) \quad , \tag{9.17}$$

a representation of the delta function in terms of any eigenfunction basis. In fact, (9.17) holds for any orthonormal basis (ϕ_n), as can easily be verified by showing that $\int [\Sigma\phi_n(x)\phi_n(y)]f(y)\,\mathrm{d}y = f(x)$ for any function f.

Of course, we have not *proved* (9.17); we derived it by a logically unsound mixture of classical analysis and the generalised function δ. We shall not pause to put (9.17) on a solid foundation, but just note that we have already given a distribution-theoretic proof of a special case of (9.17): replace x by $x - y$ in equation (1.19), and use the identity $\cos(x - y) = \cos x \cos y + \sin x \sin y$, and the result has the form of (9.17) where (ϕ_n) is the basis consisting of sine and cosine functions (see Problem 9.7).

9.3 Partial Differential Equations

The methods of section 9.2 can be applied to partial differential equations of elliptic type. We shall consider the Dirichlet problem for Laplace's equation in three dimensions; other equations and other boundary conditions can be discussed in a similar way.

Consider, then, the eigenvalue problem for the operator $-\nabla^2$ on the region V inside a well-behaved surface S (for the meaning of this phrase, see the discussion following 3.32), with the boundary condition that the functions vanish on S. Write $C^2[V]$ for the space of $\mathbb{R}^3 \to \mathbb{C}$ functions twice continuously differentiable on V, with inner product

$$(u,v) = \int_V u(x)\overline{v(x)}\,d^3x \quad . \tag{9.18}$$

$C^2[V]$ is incomplete; the corresponding Hilbert space is the space $L_2[V]$ of functions square-integrable over V. Corresponding to Definition 9.10, we have

Definition 9.19 The domain for the operator $-\nabla^2$ with Dirichlet conditions on a surface S is $\mathscr{D} = \{f \in C^2[V] : f(x) = 0 \text{ for } x \text{ on } S\}$.

Proposition 9.20 The domain $\mathscr{D}$ defined in 9.19 is a dense vector subspace of $L_2[V]$.

Proof If $f,g \in \mathscr{D}$, then clearly $af + bg \in \mathscr{D}$ for any numbers a and b, so $\mathscr{D}$ is a vector subspace. The fact that it is dense follows at once from 9.23 below, which shows that a certain subset of $\mathscr{D}$ is dense. □

The proof of 9.20 depends upon 9.23, which is a three-dimensional version of 4.51. The details are more complicated, because of the richer geometry of three-dimensional space, and if you are willing to believe 9.20 without a complete proof, you may omit 9.21–9.23. However, 9.23 will be used again in 9.27 and in Chapter 12.

Definition 9.21 The **distance of a point x from a set $S \subset \mathbb{R}^3$** is $d(x) = \inf\{|x - y| : y \in S\}$. A function f: $\mathbb{R}^3 \to \mathbb{C}$ is said to **vanish near** S if there is a $\delta > 0$ such that $f(x) = 0$ whenever $d(x) < \delta$. □

Lemma 9.22 The distance function $d(x)$ defined in 9.21 is continuous.

Proof By Problem 4.6, for any x, y, ξ we have

$$\big| |x-y| - |\xi - y| \big| \leqslant |x - \xi| \quad ,$$

$$\therefore \big| \inf\{|x-y| : y \in S\} - \inf\{|\xi - y| : y \in S\} \big| \leqslant |x - \xi| \quad ,$$

$$\therefore |d(x) - d(\xi)| \leqslant |x - y| \quad .$$

Hence the definition of continuity is satisfied with $\delta = \epsilon$. □

Proposition 9.23 Let $C_0^\infty(V)$ be the set of smooth functions on the bounded region $V \subset \mathbb{R}^3$ which vanish near the boundary S of V. Then $C_0^\infty(V)$ is dense in $L_2[V]$.

Proof We begin by defining a mollifier in three-space. For any $\alpha > 0$ let M_α be a smooth $\mathbb{R}^3 \to \mathbb{R}$ function such that $M_\alpha(x) \geqslant 0$ for all x, $M_\alpha(x) = 0$ for $|x| \geqslant \alpha$, and $\int_{\Sigma_\alpha} M_\alpha(x)\,\mathrm{d}^3x = 1$, where

$$\Sigma_\alpha = \{x : |x| \leqslant \alpha\}$$

is a sphere of radius α. For example, one can take $M_\alpha(x) = Nm_\alpha(|x|)$, where m_α is the one-dimensional mollifier defined in 4.51, and N is a suitable constant, chosen to make $\int M_\alpha \mathrm{d}^3x = 1$. Now, for any continuous function f, we define a smoothed version $(f)_\alpha$ of f as follows:

$$(f)_\alpha(x) = \int_{\Sigma_\alpha} f(x-y) M_\alpha(y)\,\mathrm{d}^3y \quad . \tag{9.19}$$

Note that when x is within a distance α of the boundary of V, this expression involves values of f at points outside V; we assume that f is defined and continuous everywhere. An argument identical with the proof of 4.51(ii) shows that $(f)_\alpha$ is smooth, and that

$$\|(f)_\alpha - f\| \to 0 \quad \text{as} \quad \alpha \to 0 \quad .$$

We shall now show how to modify a given function so as to bring it smoothly to zero near the boundary S of V. For any $\beta > 0$, define a function ν_β by

$$\nu_\beta(x) = \begin{cases} 1 \quad \text{if} \quad d(x) \geqslant 3\beta \quad , \\ [d(x) - 2\beta]/\beta \quad \text{if} \quad 2\beta \leqslant d(x) \leqslant 3\beta \quad , \\ 0 \quad \text{if} \quad d(x) \leqslant 2\beta \quad , \end{cases}$$

where $d(x)$ is the distance of x from S. Then $\nu_\beta \in C[V]$ because d is continuous by 9.22. For any $f \in C[V]$, define

$$f_\beta(x) = \nu_\beta(x) f(x) \quad .$$

Then $f_\beta \in C[V]$, and f_β is zero within a distance 2β of S. We have

$$\|f_\beta - f\|^2 = \int_V \{\nu_\beta - 1\}^2 |f(\boldsymbol{x})|^2 \,\mathrm{d}^3\boldsymbol{x}$$

$$\leqslant B^2 \int_V \{\nu_\beta - 1\}^2 \,\mathrm{d}^3\boldsymbol{x}$$

where $B = \max\{|f(\boldsymbol{x})| : \boldsymbol{x} \in V\}$. Hence

$$\|f_\beta - f\|^2 \leqslant B^2 \int_W \mathrm{d}^3x \quad ,$$

where $W = \{x : d(x) \leqslant 3\beta\}$ is a layer of thickness 3β adjacent to S. If A is the area of S, then the volume of this layer will be approximately $3A\beta$, hence less than $4A\beta$, say, for sufficiently small β. Therefore

$$\|f_\beta - f\|^2 \leqslant B^2 4\beta A \quad .$$

Now, f_β vanishes on S, but it is not in general differentiable. We now apply the mollifier defined above: let

$$\phi(\boldsymbol{x}) = \int_{\Sigma_\alpha} f_\beta(\boldsymbol{x} - \boldsymbol{y}) M_\alpha(\boldsymbol{y}) \,\mathrm{d}^3\boldsymbol{y} \quad . \tag{9.20}$$

Note that since $f_\beta(\boldsymbol{x}) = 0$ on S, we can define it to vanish outside V, and f_β is then continuous everywhere. $f_\beta(\boldsymbol{x}) = 0$ when $d(\boldsymbol{x}) < 2\beta$, hence $f_\beta(\boldsymbol{x} - \boldsymbol{y}) = 0$ whenever $d(\boldsymbol{x}) < \beta$ and $|\boldsymbol{y}| < \beta$. Hence if $\alpha < \beta$, then the integrand in (9.20) vanishes whenever $d(\boldsymbol{x}) < \beta$, and we have

$$\phi(\boldsymbol{x}) = 0 \quad \text{when} \quad d(\boldsymbol{x}) < \beta \quad ,$$

if $\alpha < \beta$. We showed above that ϕ is smooth, hence $\phi \in C_0^\infty(V)$. We showed also that $\|\phi - f_\beta\| \to 0$ as $\alpha \to 0$, hence there is an α_0 such that $\|\phi - f_\beta\| < 2B\sqrt{A\beta}$ if $\alpha < \alpha_0$. Taking $\alpha = \frac{1}{2}\min(\alpha_0, \beta)$, we therefore have

$$\|\phi - f\| < 4B\sqrt{A\beta} \quad .$$

So for any $\epsilon > 0$ we can choose $\beta = \epsilon^2/16AB^2$; then the above construction gives a $\phi \in C_0^\infty(V)$ with $\|\phi - f\| < \epsilon$, showing that $C_0^\infty(V)$ is dense in $C[V]$.

Finally, $C[V]$ is dense in $L_2[V]$, because L_2 is the completion of C, so it follows that $C_0^\infty(V)$ is dense in L_2. □

We now return to the operator $-\nabla^2 : \mathscr{D} \to C[V]$. The domain $\mathscr{D}$ is dense in $L_2[V]$ by 9.20, and for any $u, v \in \mathscr{D}$,

$$\int_V u \nabla^2 \bar{v} \,\mathrm{d}^3\boldsymbol{x} = \int u \boldsymbol{\nabla} \cdot (\boldsymbol{\nabla} \bar{v}) \,\mathrm{d}^3\boldsymbol{x}$$

$$= \int [\boldsymbol{\nabla} \cdot (u \boldsymbol{\nabla} \bar{v}) - (\boldsymbol{\nabla} u) \cdot (\boldsymbol{\nabla} \bar{v})] \,\mathrm{d}^3\boldsymbol{x}$$

$$= \int [\boldsymbol{\nabla} \cdot (u \boldsymbol{\nabla} \bar{v} - \bar{v} \boldsymbol{\nabla} u) + (\nabla^2 u) \bar{v}] \,\mathrm{d}^3\boldsymbol{x}$$

$$= \int (\nabla^2 u) \bar{v} \,\mathrm{d}^3\boldsymbol{x} \quad ,$$

using the divergence theorem and the boundary condition. Thus $-\nabla^2$ is symmetric, and therefore has real eigenvalues and orthogonal eigenfunctions.

It follows from Theorem 3.32 that zero is not an eigenvalue of $-\nabla^2$. *Proof:* the eigenvalue equation with $\lambda = 0$ is just Laplace's equation, with the boundary condition of vanishing on S; the zero function is obviously a solution, therefore it is the only solution by Theorem 3.32, so there are no eigenfunctions for eigenvalue zero. By the same reasoning as in section 9.2 it follows that the operator $-\nabla^2$ has an inverse, K, say, which is an integral operator whose kernel is Green's function. Corollary 5.21 shows that this operator extends in a natural way to the whole of $L_2[V]$, just as in section 9.2 Γ extends to $\Gamma': L_2[a,b] \to L_2[a,b]$. We shall here denote the integral operator by K, no matter what domain it acts on. K is compact by 5.20, since the compactness criterion 8.54 extends easily to three dimensions. Green's function is real and symmetric, so K is a compact self-adjoint operator. The spectral theorem says that K has an orthogonal set of eigenfunctions forming a basis for $L_2[V]$, with a sequence of eigenvalues tending to zero. To show that $-\nabla^2$ has a complete set of eigenfunctions, we must show that all eigenfunctions of K are eigenfunctions of $-\nabla^2$.

The chain of reasoning is similar to that of the last section, but the details are more complicated because of the singularity in the Green's function. The proofs of the following two results are therefore relegated to an appendix.

Lemma 9.24 Let K be the integral operator on $L_2[V]$ whose kernel is the Dirichlet-problem Green's function for the region $V \subset \mathbb{R}^3$ inside a closed surface S.

(a) If $u \in L_2[V]$, then Ku is continuous on V.

(b) If u is continuous on V, then Ku is continuously differentiable.

Proposition 9.25 In the notation of 9.24, if u is continuously differentiable on V, then the function $v = Ku$ is twice-differentiable, satisfies $\nabla^2 v = -u$, and vanishes on S.

Proofs See Appendix I. □

Proposition 9.26 In the notation of 9.24, $u = \lambda Ku$ if and only if $\nabla^2 u = -\lambda u$ and $u = 0$ on S.

Proof (a) If $u = \lambda Ku$, then $u \in C[V]$ by 9.24(a). Therefore $u \in C^1[V]$ (the space of continuously differentiable functions on V) by 9.24(b). Hence $u = \lambda Ku$ is twice-differentiable by 9.25, and $\nabla^2 u = \nabla^2(\lambda Ku) = -\lambda u$, and $u = 0$ on S, by 9.25.

(b) Suppose $\nabla^2 u = -\lambda u$ and u vanishes on S. Then the function v defined by

$$v = \lambda Ku \tag{9.21}$$

satisfies $\nabla^2 v = -\lambda u$ and vanishes on S by 9.25. Hence $\nabla^2(v - u) = 0$, and $v - u$ vanishes on S, hence $v = u$ by 3.32. Hence (9.21) becomes $u = \lambda Ku$, as required. □

This result shows that the nonzero eigenvalues of K are the reciprocals of the eigenvalues of $-\nabla^2$, and the eigenfunctions are the same. To complete the proof of the equivalence of the two eigenvalue problems we must show that zero is not an eigenvalue of K. This needs the following result.

Proposition 9.27 The operator K defined in 9.24 is invertible.

Proof Proposition 8.35 shows that K is invertible if its range is dense in $L_2[V]$. We shall show that the range contains the subspace $C_0^\infty(V)$ defined in 9.23. Take any $\phi \in C_0^\infty(V)$, and set $u = -\nabla^2\phi$. Then $u \in C_0^\infty(V)$, hence $Ku \in \mathscr{D}$ and $\nabla^2(Ku) = -u$ by 9.25. The uniqueness theorem for Laplace's equation now shows that $\phi = Ku$. Thus each $\phi \in C_0^\infty(V)$ is in the range of K, so the range is dense in $L_2[V]$ by 9.23. Hence K is invertible by 8.35. □

Theorem 9.28 (Expansion Theorem for Laplace's Equation) Let $V \subset \mathbb{R}^3$ be the interior of a well-behaved closed surface S. The equation

$$-\nabla^2 u = \lambda u \text{ in } V, \text{ with } u = 0 \text{ on } S \quad ,$$

has an infinite sequence of eigenfunctions forming an orthogonal basis for $L_2[V]$, and the corresponding eigenvalues λ_n satisfy $|\lambda_n| \to \infty$ as $n \to \infty$.

Remark: for the meaning of 'well-behaved' see the discussion following 3.32. Some such condition is needed to ensure the existence of Green's function.

Proof Proposition 9.26 shows that every eigenfunction of $-\nabla^2$ is an eigenfunction of the integral operator K, and that every eigenfunction of K corresponding to a nonzero eigenvalue (that is, $\lambda^{-1} \neq 0$ in the notation of 9.26) is an eigenfunction of $-\nabla^2$. Proposition 9.27 shows that zero is not an eigenvalue of K, so in fact all eigenfunctions of K are eigenfunctions of $-\nabla^2$.

The kernel g of the operator K is real and symmetric (Theorem 3.33), hence K is self-adjoint. We have shown that g is square-integrable, hence K is compact by the three-dimensional version of Theorem 8.54. The spectral theorem now shows that the eigenfunctions of K form an orthogonal basis, and we showed above that they are the same as the eigenfunctions of $-\nabla^2$, hence the latter are complete. Finally, the λ_n are the reciprocals of the eigenvalues of K, which tend to zero by Theorem 9.4, hence $|\lambda_n| \to \infty$. □

Results of this type for ordinary and partial differential equations form the basis of the method of solving partial differential equations by separation of variables. For example, given the wave equation

$$u_{tt} = \nabla^2 u$$

on the region $V \subset \mathbb{R}^3$, we can use Theorem 9.28 to write

$$u(x,t) = \Sigma c_n(t)\phi_n(x)$$

where ϕ_n are the eigenfunctions of $-\nabla^2$. The wave equation then gives

$$d^2 c_n/dt^2 + \lambda_n c_n = 0 \quad ,$$

and the problem has been separated into an ordinary differential equation in the time variable, and a partial differential equation in the space variables. Similarly, the space problem can sometimes be separated into three one-dimensional problems, using the Sturm-Liouville expansion theorem. For detailed accounts of this technique, see Kreider *et al.*, or books on partial differential equations.

9.4 The Fredholm Alternative

The spectral theorem of section 9.1 shows that self-adjoint compact operators in Hilbert space are in some ways similar to symmetric matrices in $\mathbb{R}^n$. This section discusses another aspect of this similarity. Theorem 9.29 is a simple generalisation of facts about sets of linear algebraic equations: an inhomogeneous system has a unique solution if and only if the determinant of the coefficients is nonzero, that is, if the corresponding homogeneous system has no nonzero solution. This statement was extended to integral equations by I. Fredholm, and was later generalised to Hilbert-space operators. It is called 'The Fredholm Alternative' because of the two alternative possibilities (i) and (ii) below.

Theorem 9.29 (Fredholm Alternative for Self-Adjoint Operators) Let A be a self-adjoint compact linear operator on a Hilbert space H. Consider the two equations

$$x = Ax + f \quad , \tag{I}$$

$$x = Ax \quad , \tag{H}$$

where f is a given element of H; (I) and (H) denote the inhomogeneous and homogeneous equations respectively.

(i) If the only solution of (H) is $x = 0$, then (I) has a unique solution x for each $f \in H$.

(ii) If (H) has nonzero solutions, then (I) has a solution only if f is orthogonal to every solution of (H), in which case (I) has infinitely many solutions, the difference between any two of them being a solution of (H).

Proof (i) By Theorem 9.4, H has an orthonormal basis (e_n) consisting of eigenvectors of A with eigenvalues λ_n. Let

$$f = \Sigma c_n e_n \quad .$$

Look for a solution of (I) in the form $x = \Sigma a_n e_n$; then

$$\Sigma a_n e_n = \Sigma a_n \lambda_n e_n + \Sigma c_n e_n$$

$$\therefore a_n = c_n/(1 - \lambda_n) \tag{9.22}$$

for all n, provided that $\lambda_n \neq 1$. We are given that the equation $Ax = x$ has no nonzero solution, therefore 1 is not an eigenvalue of A, and (9.22) is justified. Hence if (I) has a solution, it must be of the form

$$\sum \frac{c_n e_n}{1 - \lambda_n} \quad . \tag{9.23}$$

This proves that if (I) has a solution, it is unique. We shall now show that the series (9.23) always converges and gives a solution of (I).

If 1 were a limit point (Definition 4.41) of the set of eigenvalues of A, then (9.23) would not in general converge because its terms would be very large when λ_n was close to 1. However, we showed in section 9.1 that $\lambda_n \to 0$ as $n \to \infty$, hence no subsequence of (λ_n) can tend to 1, so 1 is not a limit point of $\{\lambda_n\}$ and there is an eigenvalue λ_p which is closer to 1 than any other eigenvalue. Then

$$\sum \left| \frac{c_n}{1 - \lambda_n} \right|^2 \leqslant \frac{1}{|1 - \lambda_p|^2} \Sigma |c_n|^2 \quad ,$$

and $\Sigma |c_n|^2$ converges, hence $\Sigma |c_n/(1 - \lambda_n)|^2$ converges, hence (9.23) converges by the Riesz-Fischer theorem. Using the last clause of Theorem 9.4 it is easy to verify that (9.23) is a solution of equation (I).

(ii) Suppose now that x satisfies (I) and $u \neq 0$ is a solution of (H). Then the inner product of (I) with u gives

$$\begin{aligned}(x,u) &= (Ax,u) + (f,u) \\ &= (x,Au) + (f,u)\end{aligned}$$

since A is self-adjoint. But $u = Au$; hence $(f,u) = 0$. Thus if (I) has a solution, then f must be orthogonal to every solution of (H). If x satisfies (I), then $x + u$ also satisfies (I) for any u satisfying (H); since cu also satisfies (H) for any scalar c, there are infinitely many such solutions. Finally, it is simple to verify that if x and y satisfy (I), then $(x - y)$ satisfies (H). □

Remark 9.30 The form in which the equations (I) and (H) are written should be noted. In the theory of algebraic equations it is usual and natural to write the equations in the form

$$Bx = f \quad , \tag{I'}$$

$$Bx = 0 \quad , \tag{H'}$$

and the Fredholm alternative applies to (I′) and (H′) if B is a symmetric matrix. But we shall show below that the alternative does not apply to (I′) and (H′) when B is a self-adjoint compact operator in Hilbert space. It applies only when B differs from the identity by a compact operator, that is, when $B = I - A$ where A is compact, in which case (I′) and (H′) become (I) and (H). Note that if A is compact, then $B = I - A$ is not compact, unless the space is finite-dimensional.

Example 9.31 Let B be the integral operator inverse to the Sturm-Liouville operator $-D^2$ with boundary conditions $u(0) = u(\pi) = 0$. Then B is compact and self-adjoint, and its eigenvalues are n^{-2} for $n = 1,2,\ldots,$ the reciprocals of the eigenvalues of $-u'' = \lambda u$. We shall show that the Fredholm alternative does not apply to (I′), (H′) for this operator.

Zero is not an eigenvalue of B, hence (H′) has no non-zero solution. We shall produce an f such that (I′) has no solution either, thus violating clause (ii) of the Fredholm alternative. Let (e_n) be an orthonormal sequence of eigenfunctions of B, and take $f = \Sigma n^{-1} e_n$; the series converges by the Riesz-Fischer theorem. Suppose $x = \Sigma c_n e_n$ is a solution of (I′); substituting in (I′) gives $c_n = 1/(nn^{-2}) = n$. Hence $\Sigma |c_n|^2$ diverges, contradicting Parseval's relation. Thus there is no solution of (I′). □

The following example shows that the compactness of A in Theorem 9.29 is essential.

Example 9.32 Let $A : L_2[0,1] \to L_2[0,1]$ be the operator which multiplies any function $x(t)$ by $(1 - t)$. It is easily seen to be self-adjoint, but not compact. The homogeneous equation $x(t) = (1 - t)x(t)$ has no solution other than $x(t) = 0$. Yet the inhomogeneous equation has no solution in $L_2[0,1]$ when f is a constant function; the solution of $x = Ax + f$ is then proportional to $1/t$, which is not square-integrable, and therefore not a member of our space. □

The burden of the above discussion is that although the Fredholm Alternative is so simple to state, being word for word the same as in elementary linear algebra, yet the compactness condition is essential, and the theorem therefore necessarily involves all the subtle considerations which have occupied us in this and the previous chapters. The self-adjointness condition, however, is not essential. The following version of the theorem applies to non-self-adjoint operators.

Theorem 9.33 (Fredholm Alternative for Non-Self-Adjoint Operators) Let A be a compact operator on a Hilbert space H. Consider the four equations

$$x = Ax + f \quad , \tag{I}$$

$$y = A^* y + g \quad , \tag{I*}$$

$$x = Ax \quad , \tag{H}$$

$$y = A^*y \quad , \tag{H*}$$

where f,g are given elements of H. Then *either*

(i) the only solutions of (H) and (H*) are zero, in which case (I) and (I*) have unique solutions for any $f,g \in H$; *or*

(ii) there are nonzero solutions of (H) and of (H*), in which case (I) has a solution if and only if f is orthogonal to every solution of (H*), and (I*) has a solution if and only if g is orthogonal to every solution of (H). □

We are not in a position to prove this theorem, because the spectral theorem which we used to prove 9.29 does not apply to non-self-adjoint operators. The proof of 9.33 needs a different approach; see Kolmogorov & Fomin (1957), for example.

Some parts of the theorem can be proved easily, however. In particular, the statement that the equation (I) has a solution only if f is orthogonal to every solution of (H*) is easily proved by taking the inner product of (I) with a solution of (H*). This orthogonality condition is sometimes called a 'solubility condition' for (I), for obvious reasons; the argument is independent of compactness considerations, and the condition applies also to non-compact operators such as differential operators. For a physical interpretation, see Problem 9.12. For an example of the use of the Fredholm Alternative in applied mathematics, see section 11.7.

9.5 Projection Operators

We shall now look at the spectral theorem from a more geometrical point of view, using the idea of orthogonal projection discussed in Chapter 7. The beginning of this section is really a continuation of section 7.5; we define projection operators and discuss their properties. We then use them to reformulate the spectral theorem, and we end the chapter by using projections to define functions of an operator. We shall use all this machinery in the next chapter, when we discuss the theory of the Rayleigh-Ritz approximation method.

We begin with a reminder of the result 7.47: if S is a subspace of H, then for any $x \in H$ there is a unique $y \in S$ and $z \perp S$ such that $x = y + z$. The vector y is called the projection of x on to S.

Definition 9.34 Let S be a subspace of the Hilbert space H. The operator $P: H \to S$ defined by $Px = y$, where y is the projection of x on to S, is called the **projection operator on to** S. S is the **range** of the projection P (cf. 5.2). □

Theorem 7.47 ensures the existence and uniqueness of y, and thus justifies this definition.

Examples 9.35 (a) Let (e_n) be an orthonormal basis for H. For any fixed r, the set of all scalar multiples of e_r is a one-dimensional subspace of H; call this subspace S_r. For any $x \in H$,

$$x = (x,e_r)e_r + \sum_{n \neq r}(x,e_n)e_n$$

with $(x,e_r)e_r \in S_r$ and $\Sigma_{n \neq r}(x,e_n)e_n \perp S_r$. Hence $(x,e_r)e_r$ is the projection of x on to S_r, and the projection operator P_r on to S_r is given by

$$P_r : x \mapsto (x,e_r)e_r \quad .$$

Since the range of P_r is one-dimensional, P_r has rank 1.

(b) If S is the $(m-n+1)$-dimensional subspace generated by $\{e_n, e_{n+1}, \ldots, e_m\}$, then for any $x \in H$ we write

$$x = \sum_{r=n}^{m}(x,e_r)e_r + \left\{\sum_{r=1}^{n-1}(x,e_r)e_r + \sum_{r=m+1}^{\infty}(x,e_r)e_r\right\} \quad .$$

We deduce as above that $\Sigma_n^m(x,e_r)e_r$ is the projection of x on to S, so the projection operator $P_{n,m}$ on to S is given by

$$P_{n,m} : x \mapsto \sum_{n}^{m}(x,e_r)e_r \quad .$$

We clearly have

$$P_{n,m} = \sum_{n}^{m} P_r \quad .$$

(c) The zero operator is a trivial example of a projection. It is the projection on to the trivial subspace $\{0\}$.

Lemma 9.36 If P is a projection operator with range $R(P)$, and $x \perp R(P)$, then $Px = 0$.

Proof We have $x = 0 + x$ where $0 \in R(P)$ and $x \perp R(P)$. Hence the projection of x on to $R(P)$ is zero, so $Px = 0$. □

Lemma 9.36 shows that we can describe the projection operator on to S as a linear operator which is the identity when acting on S and the zero operator when acting on $S^{\perp}$. The following result lists some other properties of projections.

Proposition 9.37 A projection operator P is linear, bounded, self-adjoint, and satisfies $P^2 = P$. If $P \neq 0$, then $\|P\| = 1$.

Proof (a) Linearity: We write $x_i = y_i + z_i$, with $y_i \in S$ and $z_i \perp S$, for $i = 1,2$, where S is the range of P. For any x_1 and x_2,

$$x_1 + x_2 = (y_1 + y_2) + (z_1 + z_2)$$

is the orthogonal decomposition of $x_1 + x_2$; thus

$$P(x_1 + x_2) = y_1 + y_2 = Px_1 + Px_2 \quad .$$

Again, for any scalar a, $ax_1 = ay_1 + az_1$, hence

$$P(ax_1) = aPx_1 \quad .$$

We have now shown that P is linear.

(b) Boundedness: If $x = y + z$ with $y \in S$, $z \perp S$, then

$$\|x\|^2 = \|y\|^2 + \|z\|^2$$

by 7.12,

$$\therefore \|Px\|^2 = \|y\|^2 = \|x\|^2 - \|z\|^2$$
$$\leqslant \|x\|^2 \quad .$$

Thus P is bounded, and

$$\|P\| \leqslant 1 \quad . \tag{9.24}$$

(c) Self-adjointness: With the notation of (a) above,

$$\begin{aligned}(x_1, Px_2) &= (y_1 + z_1, y_2)\\ &= (y_1, y_2)\\ &= (y_1, y_2 + z_2)\\ &= (Px_1, x_2)\end{aligned}$$

for any x_1, x_2. Hence P is self-adjoint.

(d) For any x, $Px \in S$, therefore $P(Px) = Px$. Thus $P^2 = P$. Now, $\|P\| = \|P^2\| \leqslant \|P\|^2$ by Proposition 8.15. Hence either $\|P\| = 0$ or

$$1 \leqslant \|P\|$$

which, combined with (9.24), gives $\|P\| = 1$. □

The following is a converse to Proposition 9.37.

Proposition 9.38 If P is a bounded self-adjoint linear operator with $P^2 = P$, then P is a projection.

Proof Set $S = R(P) = \{Px : x \in H\}$. We shall show that P acting on S is the identity and P acting on $S^\perp$ is the zero operator, and thus that P is the projection on to S.

If $y \in S$, then $y = Pu$ for some $u \in H$. Therefore

$$Py = P^2u = Pu \quad .$$

$$\therefore Py = y \quad \text{if} \quad y \in S \quad . \tag{9.25}$$

If $z \in S^{\perp}$, then for any $v \in H$,

$$(v,Pz) = (Pv,z) = 0$$

because $Pv \in S$ and $z \perp S$. Thus

$$(v,Pz) = 0 \quad \text{for all} \quad v \in H \quad ,$$

which implies that

$$Pz = 0 \quad \text{if} \quad z \in S^{\perp} \quad . \tag{9.26}$$

Now, for any $x \in H$, write

$$x = y + z$$

with $y \in S$, $z \in S^{\perp}$; then

$$Px = Py + Pz = y$$

using (9.25) and (9.26). Thus P is the projection on to S. □

It is easy to see that the sum of two projection operators is not in general a projection. Indeed, it follows from Proposition 9.37 that $\|P + P\| = 2\|P\| = 2$, so that $P + P$ cannot be a projection. However, if P and Q are orthogonal projections, in a sense to be defined below, then $P + Q$ will be a projection operator. We first define the idea of orthogonality of subspaces.

Definition 9.39 Two subspaces S and T of a Hilbert space are said to be **orthogonal** to each other if $(s,t) = 0$ for all $s \in S$ and all $t \in T$. □

This is equivalent to the statement $s \perp T$ for all $s \in S$, or $t \perp S$ for all $t \in T$. We think of subspaces as generalisations of lines and planes in $\mathbb{R}^3$; lines and planes which are perpendicular are orthogonal in the sense of 9.39.

Definition 9.40 Two projections P and Q in a Hilbert space H are **orthogonal** to each other if their ranges are orthogonal subspaces of H. A set $\{P_n\}$ of projections is called an **orthogonal set of projections** if (a) P_i is orthogonal to P_j for all $i \neq j$, and (b) $P_i \neq 0$ for each i.

Example 9.41 As in Example 9.35, let (e_n) be an orthonormal basis for H, and for each r let P_r be the projection on to the one-dimensional subspace E_r spanned by e_r. If $r \neq n$, the spaces E_n and E_r are orthogonal, and the operators P_n and P_r are orthogonal. $\{P_n\}$ is an orthogonal set of projections.

Lemma 9.42 If P and Q are orthogonal projections, then $PQ = 0$.

Proof For any $x \in H$, $Qx \in R(Q)$, therefore $Qx \perp R(P)$, hence $PQx = 0$ by Lemma 9.36. Thus PQ maps every vector into zero. □

This result bears a striking resemblance to the condition $(p,q) = 0$ for orthogonality of two vectors.

Proposition 9.43 (Sums of Projections) If $\{P_1, \ldots P_n\}$ is a finite orthogonal set of projections, then $\Sigma_{r=1}^{n} P_r$ is a projection.

Proof ΣP_r is linear, bounded, and self-adjoint because each P_r is. Using 9.42 we have

$$(\sum_r P_r)^2 = \sum_{r,s} P_r P_s = \sum_r P_r^2 = \sum_r P_r \quad .$$

Hence ΣP_r is a projection by Proposition 9.38. □

Example 9.35(b) illustrates this result.

The main idea of this section is that orthogonal sets of projections can be used to construct other operators, rather as orthogonal basis vectors are used to construct other vectors. We begin by expressing in the language of projections the familiar fact that any vector in a Hilbert space H can be expressed as a series of orthonormal basis vectors (e_n).

For any fixed n, let P_n be the projection on to the (one-dimensional) subspace spanned by e_n. Then, as we have seen in Example 9.35,

$$P_n x = (x, e_n) e_n \quad .$$

Hence for any $x \in H$ we have

$$\sum_{n=1}^{\infty} P_n x = x \quad . \tag{9.27}$$

This equation does not imply that $\Sigma_1^\infty P_n = I$. In fact, the series of operators ΣP_n does not converge according to our definition of operator convergence, because $\| P_n \| = 1$ for all n (by 9.37), whereas if ΣP_n converged we would have $\| P_n \| \to 0$ as $n \to \infty$.† However, although ΣP_n does not converge, we shall now show that $\Sigma c_n P_n$ converges if (c_n) is a sequence of scalars which tends to zero. This result is a stepping-stone to our real objective, Corollary 9.45, which says that every compact self-adjoint operator can be expressed in the form $\Sigma c_n P_n$, where c_n are its eigenvalues.

† Many authors define three kinds of operator convergence, called uniform, strong, and weak operator convergence, and say that ΣP_n converges strongly but not uniformly to I.

Theorem 9.44 (Weighted Sums of Projections) Let (P_n) be an orthogonal sequence of projection operators, and (c_n) a sequence of numbers with $c_n \to 0$ as $n \to \infty$. Then

(a) $\Sigma_1^\infty c_n P_n$ converges;

(b) For each n, c_n is an eigenvalue of the operator $A = \Sigma c_n P_n$, and the only other possible eigenvalue is zero;

(c) If all the c_n are real, then A is self-adjoint;

(d) If each P_n has finite rank, then A is compact.

Proof (a) We use the Cauchy criterion. We have

$$\left\| \sum_{r=n}^{m} c_r P_r \right\| = \sup \left\{ \left\| \sum_n^m c_r P_r x \right\| : \|x\| = 1 \right\} \quad , \tag{9.28}$$

and because the vectors $c_r P_r x$ are orthogonal,

$$\left\| \sum_n^m c_r P_r x \right\|^2 = \sum_n^m \| c_r P_r x \|^2 = \sum_n^m |c_r|^2 \, \| P_r x \|^2 \quad .$$

Now, $c_n \to 0$, hence for any $\epsilon > 0$ there is an N such that $|c_r| < \epsilon$ for $r > N$. Hence for $n > N$ we have

$$\left\| \sum_n^m c_r P_r x \right\|^2 < \epsilon^2 \sum_n^m \| P_r x \|^2 = \epsilon^2 \left\| \sum_n^m P_r x \right\|^2$$

because the vectors $P_r x$ are orthogonal,

$$\leqslant \epsilon^2 \left\| \sum_n^m P_r \right\|^2 \| x \|^2 \quad .$$

Now, $\Sigma_n^m P_r$ is a projection by 9.43, and therefore has norm 1. Hence

$$\left\| \sum_n^m c_r P_r x \right\|^2 < \epsilon^2 \| x \|^2 \quad .$$

It now follows from (9.28) that

$$\left\| \sum_n^m c_r P_r \right\| < \epsilon \quad \text{for} \quad n > N \quad ;$$

thus the sequence of partial sums $\Sigma_1^n c_r P_r$ is a Cauchy sequence, and therefore converges.

(b) We first remark that A is a bounded operator. This follows from Theorem 8.14, since A is the limit of the sequence of partial sums of $\Sigma c_n P_n$.

Now, if $x \in R(P_n)$, then $P_n x = x$, and $P_r x = 0$ if $r \neq n$ because the P_n are orthogonal. Hence $Ax = c_n x$, and c_n is an eigenvalue of A.

To show that there can be no other eigenvalue apart from zero, let x be any eigenvector with eigenvalue λ. Set $y_r = P_r x$ for $r = 1, 2, \ldots$, and $z = Qx$, where Q is the projection on to the subspace of vectors orthogonal to all the $R(P_r)$. Then

$$x = \sum_1^{\infty} y_r + z \tag{9.29}$$

with $z \perp R(P_r)$ for all r. Now,

$$A(\Sigma y_r + z) = A\Sigma y_r$$

because $P_r z = 0$ for all r,

$$= \Sigma A y_r$$

because A is bounded, as proved above, and therefore continuous. Hence the eigenvalue equation reads

$$Ax = \Sigma c_r y_r = \lambda(\Sigma y_r + z) \quad ,$$

or

$$\Sigma(\lambda - c_r)y_r + \lambda z = 0 \quad . \tag{9.30}$$

But the sum of an orthogonal set of vectors can vanish only if each term vanishes; hence $\lambda z = 0$, and for each r, either $c_r = \lambda$ or $y_r = 0$. But if (9.29) is a nonzero eigenvector we must have either $z \neq 0$ or $y_r \neq 0$ for some r, hence either $\lambda = 0$ or $\lambda = c_r$ for some r.

(c) For any r, P_r is self-adjoint by Proposition 9.37, hence $c_r P_r$ and $\Sigma c_r P_r$ are self-adjoint.

(d) If $R(P_r)$ is finite-dimensional, then P_r is compact by 8.50, hence so is $c_r P_r$, hence so is A by Theorem 8.51. □

Corollary 9.45 (Spectral Theorem Rewritten) Let A be a compact self-adjoint operator on a Hilbert space H, with orthogonal eigenvectors $e_1, e_2, \ldots$ and eigenvalues $\lambda_1, \lambda_2, \ldots$. Let P_r be the projection on to the subspace spanned by e_r. Then

$$x = \Sigma P_r x \quad \text{for all} \quad x \in H \quad , \tag{9.31}$$

and

$$A = \Sigma \lambda_r P_r \quad . \tag{9.32}$$

Proof From the spectral theorem,

$$x = \Sigma(x,e_r)e_r$$

$$\therefore x = \Sigma P_r x \quad ,$$

using Example 9.35. Again from Theorem 9.4,

$$Ax = \Sigma(x,e_r)\lambda_r e_r$$

$$= \Sigma(\lambda_r P_r x)$$

$$\therefore Ax = (\Sigma \lambda_r P_r)x \quad ; \tag{9.33}$$

the convergence of $\Sigma\lambda_r P_r$ (quite a different matter from the convergence of $\Sigma(\lambda_r P_r x)$ follows from Theorem 9.44. But (9.33) holds for all x, so (9.32) follows. □

This corollary can be regarded as another form of the spectral theorem. The statement that (e_n) is a basis for H is equivalent to the statement (9.31) that any vector can be reconstructed from its projections on to the subspaces generated by the e_r, and equation (9.32) is equivalent to the last clause of Theorem 9.4.

It follows from 9.45 that a compact self-adjoint operator is a sum (possibly infinite) of extremely simple operators: $\lambda_r P_r$ is a multiple of the identity operator on the subspace $R(P_r)$. Thus rank-one projections are not only the simplest self-adjoint compact operators, they are the basic ones, in the sense that any self-adjoint compact operator is a (possibly infinite) linear combination of them.

This version 9.45 of the spectral theorem is theoretically important because it generalises to noncompact operators more easily than Theorem 9.4. It is useful also because it leads to an elegant expression for powers and more general functions of an operator, as we shall now see.

In the notation of 9.45, we have

$$A = \Sigma\lambda_r P_r$$

$$\therefore A^2 = A\Sigma\lambda_r P_r = \Sigma\lambda_r A P_r \quad .$$

Now, for any $x \in H$, $P_r x$ is an eigenvector of A with eigenvalue λ_r, hence $AP_r = \lambda_r P_r$ and

$$A^2 = \Sigma\lambda_r^2 P_r \quad .$$

Similarly,

$$A^n = \sum_r \lambda_r^n P_r \quad \text{for} \quad n \geqslant 1 \quad ,$$

and for a polynomial $p(x) = \Sigma_{s=1}^n a_s x^s$,

$$p(A) = \sum_r p(\lambda_r) P_r \quad . \tag{9.34}$$

The constant term in p must be zero, that is, we must have $p(0) = 0$, otherwise the coefficient of P_r in (9.34) would not tend to zero as $r \to \infty$, and the series would not converge. To deal with a polynomial with a nonzero constant term a_0, one simply adds $a_0 I$ to (9.34), where I is the identity.

It is not difficult (see Problem 9.20) to show that (9.34) holds also when p is a power series, provided again that $p(0) = 0$; $p(A)$ is then an operator power series as defined in section 8.2. We shall now extend (9.34) to functions other than power series. Since we have not yet assigned a meaning to such functions of an operator, we are free to do so now: we define $f(A)$ to be the right hand side of (9.34), that is, $\Sigma_r f(\lambda_r) P_r$.

Definition 9.46 (Functions of an Operator) Let $A = \Sigma_r \lambda_r P_r$ be a self-adjoint compact operator with eigenvalues λ_r. Suppose the function f maps each eigenvalue of A into a real number, and $f(\lambda) \to 0$ as $\lambda \to 0$. Then we define the function f of the operator A by

$$f(A) = \Sigma f(\lambda_r) P_r \quad . \qquad \square \qquad (9.35)$$

Theorem 9.44 shows that the series in (9.35) converges; $f(A)$ is compact and self-adjoint. The discussion above shows that Definition 9.46 agrees with the usual definition of polynomial and power-series functions. We now consider an example which is not expressible as a power series.

Example 9.47 (Square Root of an Operator) Consider the function $f(x) = \sqrt{x}$. To use Definition 9.46 we must restrict ourselves to operators A with no negative eigenvalues (these operators play a central role in the next chapter; see section 10.1). Definition 9.46 then gives

$$\sqrt{A} = \Sigma \sqrt{\lambda_r} P_r \quad . \qquad (9.36)$$

To justify calling this the square root of A, we must show that its square is A.

For any x, we have, using (9.36)

$$\begin{aligned} \sqrt{A}(\sqrt{A}x) &= \sum_r \sqrt{\lambda_r} P_r \sqrt{A} x \\ &= \sum_r \sqrt{\lambda_r} P_r \sum_s \sqrt{\lambda_s} P_s x \\ &= \sum_r \sqrt{\lambda_r} \sum_s \sqrt{\lambda_s} P_r P_s x \\ &= \sum_r \sqrt{\lambda_r} \sqrt{\lambda_r} P_r x \end{aligned}$$

because the projections P_r are orthogonal. Thus, for all x,

$$\sqrt{A}\sqrt{A}x = Ax$$

so $\sqrt{A} \cdot \sqrt{A} = A$, and we have justified calling the operator in (9.36) the square root of A.

We have now shown that every self-adjoint compact operator has a square root, provided that it has no negative eigenvalues. We note for future reference that if B is a self-adjoint compact operator commuting with A, then B commutes with $\sqrt{A}$. This follows from Theorem 9.7: we can take the P_n to be projections on to a complete set of common eigenvectors of A and B, then it is easy to verify that B commutes with each P_n, and therefore with $\sqrt{A} = \Sigma \sqrt{\lambda_n} P_n$. $\square$

9.6 Summary and References

In section 9.1 we proved the spectral theorem for compact self-adjoint operators, along the lines of Kolmogorov & Fomin (1970). This is a standard piece of theory, and can be found in Helmberg, Pryce, and Stakgold, for example. However, many textbooks do not treat compact self-adjoint operators separately, but regard them as a special case of the theories of noncompact and non-self-adjoint operators, which are beyond the scope of this book (see Kreyszig, for example).

In section 9.2 we applied the spectral theorem to prove that the eigenfunctions of Sturm-Liouville systems are complete; see Pryce or Vulikh. In section 9.3 we proved a similar result for Laplace's equation. In section 9.4 we discussed the Fredholm alternative governing the existence and uniqueness of solutions of operator equations; this is discussed in most treatments of operator theory or integral equations. In section 9.5 we defined projection operators, and discussed the representation of operators and functions of operators as infinite linear combinations of projection operators; see Helmberg, Hutson & Pym, Kreyszig, or Naylor & Sell.

PROBLEMS 9

Section 9.1

9.1 A bounded operator A on a Hilbert space is called **normal** if it commutes with its adjoint, $AA^* = A^*A$. Every self-adjoint operator is obviously normal.

(a) Show that if the function $K(x,y)$ satisfies $K(x,y) = \overline{K(y,x)}$, then for any real d, the operator $u \mapsto du + i\int_a^b K(x,y)u(y)\,dy$ on the complex Hilbert space $L_2[a,b]$ is normal.

(b) Show that if B,C are commuting self-adjoint operators, then $B + iC$ is normal.

(c) Prove the converse of (b), that is, for any normal operator A, there are self-adjoint commuting operators B,C such that $A = B + iC$.

9.2 Use the result of Problem 9.1(c), together with Theorem 9.7, to show that a compact normal operator has a complete set of orthogonal eigenvectors.

9.3 Given an array (infinite matrix) of numbers k_{ij}, $i,j = 1,2,\ldots$, we say the double series $\Sigma_{ij}|k_{ij}|^2$ converges if for each i the series $\Sigma_j|k_{ij}|^2$ converges to a number L_i such that $\Sigma_i L_i$ converges, and for each j the series $\Sigma_i|k_{ij}|^2$ converges to a number M_j such that $\Sigma_j M_j$ converges. If $\Sigma_{ij}|k_{ij}|^2$ converges and $\overline{k_{ij}} = k_{ji}$ for all i,j, we define an operator K on the space l_2 (defined in 7.4) by $(Kx)_i = \Sigma_{j=1}^{\infty} k_{ij}x_j$. Show that K is a compact self-adjoint operator $l_2 \to l_2$, and write out what the spectral theorem says in this case.

9.4 Given a function K such that $\overline{K(x,y)} = K(y,x)$ and $\int_a^b\int_a^b |K(x,y)|^2\,dx\,dy$ exists, let λ_i and ϕ_i be the eigenvalues and orthonormal eigenfunctions of the integral operator on $L_2[a,b]$ whose kernel is K. Show that

$$K(x,y) = \sum_i \lambda_i \phi_i(x)\overline{\phi_i(y)} \quad ,$$

convergence being with respect to the norm in the space H of Problem 8.28. Show also that

$$\iint |K(x,y)|^2 \, dx \, dy = \sum_i |\lambda_i|^2 \quad .$$

9.5 $K: \mathbb{R}^2 \to \mathbb{C}$ is a piecewise continuous function, and $K(x,y) = \overline{K(y,x)}$. The integral operator A on $L_2[a,b]$ with kernel K has eigenvalues λ_i and orthonormal eigenfunctions ϕ_i.

(a) Show that the series $\Sigma c_n \phi_n(x)$ converges absolutely and uniformly if the constants c_n satisfy $\Sigma |c_n/\lambda_n|^2 < \infty$.

(b) Show that if f is in the range of A (that is, if $f(x) = \int_a^b K(x,y)g(y)\,dy$ for some $g \in L_2[a,b]$), then the series $\Sigma(f,\phi_n)\phi_n(x)$ converges absolutely and uniformly to f on $[a,b]$. Is this still true it we remove the condition that f lie in the range of A?

9.6 In Problem 9.4 it is shown that the eigenvalues λ_i of an integral operator with square-integrable kernel are such that $\Sigma |\lambda_i|^2$ converges. Is this true for compact self-adjoint operators in general?

Sections 9.2 and 9.3

9.7 Verify that the eigenfunctions of the system $u'' + \lambda u = 0$, $u(0) = u(a)$, $u'(0) = u'(a)$, are the sine and cosine functions which appear in the full-range Fourier series on the interval $(0,a)$. Use the result of Problem 9.5(b) to show that if f is a twice-differentiable function such that $f(0) = f(a)$ and $f'(0) = f'(a)$, then its Fourier series converges absolutely and uniformly to f on the interval $[0,a]$. (This result holds under weaker conditions on f, of course, but the proof is not as simple.)

9.8 Solving partial differential equations in circular regions leads to the eigenvalue problem

$$r^2u'' + ru' + (\lambda r^2 - p^2)u = 0 \quad \text{for} \quad 0 < r < 1 \quad ,$$

with $u(1) = 0$ and u bounded as $r \to 0$. Here p is a given integer. The theorem of section 9.2 does not apply because the coefficient of u'' vanishes at $r = 0$. However, the eigenfunctions can be shown to be complete as follows.

Show that after dividing by r the equation can be written $Lu = \lambda u$ where L has the form of (9.8). Find Green's function g for L, and show that $\int_0^1 \int_0^1 |g(r,s)|^2 s\,ds\, r\,dr$ exists. Deduce that the eigenfunctions are complete and orthogonal with respect to the inner product $(f,g) = \int_0^1 f(r)\overline{g(r)}\, r\,dr$.

(Note: the cases $p > 0$ and $p = 0$ need separate treatment.)

9.9 If you are familiar with the method of separation of variables, use it to find the eigenvalues and eigenfunctions of the operator $-\nabla^2$ in two dimensions, on the region $0 < x < a$, $0 < y < a$, with Dirichlet boundary conditions. Hence write down a series representation of Green's function for this problem.

9.10 Use the method outlined at the end of section 9.3 to obtain a solution of the two-dimensional heat equation $u_t = u_{xx} + u_{yy}$ for $(x,y) \in R$ and $t > 0$, given that $u(x,y,0) = \phi(x,y)$, and $u(x,y,t) = 0$ on the boundary of R for all t; here R is the square $\{(x,y): 0 < x < a, 0 < y < a\}$ and ϕ is a given element of $L_2[R]$.

Section 9.4

9.11 (Fredholm Alternative for differential operators.) Let L be a Sturm-Liouville operator. Since L is not compact, Theorem 9.29 does not directly apply. Prove that nevertheless, the Fredholm Alternative applies to the equation $Lu = f$.

9.12 The equation $\ddot{x} + \alpha^2 x = F(t)$ describes the response of an oscillator of natural (angular) frequency α to an external force $F(t)$ – see section 6.7. Suppose F is periodic, with period $T = 2\pi/\omega$. Consider the periodic solutions with frequency ω, that is, the solutions satisfying $u(0) = u(T)$, $\dot{u}(0) = \dot{u}(T)$. Show that the Fredholm Alternative of Problem 9.11 is equivalent to the following statement: there is always a periodic solution with frequency ω, unless the natural vibration frequency α is a multiple of ω, in which case there is no periodic solution if the force F has a non-zero Fourier component with frequency α. (Thus the solubility condition $(f,u) = 0$ in the Fredholm Alternative can be interpreted as saying that in order to avoid resonance, and thus ensure the existence of a periodic solution, the force must not contain any component of the natural vibration of the system, that is, it must be orthogonal to it. If there is such a nonzero component of the force, it feeds energy continuously into the system, the amplitude of oscillation increases, and a periodic solution, representing steady oscillation, is impossible.)

9.13 The equation $u'' + \lambda u = f(x)$ for $0 < x < 1$, with $u(0) = u(1) = 0$, describes the deflection of a horizontal girder with a vertical load $f(x)$ distributed along its length, and a compressive load λ; we have made the approximation $u \ll 1$, corresponding to small deflections of the beam. Eigenvalues of the equation $u'' + \lambda u = 0$ correspond to values of the compressive load at which the girder buckles, and the eigenfunction gives the shape into which it buckles. Give a physical interpretation of the Fredholm Alternative for this problem, along similar lines to Problem 9.12.

9.14 In section 8.3 we defined the adjoint of a bounded operator. The theory for unbounded operators is harder because the domain of the operator must be considered, whereas bounded operators are generally defined on the whole space.

Roughly speaking, the adjoint A^* of an operator $A: \mathscr{D} \to H$ is defined so that $(x,A^*y) = (Ax,y)$ for all $x \in \mathscr{D}$ and all y in a suitable domain $\mathscr{D}^*$. This will suffice for the purposes of this problem; for more details, see Hutson & Pym or Stakgold, for example.

(a) Find the adjoint of the differential operator $L = \mathrm{D}^2 + 2k\mathrm{D} + \omega^2$ on the domain $\mathscr{D} = \{u \in C^2[0,a] : u(0) = u(a) = 0\}$. (Hint: look at the proof of 2.15.)

(b) Given that $k = 2$, $\omega^2 = 5$ and $a = \pi$ in (a), use the method at the end of section 9.4 to write down a condition on f that will ensure that the equation $Lu = f$ has a solution which vanishes at the endpoints.

Section 9.5

9.15 If P_1 and P_2 are projections, show that P_1P_2 is a projection if and only if P_1 and P_2 commute. If this is so, prove that $P_1 + P_2$ is not a projection unless P_1 and P_2 are orthogonal, but $P_1 + P_2 - P_1P_2$ is always a projection if P_1 and P_2 commute. Determine the ranges of P_1P_2 and $P_1 + P_2 - P_1P_2$ in terms of $R(P_1)$ and $R(P_2)$. Illustrate with an example.

9.16 Let E be the one-dimensional subspace of $\mathbb{R}^2$ generated by the unit vector (α,β). Write down the matrix P representing the projection operator on to E, and verify that $P^2 = P$.

9.17 Find the eigenvalues and eigenvectors of the matrix

$$\begin{pmatrix} -7 & 24 \\ 24 & 7 \end{pmatrix} .$$

Express the matrix as a sum of projections, using the result of Problem 9.16.

9.18 Let P be the projection operator on to the subspace of $L_2[0,\pi]$ generated by $\{\sin x, \sin 2x, \ldots, \sin kx\}$. Write down an explicit formula for P as an integral operator. Verify that P is self-adjoint and satisfies $P^2 = P$.

9.19 (Non-orthogonal projections.) The theory of section 9.5 concerns orthogonal projection. It can be generalised as follows, to operators which are not self-adjoint, corresponding to decomposition into non-orthogonal subspaces. A bounded linear operator P on a Banach space B will be called a **projector** if $P^2 = P$. $R(P)$ denotes the range of P.

(a) Show that $I - P$ is a projector if P is. Show that if $u \in R(P)$ then $Pu = u$, and if $u \in R(I - P)$ then $Pu = 0$.

(b) Show that for any projector P on a Banach space B, the range $R(P)$ of P is a closed subspace, and is therefore itself a Banach space.

(c) Show that any $u \in B$ can be uniquely expressed in the form $u = x + y$ with $x \in R(P)$ and $y \in R(I - P)$.

(d) What can you say about the norm of a projector?

9.20 Let $p(x) = \Sigma_1^\infty c_n x^n$ be a power series which converges absolutely whenever $|x| \leqslant \| A \|$, where A is a compact self-adjoint operator. Theorem 8.20 shows that $\Sigma c_n A^n$ converges; we call its sum $p(A)$. Show that $p(A) = \Sigma_r p(\lambda_r) P_r$, where λ_r are the eigenvalues of A and P_r the projection operators on to the one-dimensional eigenspaces.

9.21 The matrix exponential was defined in Problem 8.9. Use the result of Problem 9.20 to calculate e^{At}, where t is a real number and A is the matrix of Problem 9.17. Verify, using the projection matrices, that for any $u_0 \in \mathbb{R}^2$ the vector $u = e^{At} u_0$ satisfies the differential equation $\dot{u} = Au$.

9.22 (The Power Method for computing eigenvalues and eigenvectors.) A is a compact self-adjoint operator with eigenvalues λ_n and eigenvectors e_n. Suppose $|\lambda_1| > |\lambda_2| > \ldots$ and let P_n be the projection on to the n-th eigenspace. Take any vector x, and set $x_n = A^n x / \| A^n x \|$, $a_n = \| P_n x \|$.

(a) Show that $(x_n, Ax_n) = \dfrac{\Sigma_r \lambda_r^{2n+1} a_r^2}{\Sigma_r \lambda_r^{2n} a_r^2}$.

(b) If $a_1 \neq 0$ and if only a finite number of eigenvalues are nonzero, show that $(x_n, Ax_n) \to \lambda_1$ as $n \to \infty$.

(c) Extend the result of (b) to the case where infinitely many eigenvalues are nonzero.

(d) Show that if $a_1 \neq 0$ and $\lambda_1 > 0$, then x_n converges to an eigenvector of A.

(e) If $\lambda_1 < 0$, is the sequence (x_n) convergent? Can you use, or adapt, the method to find an eigenvector in this case?

(f) How can you find the higher eigenvalues and eigenvectors by this technique?

(g) So far we have assumed that $|\lambda_1| > |\lambda_2| > \ldots$. What if $\lambda_1 = -\lambda_2$?

More problems will be found on page 378.

Chapter 10

Variational Methods

This chapter discusses some approximate methods for the solution of linear equations, using the theory of operators in Hilbert space. The methods to be discussed are called **variational methods**, because they involve maximising or minimising a functional, and 'variational' is a general term covering a maximum or minimum. We have already met a **variational principle**, or statement that something is a maximum or minimum: Theorem 9.2 states that one of the eigenvalues of an operator A is (apart from a sign) the maximum value of $|(x,Ax)|$. In this chapter we develop other variational principles, and use them to give approximate solutions of linear problems.

Section 10.1 contains an essential theoretical idea, the definition of a positive operator; most of this chapter deals with positive operators. Section 10.2 discusses a simple variational method for estimating the first eigenvalue of an operator, and section 10.3 extends it to higher eigenvalues, and to higher-order approximations. Section 10.4 is theoretical; it justifies the technique of section 10.3 by proving that the results become exact as the order of approximation tends to infinity, and also proves that the technique gives upper or lower bounds for the eigenvalues. Section 10.5 gives a method of obtaining lower bounds for the solution of inhomogeneous linear equations, and section 10.6 extends the method to give upper bounds as well.

10.1 Positive Operators

Theorem 9.2 gives a useful way of finding the largest eigenvalue of a compact self-adjoint operator: the maximum value of $|(x,Ax)|$ over all x with norm 1 is the modulus of an eigenvalue of A, and the proof of 9.3 shows that it is the eigenvalue of largest modulus. Often an operator has only positive eigenvalues, which means that the awkward modulus signs in the above statement are unnecessary; such operators are called positive operators. They can be defined without mention of their eigenvalues as follows:

Definition 10.1 An operator A on an inner product space is called
positive-definite if $(x,Ax) > 0$ for all $x \neq 0$,
negative-definite if $(x,Ax) < 0$ for all $x \neq 0$,

positive semi-definite, or simply **positive**, if $(x,Ax) \geqslant 0$ for all x,
negative semi-definite, or simply **negative**, if $(x,Ax) \leqslant 0$ for all x. □

This is a useful definition because the quantity (x,Ax) arises frequently, and it is useful to know when it is positive or negative. It is nearly the same as the definition in terms of eigenvalues, as Propositions 10.2 and 10.5 show.

Proposition 10.2 (Eigenvalues of Positive Operators) All eigenvalues of a positive-definite operator are positive, all eigenvalues of a positive operator are non-negative, etc.

Proof If A is positive-definite and $Ax = \lambda x$, with $x \neq 0$, then $(x,Ax) = \lambda(x,x) > 0$, hence λ is real and $\lambda > 0$. Similarly for the other cases. □

Examples 10.3 (a) Consider the operator $-\mathrm{D}^2$ on the space

$$S = \{u \in C^2[a,b] : u'(a) = u'(b) = 0\}$$

with inner product $(u,v) = \int_a^b u(x)\overline{v(x)}\,\mathrm{d}x$. We have

$$(u, -\mathrm{D}^2 u) = -\int_a^b u\,\overline{u}''\,\mathrm{d}x \quad ,$$

and integration by parts gives

$$(u, -\mathrm{D}^2 u) = \int_a^b |u'|^2\,\mathrm{d}x \geqslant 0$$

if $u \in S$. Thus $-\mathrm{D}^2 : S \to C[a,b]$ is a positive operator. It is not positive-definite because $(u, -\mathrm{D}^2 u) = 0$ for any constant function u.

(b) Set

$$T = \{u \in C^2[a,b] : u(a) = u(b) = 0\} \quad .$$

Then $-\mathrm{D}^2 : T \to C[a,b]$ is positive by the same argument as above. It is also positive-definite because nonzero constant functions do not belong to T: if $(u, -\mathrm{D}^2 u) = 0$ and $u \in T$, then $\int |u'|^2\,\mathrm{d}x = 0$, hence u = constant = 0 because $u(a) = 0$. Thus the positive-definiteness of an operator depends on its domain as well as on the form of the operator.

(c) The integral operator whose kernel is Green's function for $-\mathrm{D}^2$ on T is the inverse of the differential operator, as shown in section 9.2, and is also positive definite. Indeed, for any invertible positive operator A we have $(u,A^{-1}u) = \overline{(Av,v)}$ where $v = A^{-1}u$. Thus $(u,A^{-1}u) = \overline{(v,Av)} \geqslant 0$, hence A^{-1} is positive. □

Example 10.4 The operator $-\nabla^2$ with the domain $\mathscr{D}$ of 9.19 is positive-definite.

Proof:

$$(u, -\nabla^2 u) = -\int_V u\nabla\cdot(\nabla\bar{u})\,\mathrm{d}^3x$$
$$= -\int_V [\nabla\cdot(u\nabla\bar{u}) - (\nabla u)\cdot(\nabla\bar{u})]\,\mathrm{d}^3x$$
$$= -\int_S u\nabla\bar{u}\cdot\mathbf{d}S + \int_V |\nabla u|^2\,\mathrm{d}^3x$$

using the divergence theorem. But u vanishes on S, hence

$$(u, -\nabla^2 u) = \int_V |\nabla u|^2\,\mathrm{d}^3x \qquad (10.1)$$
$$\geqslant 0 \quad ,$$

so $-\nabla^2$ is positive on $\mathscr{D}$. To show that it is positive-definite, suppose that $(u, -\nabla^2 u) = 0$. Since ∇u is continuous, (10.1) shows that $\nabla u = 0$, so u is constant. But $u = 0$ on S, hence $u = 0$ everywhere. Thus $(u, -\nabla^2 u) = 0$ implies that $u = 0$, so $-\nabla^2$ is positive-definite on $\mathscr{D}$. □

The following is a partial converse to Proposition 10.2.

Proposition 10.5 (Positive Operators) If the eigenvectors of a symmetric operator A form a complete orthogonal set, and if all the eigenvalues are positive (or non-negative etc.), then A is positive-definite (or positive semi-definite etc.).

Proof Let λ_n be the eigenvalues and e_n the orthogonal eigenvectors of A. For any vector $u = \Sigma c_n e_n$ we have

$$(u,Au) = (\Sigma c_n e_n, Au)$$
$$= \Sigma c_n(e_n, Au)$$

by 7.13,
$$= \Sigma c_n(Ae_n, u)$$
$$= \Sigma c_n(\lambda_n e_n, u)$$
$$= \Sigma c_n \lambda_n \bar{c}_n \quad .$$
$$\therefore (u,Au) = \Sigma\lambda_n |c_n|^2 \quad ,$$

from which the result follows. □

Positive-definite operators have many properties in common with positive numbers. It is not hard to prove the following: if A is positive, then $-A$ is negative; if A is positive-definite, then it is invertible and its inverse is positive-definite; positive operators have square roots (Example 9.47 – but the noncompact case is difficult); the sum of two positive operators is positive. However, the product of two positive operators is not always positive, as the following example shows.

Example 10.6 On the space $\mathbb{R}^2$, the matrix operators

$$A = \begin{pmatrix} 1 & 0 \\ 0 & 0 \end{pmatrix}, \quad B = \begin{pmatrix} 1 & 1 \\ 1 & 1 \end{pmatrix}$$

are positive, since for any vector $u = (u_1, u_2)$, $(u,Au) = u_1^2 \geqslant 0$, and $(u,Bu) = (u_1 + u_2)^2 \geqslant 0$. But

$$AB = \begin{pmatrix} 1 & 1 \\ 0 & 0 \end{pmatrix}$$

and $(u,ABu) = u_1(u_1 + u_2)$ which is negative for $u = (1,-2)$. Thus the product of positive operators need not be positive. □

What has gone wrong, so to speak, in this example is that AB is not self-adjoint (symmetric), although A and B are. However, if A and B commute, that is, if $AB = BA$, then AB will be symmetric and positive if A and B are.

Theorem 10.7 (Product of Commuting Operators) If A and B are symmetric positive operators, and $AB = BA$, then AB is symmetric and positive.

Proof $(ABu,v) = (Bu,Av) = (u,BAv) = (u,ABv)$, hence AB is symmetric. To prove that it is positive, we begin with the case where A and B are compact. Then from 9.47 we see that A has a square root $\sqrt{A}$, and $\sqrt{A}B = B\sqrt{A}$. Hence $AB = \sqrt{A}\,B\sqrt{A}$, and

$$\begin{aligned}(x,ABx) &= (x,\sqrt{A}\,B\sqrt{A}x) \\ &= (\sqrt{A}x,\ B\sqrt{A}x) \\ &\geqslant 0 \quad \text{for all} \quad x\end{aligned}$$

because B is positive. This proves the theorem for compact operators.

If A and B are not compact but have compact inverses, then $B^{-1}A^{-1}$ is positive by the above argument, hence so is AB. In this chapter we shall only need the case where A^{-1} and B^{-1} are compact, and we will not complete the proof of the general case. In fact, the proof given above for compact operators applies in the general case; the only missing link is a proof that all positive operators have square roots. This is true, but the proof is more difficult than the compact case discussed in 9.47; see, for example, Riesz & Nagy, or Kreyszig. □

10.2 Approximation to the First Eigenvalue

We now return to the idea with which this chapter began: the modulus of the largest eigenvalue λ of a compact operator A is the maximum of $|(x,Ax)|$ with $\|x\| = 1$. If A is positive, we simply have

$$\lambda = \max\{(x,Ax) : \|x\| = 1\} \quad .$$

But maximising (x,Ax) over all x with $\|x\| = 1$ is the same as maximising $(x/\|x\|, Ax/\|x\|)$ over all x. This proves

Proposition 10.8 (Rayleigh's Principle for Compact Operators) If λ is the largest eigenvalue of a positive self-adjoint compact operator A on a Hilbert space H, then

$$\lambda = \max\{R(x) : x \in H\} \quad ,$$

where $$R(x) = \frac{(x,Ax)}{(x,x)} \quad . \tag{10.2}$$

Any vector which maximises $R(x)$ is an eigenvector of A belonging to eigenvalue λ. □

$R(x)$ is called the **Rayleigh quotient.** Since λ is the maximum value of $R(x)$, if we take any particular $x_1 \in H$, then $R(x_1) \leqslant \lambda$ and we thus obtain a **lower bound** for the eigenvalue, that is, a number below which we know the eigenvalue cannot lie. The vector x_1 used in this way is called a **trial vector,** or a **trial function** if H is a function space. Since $R(x) = \lambda$ if x is the eigenvector corresponding to λ, we can expect to get the best approximation to λ by taking the trial vector as close as possible to the eigenvector. Of course, we do not usually know the eigenvector exactly – if we did, there would be no need for approximate methods. But we often know something about the eigenvector from general considerations.

Example 10.9 The smallest eigenvalue of

$$u'' + \lambda a(x)u = 0, \; u(0) = u(1) = 0 \quad , \tag{10.3}$$

where a is a given positive function, can be interpreted as the square of the lowest frequency of vibration of a rod of nonuniform cross-section given by $a(x)$. The problem is equivalent to the integral eigenvalue problem

$$u(x) = \lambda \int_0^1 g(x;y)u(y)a(y)\,\mathrm{d}y \quad ,$$

where $$g(x;y) = \begin{cases} y(1-x) & \text{if} \quad x > y \\ x(1-y) & \text{if} \quad y > x \end{cases}$$

is Green's function for the operator $-\mathrm{D}^2$, given in 5.18. The function g is continuous and positive. As in section 9.2, we use a weighted inner product

$$(u,v) = \int_0^1 u(x)\overline{v(x)}a(x)\,\mathrm{d}x \quad . \tag{10.4}$$

Then the integral operator K defined by

$$(Ku)(x) = \int_0^1 g(x;y)u(y)a(y)\,\mathrm{d}y$$

is a self-adjoint compact operator on $L_2[0,1]$ with inner product (10.4). The method of Example 10.3 shows that $-[1/a(x)]D^2$ is a positive operator, hence its inverse K is positive. The eigenvalues of K are the reciprocals of those of the differential operator. Therefore Rayleigh's principle shows that the lowest eigenvalue λ_1 of (10.3) satisfies

$$\lambda_1 \leqslant \frac{\int_0^1 dx\, a(x)\,|v(x)|^2}{\int_0^1 dx \int_0^1 dy\, a(x)a(y)g(x;y)v(x)\overline{v(y)}} \tag{10.5}$$

for any function v.

If a is constant, then the eigenfunction corresponding to the lowest eigenvalue is $\sin(\pi x)$. If a varies reasonably smoothly, we may expect the eigenfunction to be reasonably similar to $\sin(\pi x)$; at any rate, it should vanish at the end-points and be positive in between. Taking $v(x) = \sin(\pi x)$ in (10.5) should therefore give a reasonable approximation to the eigenvalue.

The integrals are easier to evaluate if we use the very crude trial function $v(x) = 1$. In the simple case where $a(x) = 1$, this gives $\lambda_1 \leqslant 12$, which may be compared with the exact value $\lambda_1 = \pi^2$ when $a(x) = 1$. Using the trial function $\sin(\pi x)$ will give a better bound, that is, one closer to the exact value. The integrals are tedious to evaluate by hand, but there is no difficulty in evaluating them numerically. In later sections of this chapter we shall consider how to estimate and improve the accuracy of such approximations when there is no exact solution available to compare them with. □

The method used above for computing the lowest eigenvalue of a differential system is convenient for numerical computation because it involves evaluating integrals, which is quite straightforward from the point of view of numerical analysis (while numerical differentiation is not). But it is rather clumsy for analytic work, involving double integration of Green's function. Can we work with the differential operator directly, instead of first transforming into an integral operator?

Proposition 10.10 (Rayleigh's Principle for Differential Operators) Let L be a positive symmetric differential operator on a domain $\mathscr{D}$, with an inverse which is a compact integral operator. Then its smallest eigenvalue λ_1 satisfies

$$\lambda_1 = \min\{R(u) : u \in \mathscr{D}\} \quad , \tag{10.6}$$

where $R(u) = (u,Lu)/(u,u)$. Any function which minimises $R(u)$ is an eigenfunction belonging to λ_1.

Proof The theory of Chapter 9 shows that L has a complete set of eigenfunctions (e_n), with eigenvalues λ_n such that $\lambda_n \to \infty$. There must therefore be a smallest eigenvalue λ_1 (for otherwise there would be an infinite sequence of eigenvalues converging to $\inf\{\lambda_n\}$, which is impossible because all subsequences of (λ_n) tend to infinity). Now, for any $u \in \mathscr{D}$, $u = \Sigma c_n e_n$ for some (c_n). Hence

$$R(u) = \frac{(\Sigma c_n e_n, \Sigma \lambda_n c_n e_n)}{(\Sigma c_n e_n, \Sigma c_n e_n)}$$

$$\therefore R(u) = \frac{\Sigma \lambda_n |c_n|^2}{\Sigma |c_n|^2} \tag{10.7}$$

$$\geqslant \frac{\lambda_1 \Sigma |c_n|^2}{\Sigma |c_n|^2} = \lambda_1$$

since $\lambda_1 = \min\{\lambda_n\}$. Furthermore, $R(e_1) = \lambda_1$; and if the coefficient in (10.7) of any eigenvalue greater than λ_1 is nonzero, then $R(u) > \lambda_1$. Hence $R(u) = \lambda_1$ if and only if u is an eigenfunction belonging to λ_1. □

Example 10.11 We can now return to Example 10.9, and regard it as an eigenvalue problem for the operator $-[1/a(x)]\mathrm{D}^2$, on the space of twice-differentiable functions vanishing at 0 and 1, with inner product (10.4). It satisfies the conditions of Proposition 10.10, so its first eigenvalue λ_1 satisfies

$$\lambda_1 \leqslant \frac{(u, -u''/a)}{(u,u)} = \frac{-\int_0^1 u\bar{u}''\,\mathrm{d}x}{\int_0^1 a|u|^2\,\mathrm{d}x}$$

for any $u \in \mathscr{D}$. Integration by parts gives the more symmetrical form

$$\lambda_1 \leqslant \frac{\int_0^1 |u'|^2\,\mathrm{d}x}{\int_0^1 a|u|^2\,\mathrm{d}x} \quad . \tag{10.8}$$

It is quite easy to calculate approximate lowest eigenvalues from (10.8), using a trial function such as $\sin(\pi x)$ or $x(1-x)$ which satisfies the boundary conditions.

As a straightforward example, consider the case

$$a(x) = \alpha + \beta x \quad ,$$

corresponding to a uniformly tapering rod. Using the trial function $\sin(\pi x)$ gives $\lambda_1 \leqslant \pi^2/(\alpha + \frac{1}{2}\beta)$; using the trial function $x(1-x)$ gives $\lambda_1 \leqslant 10/(\alpha + \frac{1}{2}\beta)$. We can compare these results with the exact value, because when $a(x) = \alpha + \beta x$, our equation can be reduced to an equation (Bessel's equation) whose solutions are known and have been tabulated. Using these tables one can calculate the eigenvalues accurately. Taking the case $\alpha = 0$, $\beta = 1$, for example, one finds the following:

Variational method using $\sin(\pi x)$	:	$\lambda_1 \leqslant 19.74$,
Variational method using $x(1-x)$	:	$\lambda_1 \leqslant 20.00$,
Exact value	:	$\lambda_1 = 18.956$.

Thus our very simple variational method gives an accuracy of a few per cent. In practical problems, of course, there is no exact solution available to compare with the approximate result, and it is not easy to assess its accuracy; one way is described in section 10.6.

We can improve the approximation by taking a trial function with an adjustable constant, such as

$$u_c(x) = \sin(\pi x) + c\sin(2\pi x) \quad , \tag{10.9}$$

where c is constant. Taking $c = 0$ gives the result above. A better approximation is obtained by calculating $R(u_c)$ in terms of the unknown constant c, and then choosing c so as to minimise R. This will give the least possible upper bound obtainable from the trial function (10.9), and therefore the best approximation. We will carry out this procedure in the next section, where we shall see that it not only improves the approximation to the lowest eigenvalue, but gives a bound for the next, and can be extended to give all the eigenvalues. □

Finally, we note that the restriction to positive operators is not essential. If A is positive and compact, say, then there must be a largest eigenvalue. If A is not positive, there may not be a largest eigenvalue. But if the set of eigenvalues is either bounded above or bounded below, then the following result applies.

Proposition 10.12 (Rayleigh's Principle for Non-Positive Operators) Let A be a symmetric operator on a Hilbert space H, such that either A or A^{-1} is compact. If the set $\{\lambda_n\}$ of eigenvalues is bounded above, then

$$(x,Ax)/(x,x) \leqslant \sup\{\lambda_n\} \quad \text{for all} \quad x \in H \quad .$$

If the set of eigenvalues is bounded below, then

$$\inf\{\lambda_n\} \leqslant (x,Ax)/(x,x) \quad \text{for all} \quad x \in H \quad .$$

Proof Let (e_r) be the orthonormal eigenvectors of A, and let $x = \Sigma c_r e_r$. Then

$$\frac{(x,Ax)}{(x,x)} = \frac{\Sigma\lambda_r|c_r|^2}{\Sigma|c_r|^2} \quad ,$$

from which the result follows at once. □

10.3 The Rayleigh-Ritz Method for Eigenvalues

In this section we extend the method of section 10.2 to give bounds for all the eigenvalues, not just the first. We begin by looking at the method from a different point of view. When we evaluate the Rayleigh quotient $R(u)$ for a trial vector u, we are in effect approximating the eigenvector by a vector in the one-dimensional subspace

$$S = \{cu : c \text{ a scalar}\}$$

generated by u. We are projecting the problem on to S, in the sense to be explained below.

Let A be a positive self-adjoint compact operator on a Hilbert space H, and let μ be the Rayleigh approximation to its first eigenvalue using a trial vector u; then

$$\mu = (u,Au)/(u,u) \quad .$$

Now, with S defined as above, we have for all $v \in S$, $v = cu$ for some scalar c, so $(v,Av)/(v,v) = (u,Au)/(u,u)$. Hence the statement

$$\mu = \max\{(v,Av)/(v,v) : v \in S\} \tag{10.10}$$

is trivially true.

Let P be the projection operator on to S; then $Pv = v$ if $v \in S$, hence

$$(v,Av) = (Pv,APv) = (v,PAPv) \tag{10.11}$$

if $v \in S$, using the self-adjointness of P (Proposition 9.37). So (10.10) becomes

$$\mu = \max\{(v,PAPv)/(v,v) : v \in S\} \quad . \tag{10.12}$$

Now, $PAPx \in S$ for any vector x, hence PAP maps S into S. Equation (10.11) shows that $PAP : S \to S$ is positive if A is, and it is easy to show that PAP is compact and self-adjoint if A is. Hence (10.12) shows that μ is the largest eigenvalue of PAP: $S \to S$. Regarded as an operator on H, PAP has rank one; it therefore has at most one nonzero eigenvalue, whose value is given by (10.12). It is easy to show that any eigenvector not in S must have eigenvalue zero. We may regard PAP as the projection of A on to S, and think of it as a crude approximation to A, whose eigenvalue is the Rayleigh approximation to the first eigenvalue of A.

This point of view suggests a way of improving the approximation. If we consider a two-dimensional subspace S_2, and let P_2 be the projection operator on to S_2, then P_2AP_2 should be a less crude approximation to A than PAP, and its first eigenvalue should be a better approximation to that of A. Furthermore, since P_2AP_2 is a rank-two operator, it will in general have two nonzero eigenvalues, and the second one should give an approximation to the second eigenvalue of A. In the language of the last section, we can say that taking a trial vector of the form $u = c_1\phi_1 + c_2\phi_2$, where ϕ_1, ϕ_2 are linearly independent elements of H, should (i) improve the approximation to λ_1, and (ii) give an estimate of λ_2. Similarly, using a trial vector of the form $u = c_1\phi_1 + c_2\phi_2 + \ldots + c_n\phi_n$ corresponds to projecting on to an n-dimensional subspace, and will give approximations to $\lambda_1, \lambda_2, \ldots, \lambda_n$. This approximation is called the Rayleigh-Ritz approximation of order n. If (ϕ_r) is a basis for H, and P_n is the projection operator on to the n-dimensional subspace generated by $\phi_1, \ldots, \phi_n$, then we might expect that $P_nAP_n \to A$ as $n \to \infty$, so that the Rayleigh-Ritz approximation becomes exact in the limit $n \to \infty$. We shall prove this in the next section, but first we spell the method out in detail.

Definition 10.13 The *n*-th **order Rayleigh-Ritz approximations** $\mu_1, \ldots, \mu_n$ to eigenvalues of a compact self-adjoint operator A, using trial functions $\phi_1, \ldots, \phi_n$, are

$$\mu_1 = \max\{R(u): u \in S_n\}$$

$$\mu_2 = \max\{R(u): u \in S_n,\ u \perp f_1\}$$

$$\mu_3 = \max\{R(u): u \in S_n,\ u \perp f_1,\ u \perp f_2\}$$

$$\cdots\cdots\cdots\cdots\cdots\cdots$$

$$\mu_n = \max\{R(u): u \in S_n,\ u \perp f_1, f_2, \ldots f_{n-1}\} \quad ,$$

where S_n is the subspace generated by $\phi_1, \ldots, \phi_n$, and f_i is the value of u which maximises $R(u) = (u,Au)/(u,u)$ at the i-th stage.

If A is not compact but has a compact inverse, the Rayleigh-Ritz approximations are defined in the same way, but with 'max' replaced by 'min'. □

The argument above suggests, but does not prove, that μ_i is a reasonable approximation to λ_i for $1 \leqslant i \leqslant n$, and the approximation improves as $n \to \infty$. In the next section we shall prove these facts, but first we work an example.

Example 10.14 In Example 10.11 we applied Rayleigh's principle, which is the same as the first-order Rayleigh-Ritz method, to the first eigenvalue of the system

$$u'' + \lambda a(x)u = 0, \quad u(0) = u(1) = 0 \quad .$$

A natural sequence of trial functions for this problem is $\sin(n\pi x)$, $n = 1,2,\ldots$. We shall compute the second-order Rayleigh-Ritz approximation, using $u(x) = c_1\sin(\pi x) + c_2\sin(2\pi x) = c_1[\sin(\pi x) + c\sin(2\pi x)]$ where $c = c_2/c_1$. Since c_1 cancels out when we compute the Rayleigh quotient, we may as well simply take

$$u = \sin(\pi x) + c\sin(2\pi x) \quad .$$

Calculating the Rayleigh quotient

$$R(u) = \frac{\int |u'|^2\,dx}{\int a|u|^2\,dx} \quad ,$$

as in Example 10.11, gives

$$R(u) = \frac{2\pi^2(1 + 4c^2)}{1 + c^2 - 64c/9\pi^2} \tag{10.13}$$

in the particular case $a(x) = x$. An easy calculation shows that this takes its minimum for

$$c = -0.11386 \quad . \tag{10.14}$$

The minimum value of R is the second-order approximation to the first eigenvalue; we find

$$\mu_1 = 18.961 \quad .$$

This is a great improvement on the first-order approximation of Example 10.11.

We can now estimate the second eigenvalue. According to 10.13, we must maximise $R(u)$ over all linear combinations of $\sin(\pi x)$ and $\sin(2\pi x)$ which are orthogonal to $f_1 = \sin(\pi x) + c \sin(2\pi x)$, where c is given by (10.12); orthogonality is with respect to the inner product (10.4). A simple calculation shows that the function

$$u(x) = A\,[\sin(\pi x) + d \sin(2\pi x)] \tag{10.15}$$

is orthogonal to f_1 when

$$d = \frac{9\pi^2 c - 32}{32c - 9\pi^2} \quad , \tag{10.16}$$

$$= 2.1957$$

when c is given by (10.14). The approximation μ_2 to λ_2 is given by the minimum of $R(u)$; since $R(u)$ is independent of A, no minimisation procedure is in fact needed. Since (10.15) is of the same form as the trial function for the first eigenvalue, $R(u)$ is given by (10.13) with c replaced by d; evaluating it for d = 2.1957 gives

$$\mu_2 = 94.45 \quad .$$

An exact calculation of λ_2 in terms of Bessel functions (cf. 10.11) gives $\lambda_2 = 81.89$; we see that $\mu_2 > \lambda_2$, just as $\mu_1 > \lambda_1$. In the next section we shall show in general that the Rayleigh-Ritz method gives an upper bound for λ_2 and the higher eigenvalues. □

10.4 The Theory of the Rayleigh-Ritz Method

In this section we prove that the Rayleigh-Ritz approximations converge monotonically to the exact eigenvalues. The proof involves some new results, which are also of interest in their own right.

Our starting-point is equation (9.6). Since $Q_n = \{e_1, \ldots, e_n\}^\perp$, where e_i are the eigenvectors, it follows that for a compact self-adjoint positive operator A, the n-th largest eigenvalue is given by

$$\lambda_n = \max\{R(u) : u \perp e_1, \ldots, e_{n-1}\} \quad . \tag{10.17}$$

For a differential operator whose inverse is compact, the corresponding statement is that the n-th smallest eigenvalue is given by

$$\lambda_n = \min\{R(u) : u \perp e_1, \ldots, e_{n-1}\} \quad . \tag{10.18}$$

For compact operators, $\mu_1 \leqslant \lambda_1$, where μ_1 is the Rayleigh-Ritz approximation, because a constrained maximum cannot be greater than the absolute maximum λ_1. But there is no such obvious inequality relating μ_2 and λ_2; maximising $R(u)$ subject to $u \in S$ and $u \perp f_1$ has no obvious relation to maximising subject to $u \perp e_1$. Since e_1 is unknown, we cannot directly use (10.17) for $n \geqslant 2$ to estimate λ_n, in the way that we can for $n = 1$. But there is an alternative variational formula for λ_n; it differs from (10.17) and (10.18) in that it does not involve the (unknown) eigenvectors, and therefore gives useful bounds on all the eigenvalues. It can be approached by the following physical argument.

The eigenvalues of a differential operator can often be interpreted as frequencies of vibration of a physical system, which may be imagined as a set of particles held together by springs. If a constraint is imposed on a system, as, for instance, if two of its particles are rigidly connected together instead of being allowed to vibrate relative to one another, then it is plausible that the system will become stiffer, and will therefore vibrate faster. Hence we expect that imposing a constraint will increase the eigenvalues.

Now, for a linear system, represented by a vector in Hilbert space, a linear equation of constraint can be written $f(u) = 0$, where f is a linear functional. If f is continuous, then the Riesz representation theorem 7.60 shows that there is a vector k, called the **constraint vector**, such that the constraint equation is

$$(u,k) = 0 \quad . \tag{10.19}$$

The eigenvalues of the constrained system are just the eigenvalues of the differential operator L of the original system, but regarded as an operator on the subspace K consisting of vectors orthogonal to k. Hence the lowest eigenvalue of the constrained system is the minimum value over K of the Rayleigh quotient $R(u)$, which cannot be less than the minimum of $R(u)$ over the whole space, which equals the unconstrained first eigenvalue. This argument confirms the physical idea above, that constraints raise eigenvalues.

We now ask how much the lowest eigenvalue is raised by the constraint. If e_1 is the first eigenvector of the unconstrained system, then in the lowest mode of vibration the system is in the state e_1. Imposing a constraint will prevent the system from adopting e_1 as a mode of vibration, and it is plausible that the farther the system is forced to depart from its natural lowest mode e_1, the more the frequency will be changed. This suggests that the greatest change will occur when the constraint prevents there being any component of e_1 in the lowest vibration mode, that is, when the vector describing the system is constrained to be orthogonal to e_1, that is, when the constraint vector k equals e_1. But in this case the lowest eigenvalue of the constrained system is

$$\min\{R(u): u \perp e_1\}$$

which is just the second eigenvalue of the original system according to (10.18). Thus we are led to expect that if a single constraint is applied to a system, the first eigenvalue of the constrained system lies between the first and second eigenvalues of the original system. We shall now prove this.

Theorem 10.15 (Rayleigh's Theorem of Constraint) The first eigenvalue of a system subject to a single constraint lies between the first two eigenvalues of the unconstrained system. (For 'system', the pure mathematician can read 'self-adjoint positive compact operator or symmetric positive operator with compact inverse'; for 'subject to a constraint' he can read 'restricted to a subspace $\{u: (u,k) = 0\}$'.)

Proof We write out the proof for the case of a differential operator L on a domain $\mathscr{D}$, where L^{-1} is compact; the proof for a compact operator is the same but with the inequalities reversed. Let k be the constraint vector and set

$$K = \{u \in \mathscr{D}: u \perp k\} \quad .$$

Let λ_1, λ_2 be the first two eigenvalues of L, with eigenvectors e_1, e_2, and λ_c the first eigenvalue of the constrained system. Then

$$\begin{aligned} \lambda_c &= \min\{R(u): u \in K\} \qquad (10.20) \\ &\geqslant \min\{R(u): u \in \mathscr{D}\} = \lambda_1 \quad , \end{aligned}$$

since a constrained minimum cannot be less than an absolute minimum. To prove that $\lambda_c \leqslant \lambda_2$, we shall produce a $v \in K$ with $R(v) \leqslant \lambda_2$. Take

$$v = c_1 e_1 + c_2 e_2 \quad ;$$

c_1 and c_2 can easily be chosen so that $(v,k) = 0$. Then

$$R(v) = \frac{(v,Lv)}{(v,v)} = \frac{\lambda_1 |c_1|^2 + \lambda_2 |c_2|^2}{|c_1|^2 + |c_2|^2} \quad ,$$

which is a weighted average of λ_1 and λ_2 and therefore lies between them. Hence $R(v) \leqslant \lambda_2$, and (10.20) shows that $\lambda_c \leqslant R(v)$ for all $v \in K$, so we have $\lambda_c \leqslant \lambda_2$. □

Rayleigh's theorem is an interesting result, but here we are only using it as a stepping-stone towards a characterisation of the second eigenvalue which does not involve the first eigenvector. If we focus on the case of differential operators, Rayleigh's theorem gives

$$\lambda_1 \leqslant \lambda_c \leqslant \lambda_2 \quad .$$

The second half of this inequality says that

$$\lambda_2 \geqslant \min\{R(u): u \perp k\} \tag{10.21}$$

for any k in the domain of L. Hence λ_2 is greater than or equal to the maximum, over all k, of the expression on the right of (10.21). We shall now show that λ_2 equals that maximum.

Proposition 10.16 (a) If L is a positive symmetric differential operator, with domain $\mathscr{D}$, and with a compact inverse, then its second smallest eigenvalue is given by

$$\lambda_2 = \max_{k \in \mathscr{D}}\{\min\{R(u): u \perp k\}\} \quad . \tag{10.22}$$

(b) If A is a positive compact self-adjoint operator, then its second largest eigenvalue is given by

$$\lambda_2 = \min_{k \in \mathscr{D}}\{\max\{R(u): u \perp k\}\} \quad . \tag{10.23}$$

Proof (a) If $k = e_1$, the first eigenvector of L, we have $\min\{R(u): u \perp k\} = \lambda_2$, hence the right hand side of (10.22) is $\geqslant \lambda_2$. But (10.21) shows that the right hand side of (10.22) is $\leqslant \lambda_2$. This proves (10.22). The same argument applies to (b). □

Equation (10.22) is a variational expression for λ_2 which does not involve e_1; it is sometimes called the Courant, or Courant-Weyl, maximin principle, and (10.23) is called the minimax principle. They can be extended to all the eigenvalues by means of the following generalisation of 10.15.

Theorem 10.17 (Rayleigh's Theorem for Several Constraints) The first eigenvalue of a system subject to c constraints lies between the first and the $(1 + c)$-th eigenvalues of the unconstrained system. □

The proof of this theorem is a straightforward extension of that of Theorem 10.15, and is left as an exercise. A generalisation of 10.16 now follows easily.

Theorem 10.18 (Minimax and Maximin Principles) (a) If A is a positive self-adjoint compact operator on a Hilbert space H, with eigenvalues $\lambda_1, \lambda_2, \ldots$, then

$$\lambda_{n+1} = \min_{k_1, \ldots, k_n \in \mathscr{D}}\{\max\{R(u): u \perp k_1, \ldots, k_n\}\} \quad . \tag{10.24}$$

(b) If A is a positive symmetric differential operator with domain $\mathscr{D}$ and a compact inverse, and with eigenvalues $\lambda_1, \lambda_2, \ldots$, then

$$\lambda_{n+1} = \max_{k_1, \ldots, k_n \in \mathscr{D}}\{\min\{R(u): u \perp k_1, \ldots, k_n\}\} \quad . \tag{10.25}$$

Proof (a) If $k_i = e_i$ for $i = 1, \ldots, n$, where e_i are the eigenvectors of A, then

$$\max\{R(u): u \perp k_1, \ldots, k_n\} = \lambda_{n+1} \quad ,$$

hence the right hand side of (10.24) is $\leqslant \lambda_{n+1}$. But it is $\geqslant \lambda_{n+1}$ by Theorem 10.17. This proves (10.24). Part (b) is proved in the same way, but with the inequalities reversed. □

We shall need the following result, generalising the ideas discussed at the beginning of section 10.3.

Lemma 10.19 Let μ_r be the n-th order approximation to the eigenvalue λ_r of a symmetric positive operator A, using trial functions $\phi_1, \ldots, \phi_n$. Then μ_r is an eigenvalue of the operator PAP, where P is the projection operator on to the subspace S_n spanned by $\{\phi_1, \ldots, \phi_n\}$.

Proof We write out the proof assuming that A is compact; an entirely similar argument applies if A has a compact inverse. Write $R_A(u) = (u,Au)/(u,u)$. Then

$$R_{PAP}(u) = (u,PAPu)/(u,u) = (Pu,APu)/(u,u) \quad ,$$

$$\therefore R_{PAP}(u) = R_A(u) \quad \text{if} \quad u \in S_n \quad . \tag{10.26}$$

Now, $PAPx \in S_n$ for any x, hence PAP maps S_n into itself. The operator PAP: $S_n \to S_n$ is compact and self-adjoint; it is positive because (10.26) shows that $(u,PAPu) = (u,Au) \geqslant 0$ for $u \in S_n$. Hence the eigenvalues Λ_r of $PAP: S_n \to S_n$ are given by (10.17):

$$\Lambda_r = \max\{R_{PAP}(u): u \in S_n, \ u \perp f_1, \ldots f_{r-1}\}$$

for $r = 1, \ldots, n$, where f_i is an eigenvector of PAP and gives the maximum of R_{PAP} at the i-th stage. Using (10.26), we have

$$\Lambda_r = \max\{R_A(u): u \in S_n, \ u \perp f_1, \ldots, f_{r-1}\}$$

for $r = 1, \ldots, n$, where f_i is the maximising vector at the i-th stage. But this sequence of maximisation problems is precisely the Rayleigh-Ritz procedure. Hence $\Lambda_r = \mu_r$. □

We can now, at last, prove the main facts about the Rayleigh-Ritz method. The following theorem says that (i) the n-th order Rayleigh-Ritz method gives a bound for the r-th eigenvalue, if $r \leqslant n$, and (ii) increasing the order of the method improves, or at least does not worsen, the approximation.

Theorem 10.20 (Rayleigh-Ritz Bounds for Eigenvalues) (a) Let A be a positive compact self-adjoint operator on a Hilbert space H, with eigenvalues λ_r, and let $\mu_r^{(n)}$ be the approximation to λ_r computed by the Rayleigh-Ritz method of order n. Then

$$\mu_r^{(n)} \leqslant \mu_r^{(n+1)} \leqslant \lambda_r \quad .$$

(b) Let A be a positive symmetric differential operator on a domain $\mathscr{D}$, with a compact inverse and with eigenvalues λ_r. Let $\mu_r^{(n)}$ be the approximation to λ_r computed by the Rayleigh-Ritz method of order n. Then

$$\mu_r^{(n)} \geqslant \mu_r^{(n+1)} \geqslant \lambda_r \quad .$$

Proof We write out the proof of case (a); case (b) is the same but with the inequalities reversed. Using the notation of Definition 10.13, we have

$$\lambda_r = \max\{R(u): u \in H,\ u \perp e_1, \ldots, e_{r-1}\} \quad , \tag{10.27}$$

$$\mu_r^{(n)} = \max\{R(u): u \in S_n,\ u \perp f_1, \ldots, f_{r-1}\} \quad , \tag{10.28}$$

where e_i are the eigenvectors of A. The variational problems for λ_r and $\mu_r^{(n)}$ cannot be compared directly because they involve maximising over different spaces and with different orthogonality constraints. We therefore introduce a third variational problem, which can be compared directly with each of the other two because it involves the same space S_n as in (10.28) and the same constraints as in (10.27). Thus we define

$$\nu_r = \max\{R(u): u \in S_n,\ u \perp e_1, \ldots, e_{r-1}\} \quad . \tag{10.29}$$

Then

$$\nu_r \leqslant \lambda_r \quad , \tag{10.30}$$

because ν_r is the maximum of $R(u)$ over a subset of the vectors for which λ_r is the maximum.

Now, $\mu_r^{(n)}$ is the r-th eigenvalue of $PAP: S_n \to S_n$ by Lemma 10.19. By the minimax principle, 10.18(a), we therefore have

$$\mu_r^{(n)} = \min_{k_1, \ldots, k_{r-1} \in S_n} \left\{ \max \left\{ \frac{(u, PAPu)}{(u,u)} : u \in S_n,\ u \perp k_1, \ldots, k_{r-1} \right\} \right\}$$

$$\therefore \mu_r^{(n)} = \min_{k_1, \ldots, k_{r-1} \in S_n} \{\max\{R(u): u \in S_n,\ u \perp k_1, \ldots, k_{r-1}\}\} \quad , \tag{10.31}$$

using (10.26). Comparing (10.31) and (10.29), we see that if $e_1, \ldots, e_n$ belonged to S_n, then ν_r would be a member of the set of numbers $\max\{R(u): u \in S_n, u \perp k_1, \ldots, k_{r-1}\}$ for which $\mu_r^{(n)}$ is the minimum. In general, e_i will not belong to S_n. But if we define

$$g_i = Pe_i \quad \text{for} \quad i = 1, 2, \ldots$$

then $g_i \in S_n$, and for any $u \in S_n$ we have

$$(u,g_i) = (u,Pe_i) = (Pu,e_i) = (u,e_i) \quad ,$$

hence $\{u \in S_n : u \perp e_1, \ldots, e_{r-1}\} = \{u \in S_n : u \perp g_1, \ldots, g_{r-1}\}$.

We can now rewrite (10.29) as

$$\nu_r = \max\{R(u) : u \in S_n,\ u \perp g_1, \ldots, g_{r-1}\} \quad .$$

Since $g_i \in S_n$ for all i, it follows from (10.31) that

$$\nu_r \geqslant \mu_r^{(n)} \quad .$$

Combining this with (10.30), we deduce that

$$\mu_r^{(n)} \leqslant \lambda_r \quad \text{for all} \quad r \leqslant n \quad .$$

Finally we must show that $\mu_r^{(n)} \leqslant \mu_r^{(n+1)}$. This follows easily from (10.31): for any vectors $k_1, \ldots k_{r-1}$,

$$\max\{R(u) : u \in S_n,\ u \perp k_1, \ldots, k_{r-1}\}$$
$$\leqslant \max\{R(u) : u \in S_{n+1},\ u \perp k_1, \ldots, k_{r-1}\}$$

because $S_n \subset S_{n+1}$, and the maximum over a subset of S_{n+1} cannot exceed the maximum over S_{n+1}. Minimising both sides of this inequality over $k_1, \ldots k_{r-1}$, and using (10.31), gives

$$\mu_r^{(n)} \leqslant \mu_r^{(n+1)} \quad .$$

□

Theorem 10.21 (Convergence of the Rayleigh-Ritz Method)

$$\mu_r^{(n)} \to \lambda_r \quad \text{as} \quad n \to \infty \quad ,$$

where λ_r is the r-th eigenvalue of a positive symmetric operator A such that either A or A^{-1} is compact, and $\mu_r^{(n)}$ is the n-th order Rayleigh-Ritz approximation to λ_r using trial functions $\phi_1, \ldots, \phi_n$ where (ϕ_i) is an orthogonal basis.

Proof We write out the case where A is compact; if A is invertible and A^{-1} is compact, the proof is the same except that one uses A^{-1}, with eigenvalues λ_r^{-1}, in place of A. We shall show that $\lambda_r - \|A - P_nAP_n\| \leqslant \mu_r^{(n)} \leqslant \lambda_r$, and then that $\|A - P_nAP_n\| \to 0$ as $n \to \infty$; P_n is the projection operator on to the subspace spanned by $\{\phi_1, \ldots, \phi_n\}$.

The minimax principle, Theorem 10.18, gives

$$\lambda_r = \min_{k_1, \ldots, k_{r-1}} \{\max\{(u,Au) : \|u\| = 1,\ u \perp k_1, \ldots, k_{r-1}\}\}$$
$$\leqslant \max\{(u,Au) : \|u\| = 1,\ u \perp f_1, \ldots, f_{r-1}\}$$

where f_i is the maximising vector at the i-th stage of the Rayleigh-Ritz procedure of order n. Therefore

$$\lambda_r \leqslant \max\{(u,[A - P_nAP_n]u) + (u,P_nAP_nu)\colon \|u\| = 1,\ u\perp f_1, \ldots, f_{r-1}\}\ ,$$

$$\therefore \lambda_r \leqslant \max\{(u,[A - P_nAP_n]u)\colon \|u\| = 1,\ u\perp f_1, \ldots, f_{r-1}\}$$

$$+ \max\{(u,P_nAP_nu)\colon \|u\| = 1,\ u\perp f_1, \ldots, f_{r-1}\}\ . \tag{10.32}$$

Now, $\max\{(u,[A - P_nAP_n]u)\colon \|u\| = 1,\ u\perp f_1, \ldots, f_{r-1}\}$

$$\leqslant \max\{(u,[A - P_nAP_n]u)\colon \|u\| = 1\} = \|A - P_nAP_n\|\ ,$$

since removing the orthogonality constraints cannot reduce the maximum. And

$$\max\{(u,P_nAP_nu)\colon \|u\| = 1,\ u\perp f_1, \ldots, f_{r-1}\}$$

$$= \max\{(u,Au)\colon u\in S_n,\ \|u\| = 1,\ u\perp f_1, \ldots, f_{r-1}\}$$

$$= \mu_r^{(n)}\ ,$$

since P_nu runs through S_n as u runs through the Hilbert space. Thus (10.32) gives

$$\lambda_r \leqslant \|A - P_nAP_n\| + \mu_r^{(n)}\ .$$

But from Theorem 10.20 we have $\lambda_r \geqslant \mu_r^{(n)}$. Hence

$$\lambda_r - \|A - P_nAP_n\| \leqslant \mu_r^{(n)} \leqslant \lambda_r\ .$$

The lemma below shows that $\|A - P_nAP_n\| \to 0$ as $n \to \infty$, and thus completes the proof. □

Lemma 10.22 Let A be a compact operator on a Hilbert space H, and (P_n) a sequence of projections such that

$$P_nx \to x \quad \text{as} \quad n \to \infty \tag{10.33}$$

for all $x \in H$. Then $P_nAP_n \to A$ as $n \to \infty$.

Proof We must prove that $\|P_nAP_n - A\| \to 0$. Suppose not; then there is an $\epsilon > 0$, and a subsequence (r) of the sequence $(n) = (1,2,3, \ldots)$, such that

$$\|P_rAP_r - A\| > \epsilon$$

for all r in the subsequence. By definition of the operator norm, it follows that for each r there is an x_r with $\|x_r\| = 1$ such that

$$\|(P_rAP_r - A)x_r\| > \epsilon\ . \tag{10.34}$$

Now, (x_r) is a sequence in the unit sphere; by 7.70 it has a weakly convergent subsequence, call it (x_s), with

$$x_s \to \xi \quad \text{weakly as} \quad s \to \infty$$

for some ξ. We shall obtain a contradiction to (10.34) by showing that Ax_s and $P_sAP_sx_s$ converge to the same limit.

Since A is compact, we have

$$Ax_s \to A\xi \quad \text{strongly as} \quad s \to \infty \tag{10.35}$$

by 8.55.

Next we show that $P_sx_s \to \xi$ weakly. We have, for any $y \in H$,

$$(y,P_sx_s) = (P_sy,x_s)$$

$$\therefore (y,P_sx_s) = (P_sy - y,x_s) + (y,x_s) \quad . \tag{10.36}$$

But $\quad |(P_sy - y,x_s)| \leqslant \|P_sy - y\| \, \|x_s\| \to 0 \quad \text{as} \quad s \to \infty$

using (10.33), and the fact that $\|x_s\| = 1$. Now (10.36) gives

$$(y,P_sx_s) \to (y,\xi) \quad \text{as} \quad s \to \infty \quad ,$$

or $\quad P_sx_s \to \xi \quad$ weakly.

Because A is compact, it follows that

$$AP_sx_s \to A\xi \quad \text{strongly.} \tag{10.37}$$

Now, $\quad P_sAP_sx_s = P_s(AP_sx_s - A\xi) + P_sA\xi \quad .$

Since $\|P_s\| = 1$, the first term here tends to zero by (10.37), and using (10.33) we have

$$P_sAP_sx_s \to A\xi \quad .$$

Finally, combining this result with (10.35) gives

$$(P_sAP_s - A)x_s \to 0$$

which contradicts (10.34), (s) being a subsequence of (r). This proves the lemma. □

10.5 Inhomogeneous Equations

Up to now we have concentrated on the eigenvalue problem for positive operators. We now turn to another important class of linear problems, the solution of

$$Au = f$$

where f is given and u is unknown. We begin by looking at two concrete examples.

Application 10.23 (Reflection of Water Waves) A train of waves on the surface of deep water approaches a barrier, idealised as a vertical flat plate immersed in the water to a depth a; see Fig. 19. Some of the incident wave is reflected, and the rest passes beneath the barrier and reappears on the far side as a transmitted wavetrain. The effectiveness of the barrier is measured by a 'reflection coefficient'

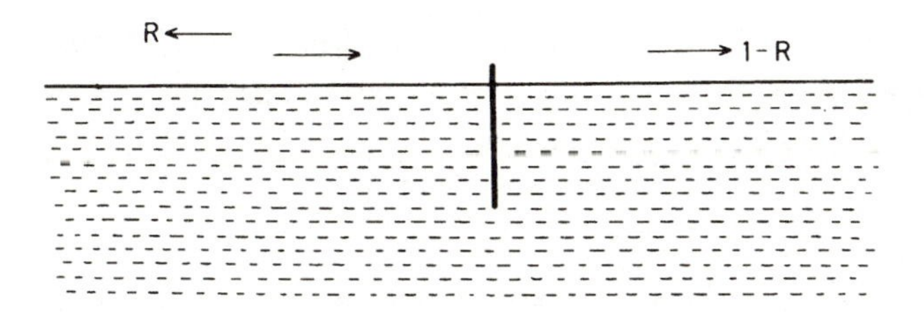

Fig. 19. Reflection of waves by a barrier.

R, which is the fraction of the incident wave which is reflected, so that $1 - R$ is the transmitted fraction. Of course, if the barrier is a solid wall from the surface to the bottom of the water, then the problem is trivial and $R = 1$. It is useful to be able to estimate R for various types of incomplete barrier, such as the one described above.

The parameters describing the problem are the depth of immersion a of the barrier, the wavelength of the incident waves, described by a parameter K, and the angle at which they approach the barrier, described by a parameter m. The fluid-mechanical equations can be reduced to an integral equation for an unknown function $u(y)$ (related to the water velocity at a depth y vertically below the barrier):

$$\int_a^\infty L(y,\eta)u(\eta)\,\mathrm{d}\eta = \mathrm{e}^{-Ky} \quad \text{for} \quad y > a \quad ,$$

where

$$L(y,\eta) = \int_0^\infty \frac{(k\cos ky - K\sin ky)(k\cos k\eta - K\sin k\eta)}{(k^2 + K^2)\sqrt{k^2 + K^2 - m^2}}\,\mathrm{d}k \quad .$$

The reflection coefficient is determined by

$$\left(\frac{-im}{\pi K}\right)\left(\frac{1-R}{R}\right) = \int_a^\infty u(y)\mathrm{e}^{-Ky}\,\mathrm{d}y \quad .$$

For the derivation of these equations, see Evans & Morris. If we work in the space of square-integrable functions on $[a,\infty)$, then the equations can be written

$$Au = f \quad ,$$

$$\left(\frac{-im}{\pi K}\right)\left(\frac{1-R}{R}\right) = (u,f) \quad ,$$

where $f(x) = \mathrm{e}^{-Kx}$ and A is the integral operator with kernel L. The problem then consists of solving $Au = f$ for u, and then calculating (u,f) in order to determine R.

Now, to solve the equation $Au = f$ takes a good deal of work. Most of it is wasted, because we are not really interested in the function u, but only in the reflection coefficient R, which depends on the single number (u,f). The variational method described below is designed for problems of this type: it gives good approximations to (u,f) directly, without having to solve $Au = f$ for u. For details of its application to this problem, see Evans & Morris.

Application 10.24 (Torsional Stiffness) Consider a rod whose cross-section is an arbitrary plane region S, the same at all points along its length. If a twisting force, or torque, is applied to the rod, it is natural to ask how easily it twists in response. This is measured by a constant k, called the modulus of torsional rigidity, proportional to the angle of twist produced by applying a unit torque to unit length of the rod. In books on elasticity theory it is shown that k can be calculated as follows. Solve the equation

$$-\nabla^2 u = 2 \quad \text{in} \quad S,$$

where ∇^2 is the two-dimensional Laplacian, with boundary conditions $u = 0$ on the boundary of S. Then

$$k = 2\iint_S u \, dx \, dy \quad .$$

Here $u(x,y)$ gives the displacement of all points in the rod; it is of relatively little practical interest, although it must in principle be calculated in order to find the important quantity k. As in the previous example, we can write

$$k = (u,2)$$

using the inner product $(f,g) = \iint_S f\bar{g} \, dx \, dy$; the method of this section allows one to estimate k without having to solve the equation $-\nabla^2 u = 2$. □

These examples are typical of a wide class of problems in applied mathematics where the centre of the problem is the solution of a linear operator equation $Au = f$, but where one is more interested in the scalar (u,f) than in the solution vector or function u. For a problem in elasticity, (u,f) may represent a measure of the stiffness of a system; in electrostatics, (u,f) may represent a capacitance; in problems of fluid flow past a body (u,f) may represent the frictional drag on the body; and so on. In this section we give an efficient method for estimating (u,f) directly. It is a variational method, very much like the Rayleigh-Ritz method, and depends on reducing the problem of solving $Au = f$ to the problem of maximising a certain functional.

Proposition 10.25 (Lower Bound for Inhomogeneous Equations) Let A be a symmetric positive linear operator on an inner product space V, and f a given member of V. Define

$$J(v) = (v,f) + (f,v) - (v,Av)$$

for all $v \in V$. If u satisfies $Au = f$, then for all $v \in V$

$$J(v) \leqslant J(u) = (u,f) \quad . \tag{10.38}$$

Proof We have

$$J(u) = (u,f) + (f,u) - (u,f)$$

and $$(u,f) = (u,Au) = (Au,u) = (f,u) \quad .$$

This proves the second part of (10.38). Now, set $v = u + w$;

$$\begin{aligned} J(u+w) &= (u+w,f) + (f,u+w) - (u+w,f+Aw) \\ &= (f,u+w) - (A[u+w],w) \\ &= (f,u) - (Aw,w) \quad . \end{aligned}$$

Hence $J(u+w) \leqslant (f,u) = (u,f) = J(u)$

since A is a positive operator. □

Thus (u,f) is the maximum value of the functional J, and one can obtain a lower bound for (u,f) by choosing a trial vector v and evaluating $J(v)$. Note how much simpler this is than the corresponding statement for eigenvalues, where the Rayleigh quotient gives either an upper or a lower bound according to whether A or A^{-1} is compact. Note also that $J(u+w) - J(u) = -(Aw,w)$ is quadratic in w; hence if w is fairly small, then $J(u+w) - (u,f)$ is much smaller, being second order in w. Thus (u,f) can be found to high accuracy by using only moderately good approximations to u. This is a general property of variational methods, and is familiar from elementary calculus: a smooth function varies very slowly in the neighbourhood of a maximum or minimum.

Example 10.26 Consider the very simple equation

$$u'' + \alpha u = -f(x), \ u(0) = u(1) = 0 \quad .$$

The eigenvalues of the operator $-D^2 - \alpha$, with these boundary conditions, are easily found; they are $n^2\pi^2 - \alpha$ for $n = 1,2,\ldots$. Hence if $\alpha < \pi^2$ the eigenvalues are all positive and Proposition 10.5 shows that $-D^2 - \alpha$ is a positive operator.

We shall work in the real vector space of real-valued functions, so $(v,f) = (f,v)$, and

$$J(v) = \int_0^1 v(x)[2f(x) + v''(x) + \alpha v(x)]\,dx \quad .$$

The domain of our operator is $\mathscr{D} = \{u \in C^2[0,1] : u(0) = u(1) = 0\}$. A natural choice of trial function satisfying the boundary conditions, and therefore belonging to $\mathscr{D}$, is

$$v(x) = c\sin(\pi x)$$

for some constant c, which we leave undetermined for the moment. Then some calculation gives

$$J(v) = c\int_0^1 2f(x)\sin(\pi x)\,dx - c^2(\pi^2 - \alpha)/2 \quad . \tag{10.39}$$

We have $(u,f) \geqslant J(v)$ for all values of c; the best estimate is given by choosing c so as to maximise J. This is easily done because (10.39) is quadratic in c; we obtain

$$(u,f) \geqslant [2/(\pi^2 - \alpha)]\left(\int_0^1 f(x)\sin(\pi x)\,dx\right)^2 \quad . \tag{10.40}$$

Taking the concrete example $f(x) = 1$ we have

$$(u,f) \geqslant \frac{8}{\pi^2(\pi^2 - \alpha)} \quad . \tag{10.41}$$

Now, this example is so simple that it is easy to solve exactly by the method mentioned in 9.18, expanding u as a series of eigenfunctions, that is, a Fourier sine series. One finds

$$(u,f) = \frac{8}{\pi^2}\left[\frac{1}{\pi^2 - \alpha} + \frac{1}{9(9\pi^2 - \alpha)} + \frac{1}{25(25\pi^2 - \alpha)} + \ldots\right] \quad . \tag{10.42}$$

Thus if $\alpha < \pi^2$, the error in (10.41) is no more than about 1%. □

One can improve the approximation by choosing a trial function with more adjustable constants; in particular one can use a Rayleigh-Ritz method, choosing a trial function $\Sigma_1^n c_r\phi_r$, where (ϕ_r) is a complete orthogonal set. As in section 10.4, one can prove that under certain conditions the n-th order approximation converges to the exact solution as $n \to \infty$. The theory is simpler than for the eigenvalue problem; see Problem 10.22. Rather than discuss the convergence question further, we shall turn to another aspect of the approximation problem, namely, the question of how to assess the accuracy of the approximation when there is no exact solution available for comparison.

10.6 Complementary Bounds

For both the eigenvalue problem and the inhomogeneous equation we have obtained one-sided bounds, that is, inequalities of the form $(u,f) > k$ where k is known or calculable, and similar inequalities for eigenvalues. But we do not know how much greater than k the value of (u,f) is. Even given a convergence theorem, that $k \to (u,f)$ as the order of approximation tends to infinity, we still do not know how large the error $|k - (u,f)|$ is for a given order of approximation. If we could find an upper bound K for (u,f), then we would have

$$k \leqslant (u,f) \leqslant K \quad ,$$

and the range of possible values of (u,f) would be a finite interval, a great improvement on the semi-infinite range of possible values left open by the one-sided bound $(u,f) \geqslant k$. If the approximations are skilfully chosen, we might even hope for k and K to be close together, in which case (u,f) would be known with considerable precision.

Such upper and lower bounds K and k are often called **complementary bounds**, or complementary approximations, for (u,f). There is a large literature on methods of finding complementary bounds for eigenvalue problems, and for linear and nonlinear inhomogeneous equations. In this section we shall discuss a simple method for the linear inhomogeneous equation $Au = f$. The method applies to operators which are not only positive but satisfy the stronger condition that $(u,Au)/(u,u)$ is never near zero, in the following sense.

Definition 10.27 A symmetric operator A on an inner product space V is called **strongly positive**, or **coercive**, if there is a number $M > 0$ such that

$$(u,Au) \geqslant M(u,u) \quad \text{for all} \quad u \in V \quad .$$

Any such M is called a **lower bound** for the operator A.

Example 10.28 If B is a bounded self-adjoint operator with $\|B\| < 1$, and I is the identity, then the operator $A = I - B$ is strongly positive, with lower bound $1 - \|B\|$. □

A positive operator has positive eigenvalues (Proposition 10.2). We now show that if the eigenvalues are positive and bounded away from zero, then the operator is strongly positive.

Proposition 10.29 (Strongly Positive Operators) If A is a symmetric operator with a complete set of eigenvectors, then A is strongly positive if there is an $M > 0$ such that $\lambda_r \geqslant M$ for all eigenvalues λ_r of A. M is then a lower bound for A.

Proof Let λ_n be the eigenvalues and e_n the orthonormal eigenvectors of A. For any vector v, we have, using Problem 7.11,

$$\begin{aligned}(v,Av) &= \Sigma(v,e_n)(e_n,Av) \\ &= \Sigma(v,e_n)(Ae_n,v) \\ &= \Sigma\lambda_n(v,e_n)(e_n,v) \\ &\geqslant M\Sigma(v,e_n)(e_n,v) = M(v,v) \quad .\end{aligned}$$

□

This result implies that compact self-adjoint operators on infinite-dimensional spaces cannot be strongly positive (cf. Theorem 9.4). But many common differential operators are strongly positive.

Example 10.30 The operator $-D^2$, with boundary conditions $u(0) = u(1) = 0$, is strongly positive by Proposition 10.29; its eigenfunctions are complete (Theorem 9.16), and its eigenvalues are $\pi^2, 4\pi^2, \ldots$, so π^2 is a lower bound for $-D^2$.

We now give another proof that this operator is strongly positive, which does not depend on knowing the eigenvalues and is therefore more easily generalised to other Sturm-Liouville systems. Integration by parts gives

$$(u, -D^2 u) = \int_0^1 |u'|^2 \, dx \tag{10.43}$$

if u satisfies the boundary conditions. Now,

$$\begin{aligned} |u(x)| &= \left|\int_0^x u'(t)\, dt\right| \\ &\leqslant \left(\int_0^x |u'|^2 \, dt \cdot \int_0^x dt\right)^{\frac{1}{2}} \end{aligned}$$

using the Cauchy-Schwarz inequality. Hence

$$\begin{aligned} \int_0^1 |u|^2 \, dx &\leqslant \int_0^1 \left(\int_0^x |u'|^2 \, dt\, x\right) dx \\ &\leqslant \int_0^1 \left(\int_0^1 |u'|^2 \, dt\right) x \, dx \\ &= \tfrac{1}{2} \int_0^1 |u'|^2 \, dt \\ &= \tfrac{1}{2} (u, -D^2 u) \end{aligned}$$

using (10.43). Hence $-D^2$ is strongly positive, with lower bound 2. As might be expected, the lower bound given by this method is less than that given by the first method, which uses more detailed information about the operator. □

We now return to the problem of finding complementary bounds for (u,f) where u satisfies $Au = f$. We must find a pair of functionals, J and K say, such that $J(v) \leqslant (u,f) \leqslant K(v)$ for all v in the space, and which give good approximations to (u,f) when v is not too far from u. We have found such a J in Proposition 10.25, and it satisfies $J(u) = (u,f)$. Let us construct a new functional G by adding an extra term to J; if we arrange for this extra term to vanish when $v = u$, then $G(u)$ will equal (u,f), and $G(v)$ may therefore be expected to give a reasonable approximation to (u,f) if v is close to u. Thus we are led to consider the functional

$$G(v) = (v,f) + (f,v) - (v,Av) + (f - Av, H[f - Av]) \quad ,$$

where H is an operator to be specified later. The last term has been chosen so as to vanish when $v = u$, where $Au = f$. Now, setting $v = u + w$, we have, after some easy algebra,

$$G(u + w) = (u,f) + (w, [AHA - A]w) \quad . \tag{10.44}$$

This is quadratic in w, showing that $G(v)$ is stationary at $v = u$, that is, a small change in v produces a much smaller change in the value of G. Thus for any operator H, the functional G can reasonably be used to estimate (u,f); special choices of H may give upper or lower bounds.

If H is chosen so that $AHA - A$ is a negative operator, then (10.44) shows that $G(v)$ gives a lower bound for (u,f). The simplest such choice is $H = 0$; then $AHA - A$ is negative because A is positive. In this case, G reduces to J, which we have discussed in the last section. Other choices are possible, however, giving better approximations (see Problem 10.20).

If, on the other hand, H is chosen so that $AHA - A$ is positive, then $G(v)$ gives an upper bound. The simplest choice is $H = cI$ where c is a positive number and I the identity. If c is large enough, then the positive term cA^2 will outweigh the negative term $-A$, and $AHA - A$ will be positive. Indeed, we have

$$AcIA - A = A(cA - I) \quad ;$$

A is positive, and if $cA - I$ is positive, then Theorem 10.7 shows that $AHA - A$ is positive. But if A is strongly positive, then (using the notation of 10.27) $A - MI$ is positive, as is easily verified. Hence $cA - I$ is positive if we take $c = M^{-1}$. Thus if

$$H = M^{-1}I$$

then $AHA - A$ is positive, and G gives an upper bound for (u,f). We have now proved the following.

Proposition 10.31 (Complementary Bounds) Let A be a strongly positive symmetric linear operator on an inner product space V, and f a given member of V. Define a functional K by

$$K(v) = J(v) + M^{-1}\,\|f - Av\|^2$$

where $J(v) = (v,f) + (f,v) - (v,Av)$, and M is a lower bound for A. If u satisfies $Au = f$, then

$$J(v) \leqslant (u,f) \leqslant K(v) \quad \text{for all} \quad v \in V \quad . \tag{10.46}$$

Example 10.32 Consider the problem of Example 10.26:

$$u'' + \alpha u = -f(x), \;\; u(0) = u(1) = 0 \quad .$$

Using the trial function $v(x) = c\sin(\pi x)$ gave us the lower bound (10.40) for (u,f). Using the same trial function in the functional K gives

$$K(v) = \frac{1}{M}\int [f(x)]^2\,dx - \left(\frac{\pi^2 - \alpha}{M} - 1\right)\left[2c\int f(x)\sin(\pi x)\,dx - c^2\left(\frac{\pi^2 - \alpha}{2}\right)\right] \quad .$$

Choosing c so as to minimise this gives

$$(u,f) \leqslant \frac{1}{M}\left[\int f^2\,dx - 2\left(\int f(x)\sin(\pi x)\,dx\right)^2\right] \quad .$$

$$+\left(\frac{2}{\pi^2 - \alpha}\right)\left[\int f(x)\sin(\pi x)\,dx\right]^2 \quad .$$

This holds for any $M \leqslant \pi^2 - \alpha$. Since the coefficient of M^{-1} here is positive (as is easily shown using the Schwarz inequality), the best bound is given by the largest possible M, namely, $\pi^2 - \alpha$. Thus

$$(u,f) \leqslant \left(\frac{1}{\pi^2 - \alpha}\right)\int f^2 \,\mathrm{d}x \quad ,$$

and combining this with (10.40) we have the complementary bounds

$$\left(\frac{2}{\pi^2 - \alpha}\right)\left(\int f(x)\sin(\pi x)\,\mathrm{d}x\right)^2 \leqslant (u,f) \leqslant \left(\frac{1}{\pi^2 - \alpha}\right)\int [f(x)]^2 \,\mathrm{d}x \quad . \tag{10.47}$$

For the simple case $f(x) \equiv 1$, (10.47) gives

$$0.81 \leqslant (u,f)(\pi^2 - \alpha) \leqslant 1 \quad .$$

Thus the crude trial function $c\sin(\pi x)$ gives (u,f) to an accuracy of 10% – that is, taking (u,f) to be in the middle of the interval (0.81,1) gives an error around 10% at most.

For the case $f(x) \equiv 1$, we saw in 10.26 that the exact value of (u,f) is only 1% away from the lower bound $0.81/(\pi^2 - \alpha)$; the Rayleigh-Ritz lower bound is thus closer than the upper bound of Proposition 10.31. The accuracy can be improved either by using better upper bounds (see Problem 10.20), or by using a trial function with more adjustable constants, as follows.

We shall use a second-order Rayleigh-Ritz method, taking v to be a linear combination of two orthogonal functions. One might at first try $v(x) = c\sin\pi x + d\sin 2\pi x$. However, for the case $f(x) \equiv 1$, one finds that $J(v)$ is greatest and $K(v)$ is least when $d = 0$, and the bounds are the same as those given by the first-order method. This is because the problem is symmetrical about the mid-point of the interval (0,1), whereas $\sin 2\pi x$ is antisymmetrical. A good trial function should be as similar as possible to the exact solution, and in particular should have the same symmetry properties; adding an antisymmetric term to the trial function for a symmetric problem will not improve the approximation.

We therefore take $v(x) = c\sin\pi x + d\sin 3\pi x$. We choose c and d first so as to maximise J, giving an upper bound, and then so as to minimise K, giving a lower bound. The calculation is a little tedious. Taking $\alpha = \pi^2/2$, so as to have concrete numbers to discuss, we obtain

$$0.1653 \leqslant (u,f) \leqslant 0.1855 \quad .$$

The first-order approximation for this value of α is

$$0.1642 \leqslant (u,f) \leqslant 0.2026 \quad ,$$

and the exact value is

$$(u,f) = 0.1655 \quad .$$

We see that the second-order upper bound is considerably better than the first-order upper bound, but the lower bounds are much better than either. This is a general feature of complementary approximations: the simple and natural Rayleigh bound is often closer than the more contrived complementary bound.

10.7 Summary and References

In section 10.1 we introduced positive operators, and in section 10.2 we discussed Rayleigh's variational principle for the first eigenvalue of a positive operator. In section 10.3 we introduced the Rayleigh-Ritz method for systematically improving the approximation and finding higher eigenvalues; in section 10.4 we discussed other variational properties of eigenvalues, and used them to prove convergence of the Rayleigh-Ritz method. In section 10.5 we described a variational principle for inhomogeneous linear equations, and section 10.6 gives a method for obtaining complementary bounds for inhomogeneous equations.

There is a rich literature on variational methods for linear equations. Elementary discussions will be found in Dettman, Hochstadt, Morse & Feshbach, and similar books. Our treatment of eigenvalue problems follows Gould; there is also an excellent treatment (along different lines) in volume I of Courant-Hilbert. Mikhlin gives many examples of variational calculations, and describes the theory in detail; Stakgold, Milne, and Nowinski are also very useful. The method given in section 10.6 for deriving complementary bounds for inhomogeneous equations is due to Cole & Pack, who give a general approach of which Proposition 10.31 is a special case. Other methods will be found in Stakgold and Mikhlin; these authors also discuss compementary bounds for eigenvalues, as do Gould and Weinberger; see also Problem 10.9. For variational methods in non-linear problems, see Bellman (1970/73), Barnsley & Robinson, or Arthurs.

PROBLEMS 10

Section 10.1

10.1 If $A = T^2$, where T is an operator on a real Hilbert space, is A positive? What if $A = T^*T$, where T^* is the adjoint of T?

10.2 Under what conditions on the functions p and q is the Sturm-Liouville operator $\mathrm{D}p\mathrm{D} + q$ positive?

10.3 Find a domain $\mathscr{D} \subset L_2[V]$ such that the biharmonic operator ∇^4 (see Problem 3.12) is positive on $\mathscr{D}$. Here V is a region in $\mathbb{R}^3$.

10.4 Let A be the linear integral operator on $L_2[a,b]$ with kernel $K(x,y)$, where K is a continuous function. Show that if A is a positive operator, then $K(x,x) \geqslant 0$ for $a \leqslant x \leqslant b$. Conversely, if $K(x,x) \geqslant 0$, does it follow that A is positive?

Section 10.2

10.5 Consider the integral operator $\int_0^1 e^{xy} f(y)\, dy$.

(a) Use the trial function $f(x) = 1$ to show that the operator has an eigenvalue greater than 1.318 (Note: you will get an integral which can easily be evaluated by expanding in a power series).

(b) Improve this estimate by using a trial function of the form $f(x) = 1 + ax$, where a is a constant.

10.6 Construct a polynomial in x and y which vanishes on the perimeter of an equilateral triangle, and is positive in the interior. (Hint: take a product of three factors, each of which vanishes on one side.)

Use this polynomial as a trial function to give an upper bound for the first eigenvalue of $-\nabla^2$ in an equilateral triangle, with the boundary condition of vanishing on the perimeter.

10.7 I wish to make a bass drum. The only shape I can build is a rectangle. I can only afford a certain area A of material to cover it. What shape of rectangle will give the deepest note, for the given area A?

(The problem is to minimise the frequency of vibration, which is proportional to the square root of the first eigenvalue of $-\nabla^2$ with the boundary condition that the function u (representing the displacement of the drumhead) vanishes on the boundary.)

10.8 Consider the operator $-\nabla^2$ in the ellipse $x^2/a^2 + y^2/b^2 \leqslant 1$, with the Dirichlet boundary condition as usual. Use a simple polynomial trial function in the Rayleigh approximation to show that for all ellipses of a given area, the circle gives the lowest value for the first eigenvalue. (This is why drums are circular: to make them any other shape would take more skin to give the same note.)

10.9 The results of Problems 10.7 and 10.8 can be summarised by saying that the most symmetrical shape (for a given area) gives the lowest eigenvalue. This is true in general: for any class of regions of the same area (or the same perimeter), the most symmetrical region gives the lowest eigenvalue. In particular, the circle gives a lower eigenvalue than any other shape. For a general discussion of results of this type, see Chapter X of Polya; for proofs, see Polya & Szegö (1951).

This principle can be used to give complementary bounds for eigenvalues. By comparing an equilateral triangle with a circle (i) of the same area, and (ii) of the same perimeter, obtain lower bounds for the eigenvalue of Problem 10.6. (Lower bounds given by this method are quite crude. Better results can be obtained by comparing the triangle with a square. But it is better still to consult the references listed in section 10.7.)

10.10 The vibration frequencies of a rod fixed at one end and carrying a weight on the other end are given by the eigenvalues of the equation $v'' + \lambda v = 0$, with $v(0) = 0$, $v'(1) + \alpha^2 v(1) = 0$, where the constant α^2 depends on the weight.

Integrate the Rayleigh quotient by parts to give a form similar to equation (10.8) (but different because of the different boundary conditions). Use a trial function of the form $x + cx^2$, for a suitable c, to estimate the first eigenvalue.

Sections 10.3 and 10.4

10.11 Use the Rayleigh-Ritz method to estimate the first two eigenvalues of the matrix

$$\begin{pmatrix} 1.0 & 0.2 & 0.1 & -0.2 \\ 0.2 & 2.0 & 0.2 & -0.1 \\ 0.1 & 0.2 & 3.0 & 0.3 \\ -0.2 & -0.1 & 0.3 & 1.0 \end{pmatrix}.$$

10.12 Apply the Rayleigh-Ritz method to one or two of the problems for section 10.2.

10.13 If $\phi_1, \ldots, \phi_n$ are orthonormal functions and A a self-adjoint operator, define an $n \times n$ matrix (a_{ij}) by $a_{ij} = (\phi_i, A\phi_j)$. Show that the eigenvalues of (a_{ij}) are the Rayleigh-Ritz approximations to the first n eigenvalues of A. (This result gives a systematic way of calculating the Rayleigh-Ritz approximations, suitable for use with a computer.)

10.14 Use the maximin principle (Theorem 10.18) to prove the following. Let L be a self-adjoint positive differential operator on a region V, with boundary condition that $u = 0$ on the boundary of V, and with a compact inverse (for example, $L = -\nabla^2$ for $V \subset \mathbb{R}^3$, or L equals the operator of Problem 10.2, with $V = [a,b]$).

(a) If $V_1 \subset V$, then the n-th eigenvalue of L on V_1 is greater than or equal to the n-th eigenvalue of L on V. (By 'L on V_1' we mean 'L with boundary conditions $u = 0$ on the boundary of V_1'). This is a generalisation of the fact that small objects vibrate faster than larger objects.

(b) If $V_1, V_2, \ldots, V_r$ are non-overlapping subregions of V, then the n-th eigenvalue of L on V is less than or equal to the n-th member of a list obtained by taking all the eigenvalues of L on V_1, L on $V_2, \ldots, L$ on V_r, and arranging them in a single increasing sequence. For further developments of these ideas, see Courant–Hilbert, vol. 1.

10.15 If A,B are self-adjoint linear operators on a Hilbert space H, we say $A \geqslant B$ if $A - B$ is a positive operator, and $A > B$ if $A - B$ is positive-definite. Show that $A > B$ & $B > C$ implies $A > C$, and that $A \geqslant B$ & $B \geqslant A$ implies $A = B$; thus the $>$ and $\geqslant$ symbols can be manipulated as in elementary algebra. Show that if A and B are not self-adjoint, then the inequalities $A \geqslant B$ & $B \geqslant A$ do not imply $A = B$.

If A and B satisfy the conditions of 10.18(b) , and $A \geqslant B$, show that $\lambda_n^A \geqslant \lambda_n^B$ where λ_n^A, λ_n^B are the n-th eigenvalues of A,B.

Use this result to give bounds for the eigenvalues of $u'' + r(x)u + \lambda u = 0$, $u(0) = u(1) = 0$, if r is a bounded positive function.

Sections 10.5 and 10.6

10.16 Work out the variation on Example 10.26 in which the boundary conditions are replaced by $u'(0) = u'(1) = 0$.

10.17 (a) Apply a Rayleigh-Ritz technique to the variational principle of Proposition 10.25: take a trial vector of the form $v = c_1\phi_1 + \ldots + c_n\phi_n$, where $\phi_1, \ldots, \phi_n$ are linearly independent vectors, and choose the scalars c_r so as to maximise the functional $J(v)$. Show that the best values of c_r are the solutions of the equations

$$\sum_{r=1}^{n} c_r(A\phi_r,\phi_s) = (f,\phi_s), \quad s = 1, \ldots, n \quad ,$$

called Galerkin's equations. (This is easy for a real vector space, where c_r are real; to prove it for a complex space, write $c_r = d_r + ie_r$, and minimise with respect to d_r and e_r.)

(b) Galerkin's equations can be obtained from a quite different point of view, as follows. Given the equation $Au = f$, we look for an approximate solution v of the form $v = \Sigma_1^n c_r\phi_r$. If v were an exact solution, we would have $Av - f = 0$. We cannot in general choose the c_r so as to make $Av - f$ vanish, but we can at least make its projection on to the subspace spanned by $\{\phi_r\}$ vanish. Show that this is achieved when the c_r satisfy Galerkin's equations.

(The advantage of this approach to the approximation problem, known as Galerkin's method, is that it applies to a wider range of equations than the variational method; A need not be positive or self-adjoint. It also suggests that if $\phi_1, \ldots, \phi_n$ are the first n elements of a basis for the space, then the approximate solution $v = \Sigma_1^n c_r\phi_r$ might converge to the exact solution as $n \to \infty$; see Mikhlin or Milne, for example.)

10.18 The vertical displacement $y(x)$ of a point at distance x from one end of a horizontal girder satisfies the equation $y'''' - \alpha y'' = cf(x)$, where $f(x)$ is the load it carries, α is related to the tension in the girder, and c to its strength (cf. Problem

2.10, which deals with the case of zero tension). If it is clamped at both ends, and has unit length, then $y(0) = y'(0) = y(1) = y'(1) = 0$. For a uniform load, $f(x) \equiv 1$, use the method of Problem 10.17 to estimate $y(x)$; take $n = 2$, and $\phi_1(x) = x^2(1-x)^2$, $\phi_2(x) = x^3(1-x)^3$ so as to satisfy the boundary conditions and the symmetry condition as discussed in Example 10.32.

If you have access to a computer, you can calculate higher-order Rayleigh-Ritz approximations, and investigate how fast they converge as n increases.

10.19 Given that $u(x) - \int_0^1 e^{-|x-y|}u(y)\,dy = f(x)$, show that

$$(e/2)\left(\int_0^1 f(x)\,dx\right)^2 \leqslant \int_0^1 f(x)u(x)\,dx \leqslant (2e^2/e^2 - 1)\int_0^1 [f(x)]^2\,dx - \left\{\int_0^1 f(x)[\alpha + \beta e^{-x} + \gamma e^x]\,dx\right\}^2 ,$$

where α, β, γ are constants, which you should find.

10.20 (A more refined upper bound for integral equations.) Consider the operator $A = I - K$ on the real Hilbert space $L_2[0,1]$ where K is an integral operator with a symmetric positive kernel $k(s,t) = k(t,s) > 0$. Assume that $\|K\| < 1$, so that A is strongly positive.

(a) Use the identity $\iint k(s,t)[u(s) - u(t)]^2\,ds\,dt \geqslant 0$ to show that $(u,Ku) \leqslant (u^2,\kappa)$, where the function κ is defined by $\kappa = K1$, and 1 is the function whose value everywhere is 1.

(b) Show that $(u,Au) \geqslant (u,\alpha u)$ for all u, where $\alpha = A1$. Deduce that $M \equiv \min\{\alpha(t)\}$ is a lower bound for A.

(c) Assume that there is a number $\lambda < 1$ such that $\int_0^1 k(s,t)\,ds < \lambda$ for all t. Show that α^{-1} and $A\alpha^{-1}A - A$ are then positive operators, where α^{-1} is the operator which multiplies any function by $1/\alpha$.

(d) Deduce from the theory of section 10.6 that

$$(u,f) \leqslant J(v) + (f - Av, \alpha^{-1}[f - Av])$$

for any v, where u is the solution of $Au = f$. Show that this upper bound for (u,f) is better, that is, smaller, than that given by Proposition 10.31.

For a discussion and generalisations of this, see Cole & Pack. Concrete examples usually involve integrals which cannot be evaluated analytically, and, as for most practical problems, numerical computation must be used. The reader with computational experience will easily be able to devise a suitable integral equation, and compare the upper bound of this problem with that of Proposition 10.31.

10.21 (Least Squares Approximation) Another approach to the approximate solution of $Au = f$ is the following. Take a set of linearly independent vectors $\phi_1, \ldots, \phi_n$, and a trial vector $v = \sum_1^n c_r\phi_r$; choose the scalars c_r so as to minimise

$\| Av - f \|^2$, which is a measure of how far v fails to satisfy $Av = f$. Show that the c_r satisfy

$$\sum_{r=1}^{n} c_r(A\phi_r, A\phi_s) = (f, A\phi_s), \quad s = 1, \ldots, n \quad .$$

Apply this method, for $n = 2$, say, to the problem of Example 10.26, and compare the results with those given by Galerkin's equations (Problem 10.17).

More problems will be found on page 378.

PART IV

FURTHER DEVELOPMENTS

Applied functional analysis is a large subject, and we can only discuss a small part of it. The last two chapters contain brief introductions to some topics too extensive to be treated in detail in a book of this size. They are intended to give a general impression of some of the main ideas, and to lead the reader to consult more complete accounts. They are not intended to cover their subjects in the more or less systematic manner of Parts I–III, and there are no difficult proofs and no Exercises. In this sense Part IV may be regarded as light reading.

Chapter 11 is of a more or less practical character. It begins with the theory of differentiation of operators in Banach spaces, and discusses applications in dynamics, that is, the theory of systems governed by differential equations involving time-derivatives. Chapter 12 is more theoretical; it discusses L_2 spaces from the point of view of the theory of generalised functions developed in Part I, and applies the theory to the existence-uniqueness problem for Laplace's equation. Chapter 11 only requires a grasp of the basic ideas about operators on Banach and Hilbert spaces. Chapter 12 is independent of Chapter 11, but needs a good understanding of Part I and of Chapter 7.

Chapter 11

The Differential Calculus of Operators and its Applications

An operator on a normed space is a generalisation of the idea of a function of a real variable, to which the methods of algebra and calculus can be applied. We have used the algebra of operators in Part II; we shall now discuss the differential calculus. The main idea is that a nonlinear operator can be approximated in small regions by a linear operator, in the same way that in ordinary calculus a curve is approximated by the straight line touching it. This leads to useful ways of dealing with nonlinear operators, by replacing them by the local linear approximations. In the first two sections of this chapter we shall outline the theory of operator differentiation, and in section 11.3 we discuss maximum–minimum problems and the calculus of variations. The rest of the chapter deals with applications of operator theory to dynamical systems, that is, differential equations involving time-derivatives. In sections 11.4 and 11.5 we discuss the theory of stability of equilibrium, and in sections 11.6 and 11.7 the closely related question of multiple solutions of nonlinear equations, that is, bifurcation theory.

11.1 The Fréchet Derivative

The derivative of a function of a real variable is defined by

$$f'(x) = \lim_{h \to 0} \left(\frac{f(x+h) - f(x)}{h} \right) \quad . \tag{11.1}$$

If we are to extend this definition to mappings of a vector space, then x and h will be vectors. But one of the first things one learns in elementary vector algebra is never to divide by a vector; it does not make sense. The definition (11.1) is therefore unsuitable for our purposes.

However, it is easy to rearrange (11.1) so as to avoid dividing by h. It is equivalent to $hf'(x) = f(x+h) - f(x) + hz(h)$, where $z(h) \to 0$ as $h \to 0$. We can now say that $f'(x)$ is the derivative of f at x if

$$f(x+h) = f(x) + f'(x)h + \phi(h) \quad , \tag{11.2}$$

where $\phi(h) = hz(h)$ is a function which vanishes faster than h as $h \to 0$, that is, $\phi(h)/h \to 0$. If we can express the idea '$\phi(h)$ vanishes faster than h' in a normed space, then we will be able to generalise (11.2). This is easily done, using the norm of a vector as a measure of how fast it is vanishing. First we need an appropriate definition of convergence.

Definition 11.1 Let f be a mapping $X \to M$, where $X \subset N$, and N and M are normed spaces. If $a \in N$, we say that f is **defined on a neighbourhood of** a if the domain X of f contains the sphere $\{x : \|x - a\| < \alpha\}$ for some $\alpha > 0$. If $b \in M$ and f is defined on a neighbourhood of a, we say $f(x) \to b$ as $x \to a$ if, given any $\epsilon > 0$, there is a $\delta > 0$ such that $\|f(x) - b\| < \epsilon$ whenever $0 < \|x - a\| < \delta$ and $x \in X$. □

This is a direct translation of the elementary definition into the language of normed spaces, just like our definition of convergence of sequences in Chapter 4. We shall now use it to define the idea 'f vanishes faster than h'.

Definition 11.2 Suppose f: $X \to M$ is defined on a neighbourhood of $0 \in X \subset N$. We say $f(h) = o(h)$ (read '$f(h)$ is little oh of h') if $\|f(h)\|/\|h\| \to 0$ as $h \to 0$. □

The letter o stands for order of magnitude, and $f = o(h)$ means that f is of a smaller order of magnitude than h.

Examples 11.3 (a) For $N = M = \mathbb{C}$, the functions z^2, $2z^{\frac{3}{2}}$, and $z \sin z$ are all $o(z)$ as $z \to 0$. The functions $2z$ and $\sin z$ are not $o(z)$.

(b) If f is a bounded linear functional on a normed space N, with $\|f\| \neq 0$, then $f(h) \neq o(h)$. *Proof:* for any $\lambda < 1$ there is a $v \in N$ with $\|v\| = 1$ and $|f(v)| \geqslant \lambda \|f\|$; then taking $h = \alpha v$ for real positive α we have $\|f(h)\|/\|h\| \geqslant \lambda \|f\| \not\to 0$ as $\alpha = \|h\| \to 0$. This result is an analogue of the statement $2z \neq o(z)$ in (a) above. It is not restricted to functionals; for any bounded linear operator f, $f(h)$ is of the same order of magnitude as h, not of smaller order, so $f(h) \neq o(h)$.

(c) Define $A : C[0,1] \to C[0,1]$ by $(Au)(t) = \int_0^1 \sin[stu(s)]u(s)\,ds$. Then $Au = o(u)$ as $u \to 0$ (*proof:* $|\sin x| \leqslant |x|$ for all x, hence $\|Au\| \leqslant \int_0^1 su^2(s)\,ds \leqslant \|u\|^2$).

Remarks 11.4 (a) The notation $f(h) = o(h)$ is very well established. But it is a peculiar use of the equality symbol, to say the least, and the usual rules governing that symbol are not obeyed. In particular, it is clear from 11.3(a) that if $f(h) = o(h)$ and $g(h) = o(h)$, it does not follow that $f(h) = g(h)$. It might be more logical to write $f(h) \in o(h)$ where $o(h)$ denotes the set of all functions satisfying the condition of Definition 11.2. However, one soon gets used to the conventional notation.

(b) In this chapter we often deal with nonlinear operators defined on subsets of a normed space N. We should write f: $X \to M$ where $X \subset N$, as in 11.1 and 11.2. But we shall often write simply $f : N \to M$, meaning that the domain of f is some unspecified subset of N. □

We now return to the problem of extending the definition of differentiation to normed spaces. We rewrite equation (11.2) as follows:

$$f(x+h) = f(x) + f'(x)h + o(h) \quad . \tag{11.3}$$

When $o(h)$ appears in an equation such as (11.3), it stands for any function which is $o(h)$; in other words, (11.3) is equivalent to $f(x+h) = f(x) + f'(x)h + \phi(h)$ where $\phi(h) = o(h)$. We obtained (11.3) by manipulating the definition of differentiation for $\mathbb{R} \to \mathbb{R}$ functions; we now extend it to general normed spaces.

Definition 11.5 A continuous linear operator $L: N \to M$ is said to be the **Fréchet derivative** of $f: N \to M$ at the point $x \in N$ if

$$f(x+h) = f(x) + Lh + o(h) \quad \text{as} \quad h \to 0 \quad . \tag{11.4}$$

We write $L = f'(x)$. □

Note that according to 11.2, f must be defined on a neighbourhood of a if 11.5 is to apply. L is called the Fréchet derivative to distinguish it from a different kind of derivative in normed spaces, the Gâteaux derivative, which we shall not discuss. The Fréchet derivative must also be distinguished from the ordinary derivative. In the case of $\mathbb{R} \to \mathbb{R}$ functions, the two things are closely related, but logically different.

Example 11.6 For a real function $f: \mathbb{R} \to \mathbb{R}$, the ordinary derivative at x, as defined in elementary calculus, is a number giving the slope of the graph of f at x. The Fréchet derivative L at x is not a number, but a linear operator $\mathbb{R} \to \mathbb{R}$. If f is a differentiable function, then $f(x+h) = f(x) + hf'(x) + o(h)$; comparing this with (11.4), we see that L is the operator which multiplies each $h \in \mathbb{R}$ by the number $f'(x)$. In equation (11.3), therefore, we can interpret $f'(x)h$ either as the product of the number h and the number $f'(x)$ (interpreted as in elementary calculus), or as the result of applying the linear operator $f'(x)$ (interpreted as in Definition 11.5) to the element h of the space $\mathbb{R}$. Equation (11.3) is true in both interpretations, and in this sense the ordinary derivative and the Fréchet derivative are really the same thing in different forms.

Remark 11.7 Elementary calculus is based on the idea of the tangent to a curve, which is the straight line giving the best approximation to the curve in the neighbourhood of a particular point. Similarly, the Fréchet derivative of an operator f gives the best local linear approximation to f, in the following sense. Consider the change in f when its argument changes from x to $x+h$. Approximate this change by a linear operator A, so that

$$f(x+h) = f(x) + Ah + e,$$

where $e = f(x + h) - f(x) - Ah$ is the error in the linear approximation. Then the error e will generally be of the same order of magnitude as h, except when A equals the Fréchet derivative of f, in which case $e = o(h)$ so that e is much smaller than h as $h \to 0$. In this sense the Fréchet derivative gives the best linear approximation near x to the nonlinear operator f.

Example 11.8 If A is any linear operator, we have $A(x + h) = Ax + Ah$. Hence the derivative of A is A itself. This agrees with Remark 11.7: the best linear approximation to a linear operator A is A itself.

Example 11.9 For a differentiable real function of n variables $f: \mathbb{R}^n \to \mathbb{R}$, we have

$$f(x + h) = f(x) + h \cdot \nabla f + o(h) \quad .$$

Hence the Fréchet derivative at x is the operator $h \longmapsto h \cdot \nabla f(x)$. It corresponds to the gradient vector in the same way that the Fréchet derivative of an $\mathbb{R} \to \mathbb{R}$ function corresponds to the ordinary derivative.

Example 11.10 For $f: \mathbb{R}^n \to \mathbb{R}^m$, we have

$$f_i(x + h) = f_i(x) + \sum_j \frac{\partial f_i}{\partial x_j} h_j + \ldots \quad \text{for} \quad i = 1, \ldots, m \quad ,$$

or

$$f(x + h) = f(x) + f'(x)h + o(h) \quad ,$$

where the Fréchet derivative $f'(x)$ is the $\mathbb{R}^n \to \mathbb{R}^m$ operator given by the matrix $(\partial f_i/\partial x_j)$ of partial derivatives of f.

Example 11.11 It is easy to find operators which are not differentiable, that is, for which there is no linear operator L satisfying the conditions of Definition 11.5. The $\mathbb{R} \to \mathbb{R}$ function $x \longmapsto |x|$, for example, has no Fréchet derivative at $x = 0$.

Example 11.12 Define a functional $J: C[0,1] \to \mathbb{R}$ by

$$J(u) = \int_0^1 g(x)[au(x) + b\{u(x)\}^2]\,\mathrm{d}x \quad ,$$

where a and b are constants and g a given function. If $b = 0$ then J is a linear functional and so $J' = J$ (see Example 11.8). If $b \neq 0$ we have

$$\begin{aligned} J(u + h) - J(u) &= \int_0^1 g(x)[ah(x) + b\{2h(x)u(x) + h^2(x)\}]\,\mathrm{d}x \\ &= \int_0^1 g(x)[a + 2bu(x)]h(x)\,\mathrm{d}x + \int_0^1 g(x)b\{h(x)\}^2\,\mathrm{d}x \quad . \end{aligned}$$

The last term here is $o(h)$; hence $J'(u)$ is the linear functional defined by

$$J'(u)h = \int_0^1 g(x)[a + 2bu(x)]h(x)\,\mathrm{d}x \quad .$$

Here $J'(u)h$ denotes the number into which the functional $J'(u)$ maps the function h.

Example 11.13 In section 6.6 we discussed the Hammerstein operator $A: C[a,b] \to C[a,b]$ defined by

$$(Au)(x) = \int_a^b K(x,y)f(y,u(y))\,\mathrm{d}y \quad ,$$

where K and f are given functions. Assuming that f is smooth, we have

$$\begin{aligned}[A(u+h)](x) &= \int_a^b K(x,y)[f(y,u) + hf_u + \tfrac{1}{2}h^2 f_{uu} + \ldots]\,\mathrm{d}y \\ &= Au + Lh + o(h) \quad ,\end{aligned}$$

where the Fréchet derivative $L = A'(u)$ is given by

$$A'(u)h = \int_a^b K(x,y)f_u(y,u(y))h(y)\,\mathrm{d}y \quad .$$

Thus the Fréchet derivative of A at u is the linear integral operator with kernel $K(x,y)f_u(y,u(y))$. □

It is easy to generalise the basic results and methods of elementary calculus. For example, the usual rules for differentiating sums and products apply to the Fréchet derivative; if an operator is differentiable at x then it is continuous at x; the mean value theorem has a straightforward analogue; and so on. We shall not discuss these matters, but proceed to consider second derivatives, a subject which is more complicated than one might think.

11.2 Higher Derivatives

In elementary calculus, the derivative of a function f is a number, giving the slope of the tangent to the graph. But in order to discuss second derivatives we must think of this number as being the value at x of a function f'; then f'' is the derivative of this function. A similar shift in perspective is needed in order to discuss repeated Fréchet differentiation.

If $f: N \to M$ is Fréchet-differentiable for all u in some region of N, then for each u there is a bounded linear operator $f'(u): N \to M$. In the language of section 8.2, $f'(u) \in B(N,M)$ for each u; we have a rule which for each u gives a member $f'(u)$ of $B(N,M)$. We can thus regard $f'(u)$ as the value at u of a mapping $f': N \to B(N,M)$.

In the $\mathbb{R} \to \mathbb{R}$ case things are simpler because a linear transformation of $\mathbb{R}$ is simply multiplication by a real number; thus there is a one-to-one correspondence between the elements of $B(\mathbb{R},\mathbb{R})$ and $\mathbb{R}$, and we can regard $B(\mathbb{R},\mathbb{R})$ as being the same as the space $\mathbb{R}$. Hence f', like f, maps $\mathbb{R}$ into $\mathbb{R}$. In general, however, $B(N,M)$ is different from M, and we must regard f' as a mapping $N \to B(N,M)$. Note that although $f'(u)$ is a linear operator $N \to M$, the mapping f' itself is generally not linear. Again, this is like elementary calculus: if $f(u) = u^3$ where $u \in \mathbb{R}$ then $f'(u) =$

$3u^2$, a nonlinear function of u, whose value at any fixed u is a number which we are now interpreting as a linear transformation on $\mathbb{R}$.

We now define f'' to be the derivative of f'. Thus $f''(u)$ is a continuous linear operator with the same domain and codomain as f', that is, $f''(u)$: $N \to B(N,M)$. Thus for each $u \in N$, $f''(u) \in B(N,B(N,M))$, and f'' is a mapping $N \to B(N,B(N,M))$. This is a complicated thing, and higher derivatives are more complicated still. We shall now show how the structure can be simplified.

Take any $\phi \in B(N,B(N,M))$. For any $h \in N$, $\phi h \in B(N,M)$, and ϕh depends linearly on h. Now, ϕh is a linear operator $N \to M$, so for any $k \in N$, $(\phi h)k \in M$, and $(\phi h)k$ depends linearly on k as well as on h. We thus have a rule which, given any two elements h, k of N, produces an element $(\phi h)k$ of M, depending linearly on h and k. Such a rule is called a bilinear mapping.

Definition 11.14 For any vector spaces N and M, a **bilinear mapping** f from N to M is a rule which, given any $h \in N$ and $k \in N$, associates with them an element of M, denoted by $f(h,k)$ or fhk, such that for any scalars a and b, and any $h,j,k \in N$,

$$f(ah + bj,k) = af(h,k) + bf(j,k) \quad ,$$

and

$$f(h,ak + bj) = af(h,k) + bf(h,j) \quad .$$

Examples 11.15 (a) A real $n \times n$ matrix gives a bilinear mapping from $\mathbb{R}^n$ to $\mathbb{R}$: for any two vectors $x,y \in \mathbb{R}^n$, $\Sigma_{ij} x_i A_{ij} y_j \in \mathbb{R}$ and depends linearly on x and y.

(b) Similarly, a continuous real function of two variables gives a bilinear map from $C[0,1]$ to $\mathbb{R}$ defined by $(f,g) \mapsto \int_0^1 \int_0^1 K(x,y)f(x)g(y)\,dx\,dy$ (here (f,g) stands for the pair of functions f,g, not for their inner product).

Example 11.16 A continuous function of three variables $L(x,y,z)$ gives a bilinear map ϕ from $C[0,1]$ to $C[0,1]$ defined by $\phi(f,g)(x) = \int_0^1 \int_0^1 L(x,y,z)f(y)g(z)\,dy\,dz$ (here $\phi(f,g)(x)$ denotes the value at x of the $\mathbb{R} \to \mathbb{R}$ function $\phi(f,g)$ obtained by applying the bilinear mapping ϕ to the pair of functions (f,g)). □

The discussion preceding Definition 11.14 shows that members of $B(N,B(N,M))$ can be regarded as bilinear mappings of N to M. In particular, if $f: N \to M$, then $f''(u)$ is a bilinear mapping of N to M, which is perhaps easier to grasp than a linear map from N to the space $B(N,M)$ of linear maps $N \to M$. Thus for any $h \in N$ and $k \in N$, $f''(u)$ maps the pair (h,k) into an element of M which we may write $f''(u)(h,k)$ or $f''(u)hk$.

We can now write down the analogue for second derivatives of the formula (11.3) defining first derivatives. It is

$$f(x + h) = f(x) + f'(x)h + \tfrac{1}{2}f''(x)h^2 + o(h^2) \quad . \tag{11.5}$$

Here $f''(x)h^2$ is an abbreviation for $f''(x)hh$ or $f''(x)(h,h)$, and $o(h^2)$ stands for a function ϕ such that $\| \phi(h)\|/\| h \|^2 \to 0$ as $h \to 0$ – an obvious extension of Definition 11.2. The formula (11.5) is a version of Taylor's theorem to second order. It is proved using a version of the mean value theorem for Fréchet derivatives; see Cartan or Dieudonné. It says that $f''(x)$ is, roughly speaking, the coefficient of the quadratic term in the expansion of $f(x+h)$ in powers of h; (11.5) is used to calculate f'' in the same way that (11.3) is used to calculate f'.

Example 11.17 For a function $f: \mathbb{R}^n \to \mathbb{R}$, we calculate $f''(x)$ by expanding $f(x+h)$ in a Taylor series:

$$f(x+h) = f(x) + \sum_i h_i \frac{\partial f}{\partial x_i}(x) + \tfrac{1}{2}\sum_{ij} h_i h_j \frac{\partial^2 f}{\partial x_i \partial x_j}(x) + \ldots \quad .$$

As we have seen in Example 11.9, $f'(x)$ is the operator $h \mapsto \Sigma h_i f_i$, where subscripts applied to f denote partial derivatives. The second Fréchet derivative $f''(x)$ is a bilinear mapping such that $(h,h) \mapsto \Sigma f_{ij}(x)h_i h_j$. It is fairly clear that for a general pair (h,k),

$$f''(x)(h,k) = \sum_{ij} f_{ij}(x)h_i k_j \quad .$$

We shall justify this by means of formula (11.6) below, which gives an expression for $f''(x)(h,k)$ in terms of $f''(x)(h,h)$. We thus see that $f''(x)$ is the bilinear operator given by the matrix of second partial derivatives of f. □

For a twice-differentiable function $\mathbb{R}^n \to \mathbb{R}$, the second derivatives are independent of the order of differentiation, that is, $f_{ij} = f_{ji}$ in the notation of 11.17. In other words, the matrix corresponding to $f''(x)$ is symmetric. This fact generalises to Fréchet derivatives as follows.

Theorem 11.18 If $f: N \to M$ is twice-differentiable at x, then $f''(x)$ is a symmetric bilinear map, that is, $f''(x)(h,k) = f''(x)(k,h)$ for all $h,k \in N$. □

The proof of this is tedious; see Cartan or Dieudonné, for example.

Remark 11.19 For a symmetric bilinear map ϕ, $\phi(h,k)$ is uniquely determined if $\phi(h,h)$ is known for all h. In fact, we have

$$\phi(h+k, h+k) = \phi(h,h) + 2\phi(h,k) + \phi(k,k) \quad ,$$

hence $$\phi(h,k) = \tfrac{1}{2}[\phi(h+k,h+k) - \phi(h,h) - \phi(k,k)] \quad . \tag{11.6}$$

It follows that an expansion of the form (11.5) completely determines $f''(x)$, even though it might seem to determine only the 'diagonal part' of $f''(x)$ in some sense (cf. Problem 7.2(b)).

Example 11.20 Consider the nonlinear integral operator $A: C[a,b] \to C[a,b]$ defined by $(Au)(t) = \int_a^b K(t,s,u(s))\,\mathrm{d}s$, where K is a continuous function of three variables, twice-differentiable with respect to its third variable u. Writing $K_u = \partial K/\partial u$, we have

$$A(x+h)(t) = \int_a^b \{K(t,s,x(s)) + h(s)K_u(t,s,x(s)) + \tfrac{1}{2}h^2(s)K_{uu}(t,s,x(s)) + \ldots\}\,\mathrm{d}s \quad .$$

Hence $A'(x)$ is the linear operator

$$A'(x): h \mapsto \int_a^b K_u(t,s,x(s))h(s)\,\mathrm{d}s$$

(cf. Example 11.13, which is a special case of this). From the quadratic term we see that

$$A''(x)(h,h)(t) = \int_a^b K_{uu}(t,s,x(s))h^2(s)\,\mathrm{d}s \quad ; \tag{11.7}$$

the left hand side here denotes the value at t of the function $A''(x)(h,h)$ obtained by applying the bilinear operator $A''(x)$ to the pair of functions (h,h). It is fairly clear from (11.7) that $A''(x)$ applied to a general pair of functions gives

$$A''(x)(h,k)(t) = \int_a^b K_{uu}(t,s,x(s))h(s)k(s)\,\mathrm{d}s \quad ;$$

this can be deduced from (11.7) by means of the identity (11.6).

Example 11.21 In this example we work with the real vector space $L_2[a,b]$ of real-valued square-integrable functions. Define a functional $J: L_2[a,b] \to \mathbb{R}$ by

$$J(u) = \int_a^b \int_a^b u(s)K(s,t)u(t)\,\mathrm{d}t\,\mathrm{d}s \quad , \tag{11.8}$$

where K is a continuous function. We can write this more neatly by defining a linear operator A on $L_2[a,b]$ by

$$(Au)(t) = \int_a^b K(t,s)u(s)\,\mathrm{d}s \quad .$$

Now make $L_2[a,b]$ an inner product space with the usual inner product $(f,g) = \int f(x)g(x)\,\mathrm{d}x$; remember that we are considering real-valued functions only. Then (11.8) reads

$$J(u) = (u,Au) \quad . \tag{11.9}$$

Now, $J(x+h) = (x+h, Ax + Ah)$

$$= J(x) + (h,Ax) + (x,Ah) + (h,Ah) \quad . \tag{11.10}$$

J'' is given by the quadratic term in (11.10); thus $J''(x)(h,h) = (h,Ah)$. Applying (11.6) now gives

$$J''(x)(h,k) = \tfrac{1}{2}\{(h,Ak) + (k,Ah)\}$$
$$= (h, \tfrac{1}{2}\{A + A^*\}k) \quad ,$$

where A^* is the adjoint of A. If $K(s,t) = K(t,s)$, then A is a self-adjoint operator, and $J''(h,k) = (h,Ak)$ which is symmetric in h and k. If K is not a symmetric function, then the formula above says that $J''(x)$ is the symmetric bilinear functional obtained from (h,Ak) by combining it with its transposed form (k,Ah).

□

We have seen that the second derivative of an operator $f: N \to M$ can be interpreted either as a linear operator from $B(N,M)$ to M, or as a bilinear map from N to M. In Example 11.21 we showed that for a functional J on $L_2[a,b]$, there is a third interpretation of J'', in terms of a linear operator A on $L_2[a,b]$. We shall now generalise this to any functional f on a real Hilbert space H. The advantage of regarding f'' as a linear operator $H \to H$ is that when we consider maxima and minima in section 11.3, we can use the theory of positive operators in Hilbert space to give a meaning to the criterion $f'' > 0$ for a minimum.

Proposition 11.22 Let $f: H \to \mathbb{R}$ be a twice-differentiable functional on a real Hilbert space H. Then for each $x \in H$ there is a self-adjoint bounded linear operator $A: H \to H$ (depending on x) such that $f''(x)(h,k) = (h,Ak)$ for all $h,k \in H$.

Proof For any fixed $k \in H$ we have a linear functional $h \mapsto f''(x)(h,k)$; it is continuous by definition of the Fréchet derivative. The Riesz representation theorem shows that there is a $u \in H$ such that $f''(x)(h,k) = (h,u)$. We thus have a map which to each $k \in H$ assigns $u \in H$; call this map A, and write $u = Ak$. Since $f''(h,k)$ is linear in k for each fixed h, it follows that A is a linear operator, and we have $f''(x)(h,k) = (h,Ak)$ as required. The boundedness and self-adjointness of A follow at once from the continuity and symmetry of $f''(x)$. □

11.3 Maxima and Minima

In elementary calculus, if a differentiable function has a maximum or a minimum at a point, then its derivative vanishes there. Similarly, if a functional on a normed space has a maximum or a minimum, then its Fréchet derivative vanishes. Note that maxima and minima of an operator only make sense if the codomain is a completely ordered space such as $\mathbb{R}$, that is, a space in which, given any two distinct elements, one can be picked out as greater than the other. We therefore limit the discussion of this section to real-valued functionals on real normed spaces.

Example 11.23 Define $J: C[0,1] \to \mathbb{R}$ by

$$J(u) = \int_0^1 b(x)u^2(x)\,\mathrm{d}x \quad ,$$

where b is a given function. The calculation of Example 11.12 gives

$$J'(u)h = 2\int_0^1 b(x)u(x)h(x)\,\mathrm{d}x \quad .$$

Thus $J'(u) = 0$ (that is, $J'(u)$ is the zero operator) when $u \equiv 0$. Hence if J has a maximum or minimum, it must be when $u \equiv 0$. Indeed, it is obvious that if $b(x) > 0$ for all x, then $J(u) \geqslant 0$ for all u, and its minimum value is zero, attained for $u \equiv 0$. Similarly, if $b(x) < 0$ for all x, then J has a maximum value of zero for $u \equiv 0$. If b changes sign in (0,1), then J has neither a maximum nor a minimum but a stationary point at $u \equiv 0$, which is the analogue of a horizontal point of inflection for an $\mathbb{R} \to \mathbb{R}$ function, or a saddle point for an $\mathbb{R}^2 \to \mathbb{R}$ function. □

In elementary calculus, maxima and minima can be identified from the sign of the second derivative. The same is true for functionals on a normed space. But what is the meaning of the 'sign' of $f''(x)$ when $f''(x)$ is a bilinear functional? The answer is easy for functionals on Hilbert space: Proposition 11.22 associates a self-adjoint operator A with $f''(x)$, and we say $f''(x) > 0$ if A is a positive-definite operator (Definition 10.1), $f''(x) \geqslant 0$ if A is positive semidefinite, and so on. We can then say that f has a minimum at x_0 when $f'(x_0) = 0$ and $f''(x_0) > 0$, and f has a maximum when $f'(x_0) = 0$ and $f''(x_0) < 0$. This is a rough approximation to the truth, sufficient for most cases, though the function $f(x) = x^4$ shows that f can have a strict minimum even when $f''(x_0) = 0$. For precise theorems, see Cartan.

Example 11.23 Continued We have shown that $J'(0) = 0$. Now the calculation in Example 11.12 shows that $J''(u)(h,h) = \int_0^1 b(x)h^2(x)\,dx$. Hence, by (11.6), $J''(u)(h,k) = \int_0^1 b(x)h(x)k(x)\,dx$, or $(Ak)(x) = b(x)k(x)$. If $b(x) > 0$ for all x, then A is a positive operator, and J has a minimum at $u = 0$; if $b(x) < 0$ for all x then A is a negative operator and J has a maximum; if $b(x)$ changes sign in the interval (0,1), then J has neither a maximum nor a minimum. Thus the second-derivative method agrees with the results obtained above from first principles.

Example 11.24 Consider the functional $J(v) = 2(v,f) - (v,Av)$, where f is a given member of a Hilbert space H and A is a linear self-adjoint operator on H. A straightforward calculation shows that $J'(u) = 2(f - Au)$, and $J''(u) = -2A$ (or, more precisely, $J'(u): v \mapsto 2(f - Au,v)$ and $J''(u): (h,k) \mapsto (h,-2Ak)$). It follows that if A is a positive operator, then J has a local maximum when u satisfies the equation $Au = f$, a fact which will be familiar to those who have read Chapter 10. Indeed, 10.25 shows that J has a global maximum (that is, $J(u) \geqslant J(v)$ for all $v \in H$, if $Au = f$), whereas the methods of the differential calculus show only that J has a local maximum. In this sense the Fréchet calculus is less powerful than the methods of Chapter 10. Its great advantage is that it can be applied to nonlinear problems.

Example 11.25 (Euler-Lagrange Equations) What is the quickest route from a point (a,α) to a point (b,β) over uneven terrain, where the speed of travel at the point (x,y) is a given function $v(x,y)$? It is easy to show that the time taken to travel the route given by the curve $y(x)$ is $J(y) = \int_a^b \{\sqrt{1 - y'^2}/v(x,y)\}\,dx$, where y is regarded as a function defined on the interval $[a,b]$. The problem then is to minimise J over all functions y such that $y(a) = \alpha$ and $y(b) = \beta$. The branch of mathematics dealing with such questions is called the calculus of variations. We shall discuss it briefly from the point of view of the Fréchet calculus; a good discussion along classical lines will be found in Courant-Hilbert, for example.

We generalise the problem, and consider the functional

$$J(y) = \int_a^b L(x,y,y')\,\mathrm{d}x \quad ,$$

where L is a given function of three variables. Then

$$\begin{aligned} J(y+h) &= \int_a^b L(x,y+h,y'+h')\,\mathrm{d}x \\ &= \int_a^b \{L(x,y,y') + hL_y + h'L_{y'} + \ldots\}\,\mathrm{d}x \\ &= J(y) + \int_a^b h[L_y - \mathrm{d}/\mathrm{d}x\,(L_{y'})]\,\mathrm{d}x + [hL_{y'}]_a^b + \ldots. , \end{aligned}$$

on integration by parts, where $[f]_a^b$ denotes $f(b) - f(a)$. Hence

$$J'(y)h = \int_a^b h(x)[L_y - \mathrm{d}/\mathrm{d}x\,(L_{y'})]\,\mathrm{d}x + [hL_{y'}]_a^b \quad . \tag{11.11}$$

The Fréchet derivative of J is thus a rather complicated operator, the sum of a linear integral operator and the operator $h \longmapsto ph(b) - qh(a)$, where p and q are numbers independent of h.

Now, we wish to minimise J over all functions y satisfying the boundary conditions $y(a) = \alpha$, $y(b) = \beta$. But if y and $y + h$ both satisfy these boundary conditions, then $h(a) = h(b) = 0$. Hence if y minimises J subject to the given boundary conditions, it follows that $J'(y)h = 0$ for all h such that $h(a) = h(b) = 0$. But the last term in (11.11) vanishes for such functions h, so we deduce from (11.11) that J is minimised when y satisfies the differential equation

$$\frac{\partial L}{\partial y} - \frac{\mathrm{d}}{\mathrm{d}x}\left(\frac{\partial L}{\partial y'}\right) = 0 \quad . \tag{11.12}$$

This is called the Euler-Lagrange equation associated with the functional J; it is to be solved with the boundary conditions $y(a) = \alpha$, $y(b) = \beta$, and the solution minimises J – or at least J is locally stationary; it may have a maximum or a minimum or neither. Our treatment has been formal: we make no attempt to state precise conditions under which the results are valid, but refer the reader to textbooks on the calculus of variations. Our aim has been to show how the classical variational procedure fits into our general scheme of differentiation in normed spaces. □

We conclude this section with two examples of the Euler-Lagrange equation.

Example 11.26 If we take

$$L(x,y,y') = \tfrac{1}{2}p(x)y'^2 - \tfrac{1}{2}q(x)y^2 + f(x)y \quad ,$$

where p,q,f are given functions, then the Euler-Lagrange equation (11.12) is $(pu')' + qu = f$, which is an inhomogeneous differential equation of Sturm-Liouville type. We thus recover from a different point of view the relation between inhomogeneous differential equations and minimisation problems discussed in section 10.5. For linear equations, the approach via the Euler-Lagrange equation is equivalent to the approach of section 10.5, but it has the advantage of leading to a more systematic treatment of complementary bounds – see, for example, the article by Robinson in Rall (1971). □

Our last example illustrates variational methods for nonlinear problems.

Example 11.27 If we take

$$L(x,y,y') = \tfrac{1}{2}y'^2 + g(x,y)$$

where g is a given function, we have the Euler-Lagrange equation

$$y'' = f(x,y) \quad ,$$

where $f = g_y$. This equation was discussed in section 5.4 in connection with non-linear diffusion; we now see that it is equivalent to a variational problem for the functional $J(y) = \int\{\tfrac{1}{2}y'^2 + g(x,y)\}\,\mathrm{d}x$, and can therefore be solved by the Rayleigh-Ritz method. See Bellman (1970) for further discussion of this problem. □

11.4 Linear Stability Theory

When discussing differential equations, we have often regarded each solution as an element of a vector space. This is a useful approach to equations describing equilibrium phenomena, the equations of statics. But the next two sections discuss systems which change with time; they are translated into vector space terms in a different way, as follows.

It is assumed that at any time t, the state of the system can be represented by an element of a normed space N. As the state of the system changes, its representative in N varies with time; we write $u(t)$ for the vector in N representing the state at time t, and regard $u(t)$ as the value at t of a mapping u: $\mathbb{R} \to N$. We say that u is differentiable at t if $[u(t+h) - u(t)]/h$ converges as $h \to 0$, and write $\dot{u}(t)$ for the limit. Note that t and h are real numbers here; we do not need the idea of a Fréchet derivative. We assume that the law governing the evolution of the system can be written

$$\dot{u} = Pu \quad , \tag{11.13}$$

where P is an operator $N \to N$. We shall consider only operators P which are independent of t; these correspond to *autonomous* systems, that is, systems whose governing laws do not change with time. It might seem at first sight that (11.13) restricts the theory to equations of the first order in time, and therefore excludes the second-order differential equations of mechanics. The following example shows that this is not so.

Example 11.28 The equation of motion of a pendulum subject to a time-dependent external force $f(t)$ is $\ddot{x} + \alpha^2 x = f(t)$. To express this in the form (11.13), define $y(t) = \dot{x}(t)$, so that we have $\dot{x} = y$, $\dot{y} = -\alpha^2 x + f(t)$. Now introduce the vector $u = (x,y) \in \mathbb{R}^2$. Then u varies with t according to (11.13), where the operator $P: \mathbb{R}^2 \to \mathbb{R}^2$ is defined by $Pu = P(x,y) = (y, f - \alpha^2 x)$. If f is zero, then P is independent of t, in fact P is the matrix $\begin{pmatrix} 0 & 1 \\ -\alpha^2 & 0 \end{pmatrix}$; the system is then autonomous in the sense defined above. If f is not constant, then P depends on t, and the system is nonautonomous. Generally speaking, systems evolving under their own constant internal laws are autonomous, while those subject to externally imposed forces are nonautonomous.

Example 11.29 The growth of two interacting species of animals or plants is often modelled by the equations

$$\dot{x} = x(ax + by + c) \quad ,$$

$$\dot{y} = y(dx + ey + f) \quad ,$$

where x,y are the sizes of the populations of the two species, and $u, \ldots, f$ are constants representing the natural growth rates of the two species and the extent to which each species helps or hinders the growth of the other. For a discussion of these equations, called Volterra's equations, see Maynard Smith, or Braun. As in the previous example, the equations can be cast in the form (11.13) by writing $u = (x,y)$ and defining $P: \mathbb{R}^2 \to \mathbb{R}^2$ by $P(x,y) = (x[ax + by + c], y[dx + ey + f])$. the operator P is nonlinear. If $a, \ldots, f$ are constant, then P is independent of t and the system is autonomous.

Example 11.30 In section 5.4 we studied a reaction-diffusion system governed by the equation $u_t = u_{xx} + h(u,x)$, with $u(0,t) = u(1,t) = 0$ for all t. To express this in the form (11.13), take $N = L_2[0,1]$, and define an operator $P: u \mapsto u'' + h(u,x)$. We must restrict P to the subspace of twice-differentiable functions in $L_2[0,1]$, so we have a slight generalisation of (11.13) in which P is defined on a subspace of N. It is useful, as in section 9.2, to incorporate the boundary conditions into the operator P by restricting it to a domain $\mathscr{D} = \{f \in C^2[0,1] : f(0) = f(1) = 0\}$. For each fixed t, we regard u as a function of x; then $u \in \mathscr{D}$, and (11.13) is satisfied for each t. □

In this section we shall consider the case where (11.13) is a linear autonomous equation. Thus we consider the equation

$$\dot{u} = Lu + f \quad , \tag{11.14}$$

where $u(t) \in N$ for each t, $L : N \to N$ is linear and independent of t, and $f \in N$ is given. If we are given u at $t = 0$, then we can in principle use (11.14) to find u for all $t > 0$, by numerical methods. But here we are more interested in the questions of existence and stability of equilibrium solutions.

An **equilibrium solution** of (11.14) is by definition a constant vector satisfying (11.14). Clearly $u_0 \in N$ is an equilibrium solution if $Lu_0 = -f$. The solution of this equation has been discussed in earlier chapters. If L is invertible, then there is just one equilibrium solution for any given f; if L is not invertible the Fredholm Alternative may apply; if L is symmetric, u can be found by the method of eigenfunction expansions, or by variational methods. We suppose that an equilibrium solution u_0 has somehow been found, and shall consider its stability.

An equilibrium state of a system is said to be stable if, when it is given a small push at some time, say $t = 0$, the disturbance is small for all $t \geqslant 0$. It is unstable if giving it a push (however small) at $t = 0$ can produce a significant disturbance at later times. This definition agrees with our intuitive idea of stability and instability, epitomised respectively by a ball resting inside a bowl and a ball resting on the top of a bowl placed upside down. We now express the definitions more precisely; 'giving a small push' is interpreted as starting the system from an initial state slightly different from the equilibrium state u_0; 'slightly different' means $\| u(0) - u_0 \| < \delta$; 'small disturbance for $t \geqslant 0$' means $\| u(t) - u_0 \| < \epsilon$ for all $t \geqslant 0$.

Definition 11.31 An equilibrium solution u_0 of the autonomous equation (11.13) is said to be **stable** if for each $\epsilon > 0$ there is a $\delta > 0$ such that $\| u(t) - u_0 \| < \epsilon$ for all $t \geqslant 0$ for all solutions $u(t)$ of (11.13) such that $\| u(0) - u_0 \| < \delta$. **Unstable** means not stable. □

This definition applies to the general nonlinear equation (11.13). In this section we shall consider the special case of linear equations; nonlinear problems are discussed in section 11.5.

Example 11.32 The very simple equation $\dot{u} = au$, where $u(t)$ is a real number for each t and a is a real constant, has the equilibrium solution $u_0(t) \equiv 0$. The general solution is $u(t) = u(0)e^{at}$. If $a \leqslant 0$, then the zero solution is stable: we can take $\delta = \epsilon$ in 11.31. If $a > 0$ it is unstable since, however small $u(0)$ is, $u(t)$ eventually becomes large. □

This very simple example is typical of the general linear autonomous system (11.14): the solutions are exponential functions of t, and stability depends on the sign of the operator. To investigate the general case we shall look for solutions of (11.14) of the form $u(t) = u_0 + e^{\lambda t}v$, where $u_0 \in N$ is an equilibrium solution, λ is some constant, and v is a fixed vector in N. Then $\dot{u}(t) = \lambda e^{\lambda t}v$ and (11.14) is satisfied if $\lambda e^{\lambda t}v = Le^{\lambda t}v$, that is, if λ is an eigenvalue of L with eigenvector v. If there is an eigenvalue λ with positive real part, and v is its normalised eigenvector, then for any $\epsilon > 0$, the function $u(t) = u_0 + \epsilon v e^{\lambda t}$ is a solution of (11.14) such that $\| u(0) - u_0 \| = \epsilon$ and $\| u(t) - u_0 \|$ becomes indefinitely large as $t \to \infty$. This shows that the equilibrium is unstable if there is an eigenvalue with positive real part, and proves

Proposition 11.33 (Stability Criterion) If the linear equation (11.14) has a stable equilibrium, then all eigenvalues λ of L satisfy $\mathcal{Re}(\lambda) \leq 0$.

The following example shows that the converse of 11.33 is false: $\mathcal{Re}(\lambda) \leq 0$ does not imply stability.

Example 11.34 Consider the equation $\dot{u} = Au$ where $u(t) \in \mathbb{R}^2$ and A is the matrix operator $\begin{pmatrix} 0 & 1 \\ 0 & 0 \end{pmatrix}$. If u_0 is an equilibrium solution, then $Au_0 = 0$; it is easy to see that $(c,0)$ is an equilibrium solution for any number c. The only eigenvalue of A is zero. If we write $u = (x,y)$, then $\dot{u} = Au$ becomes $\dot{x} = y$ and $\dot{y} = 0$; hence the general solution is $y = \alpha$, $x = \alpha t + \beta$ for some numbers α,β. The equilibrium is unstable: if α is sufficiently small, the solution $u(t) = (\alpha t + c, \alpha)$ can be made as close as we please to $u_0 = (c,0)$ at $t = 0$, yet $\| u(t) - u_0 \| \to \infty$ as $t \to \infty$. This example shows that the condition $\mathcal{Re}(\lambda) \leq 0$ for all eigenvalues is not sufficient to ensure stability. □

The converse of 11.33 is false in general. Proposition 11.35 below gives a sufficient condition for stability. It uses the idea of the adjoint of an operator, which was defined in Chapter 7 for bounded operators only. The theory for unbounded operators is not simple (see Problem 9.14). But for our present purposes the idea can be simplified by saying that L^* is the **adjoint** of L if $(u,Lv) = (L^*u,v)$ for all u,v in the domain of L.

Proposition 11.35 (Stability Criterion) If L is a linear operator and $L + L^*$ is negative semidefinite, then all equilibrium solutions of the equation $\dot{u} = Lu + f$ are stable.

Proof Write $v = u - u_0$, where u_0 is any equilibrium solution of the equation, and u is any other solution. Then $\dot{v} = Lv$. Hence

$$\begin{aligned} \mathrm{d}\| v \|^2 / \mathrm{d}t &= (v,\dot{v}) + (\dot{v},v) \\ &= (v,Lv) + (Lv,v) \\ &= (v, [L + L^*]v) \quad . \end{aligned}$$

Now, if $L + L^*$ is negative semidefinite, then $\mathrm{d}\| v \|^2/\mathrm{d}t \leq 0$. Thus $\| v \|$ is a non-increasing function, which means that if u is initially within a distance ϵ of u_0, it never moves farther than ϵ from u_0. Thus the definition of stability is satisfied with $\delta = \epsilon$. □

Example 11.36 Consider the reaction-diffusion equation of Example 11.30, and suppose that $h(u,x) = a(x)u + b(x)$. Then $u_t = Lu + b$, where $Lu = u_{xx} + au$. This is a Sturm-Liouville operator, and we have the usual boundary conditions on the interval $[0,1]$. The eigenvalues of L are easily calculated if a is constant: they are $a - n^2\pi^2$ for $n = 1,2,\ldots$. So if $a < \pi^2$, then $L + L^* = 2L$ is a negative operator by 10.5, and the equilibrium is stable; if $a > \pi^2$ then it is unstable.

If a is not constant, we cannot write down an explicit formula for the eigenvalues, but we can use the bounds for eigenvalues derived in Chapter 10 to obtain stability conditions. We have

$$\begin{aligned}(u,Lu) &= \int_0^1 (uu'' + au^2)\,dx \\ &= \int_0^1 (au^2 - u'^2)\,dx \quad , \end{aligned} \tag{11.15}$$

assuming that u is real and satisfies the boundary conditions. If $(u,Lu) > 0$ for some u, then it follows from Proposition 10.12 that L has a positive eigenvalue; Proposition 11.33 then shows that the equilibrium is unstable, under our assumption that $(u,Lu) > 0$ for some u. The simplest trial function to use is $\sin(\pi x)$; inserting this into (11.15) shows that the equilibrium is unstable if

$$\int_0^1 a(x)\sin^2(\pi x)\,dx > \pi^2/2 \quad . \tag{11.16}$$

If a is a constant, this gives the condition $a > \pi^2$ derived above. If a is not constant, (11.16) gives a sufficient but not a necessary condition for instability; if (11.16) does not hold, the equilibrium may be stable or unstable. In general, the condition (11.16) agrees with the physical idea that if the growth rate a is high enough, then the tendency to grow dominates the tendency of diffusion to discourage growth, and instability results. In the linear equations considered in this section, instability leads to unbounded growth. For the nonlinear equation $u_t = u_{xx} + au(1 - u/b)$ discussed in section 5.4, for sufficiently large values of a a second equilibrium solution exists. The instability of the zero solution for $a > \pi^2$ is associated with the appearance of a new stable nonzero solution: solutions which start close to zero will grow and approach the nonzero equilibrium. These phenomena are discussed in section 11.6.

11.5 Nonlinear Stability

We now consider the stability of the general nonlinear autonomous equation

$$\dot{u} = Pu \quad . \tag{11.17}$$

We have seen that when $Pu = Lu + f$, where L is linear, stability depends simply on the sign of the real part of the eigenvalues of L. The theory for nonlinear operators is much more difficult and subtle, and we shall describe only a crude first approach to the problem.

The question of stability of an equilibrium solution u_0 of (11.17) concerns the effects of small initial displacements of u from u_0; it only involves values of u near

u_0. Now if P is Fréchet-differentiable, then in the neighbourhood of u_0 the operator P can be approximated by the linear operator $P'(u_0)$, and linear stability theory can be used. Thus we have

$$P(u) = P(u_0) + P'(u_0)(u - u_0) + o(u - u_0) \quad ,$$

where u_0 is an equilibrium solution of (11.17), so that $P(u_0) = 0$. If we neglect the term $o(u - u_0)$, we obtain an approximate form of the equation,

$$\dot{u} = P'(u_0)(u - u_0) \quad . \tag{11.18}$$

This is a linear equation, sometimes called the linearised approximation to the nonlinear equation (11.17). Its stability can be determined by the methods of section 11.4. Since (11.18) is a good approximation to (11.17) when u is near u_0, it is natural to assume that the nonlinear equation (11.17) is stable when its linear approximation (11.18) is stable. But this need not be so.

Example 11.37 Consider the equation $\dot{x} = x^3$, where $x(t) \in \mathbb{R}$ for each t. It has an equilibrium solution $x_0 \equiv 0$, and it is easy to solve the differential equation explicitly, and show that x_0 is an unstable equilibrium. Yet the linear approximation (11.18) in this case is $\dot{x} = 0$, which is stable. We conclude that stability of the linearised equation does not imply stability of the nonlinear equation. □

The trouble with the above example is that the linear system is only marginally stable. It has eigenvalue zero, which means that an arbitrarily small perturbation can push the eigenvalue into the right half-plane, and make the system unstable. Another way of looking at it is to say that eigenvalue zero corresponds to a constant solution of the linearised equation, and an arbitrarily small perturbation can convert this constant solution into a growing solution, and thus lead to instability. But if all the eigenvalues of a linear system are negative, then its solutions approach u_0 exponentially; the small perturbations involved in going from the linearised to the nonlinear system cannot convert exponential decay of $u - u_0$ into growth, hence in this case the nonlinear system will be stable. This argument leads to the following statement.

Proposition 11.38 (Principle of Linearised Stability) For a certain class of operators P, if all eigenvalues λ of the linearised operator $P'(u_0)$ satisfy $\mathcal{Re}(\lambda) \leqslant a < 0$ (for some a independent of λ), then the equilibrium solution u_0 of the equation $\dot{u} = Pu$ is stable. □

This result has been stated rather vaguely because no precise all-encompassing theorem is known. But various special cases have been proved, and the principle of linearised stability is generally believed to be a good guide to the stability of nonlinear systems.

The principle is well established for finite systems of ordinary differential equations, corresponding to the equation $\dot{u} = P(u)$ on a finite-dimensional space; the result is known as Lyapounov's Theorem; see Coddington & Levinson, for example. In this case the condition $\mathcal{R}e(\lambda) \leqslant a < 0$ can be replaced by the simpler condition $\mathcal{R}e(\lambda) < 0$ for all λ. The two conditions are equivalent because there are a finite number of eigenvalues. If they all have negative real parts, there is an eigenvalue λ_0 with the largest real part, and we have $\mathcal{R}e(\lambda) \leqslant \mathcal{R}e(\lambda_0) < 0$ for all λ. For operators on infinite-dimensional spaces, the condition $\mathcal{R}e(\lambda) \leqslant a < 0$ is stronger than $\mathcal{R}e(\lambda) < 0$, since there may be a sequence of eigenvalues tending to zero from below; the stronger condition is needed to exclude 'marginal stability' (eigenvalues with $|\mathcal{R}e(\lambda)|$ arbitrarily small), which can be turned into instability by small perturbations. For versions of 11.38 applying to nonlinear partial differential equations of diffusion type, see Evans or Kielhofer; for the equations of fluid mechanics see Sattinger.

We shall now use the linearised stability principle to study a nonlinear problem introduced in Chapter 5.

Example 11.39 In section 5.4 we considered a reaction-diffusion problem governed by the equation

$$u_t = u_{xx} + h(u,x),\ u(0,t) = u(1,t) = 0 \quad .$$

In Example 11.30 we expressed this in the form $\dot{u} = Pu$ where P is an operator defined on a subspace $\mathscr{D}$ of $L_2[0,1]$. In section 5.4 we found that in the special case $h(u,x) = a(x)u[1 - u/b]$, the zero solution is the only equilibrium solution if $a(x) < 3\sqrt{10}$ for all x. We shall now study the stability of this equilibrium.

We have $Pu = u'' + au(1 - u/b)$. It is very easy to differentiate this:

$$\begin{aligned} P(0 + h) &= h'' + ah + ah^2/b \\ &= P(0) + Ah + o(h) \quad , \end{aligned}$$

where A is the linear operator defined by

$$Ah = h'' + ah$$

with the domain $\mathscr{D}$ defined in Example 11.30. The principle of linearised stability says that the equilibrium is stable if all the eigenvalues of A are negative. If a is a constant, the eigenvalues are $a - \pi^2, a - 2\pi^2, \ldots$. Hence we have stability if $a < \pi^2$, and instability if $a > \pi^2$. Now, in section 5.4 we found that the equilibrium is unique if $a < 3\sqrt{10}$, and were led to expect nonuniqueness for larger values of a. The closeness of the numbers π^2 and $3\sqrt{10}$ suggests a connection between loss of stability and loss of uniqueness. In fact we shall see in section 11.7 that as a increases, the zero solution becomes unstable precisely at the point where a nonzero solution appears: stability is transferred to the new nonzero solution.

If a is not constant we cannot evaluate the eigenvalues of A exactly. But it is easy to show that all the eigenvalues λ satisfy $\lambda \leqslant \sup\{a(x)\} - \pi^2$ (see the last part of Problem 10.15). Hence if $\sup\{a(x)\} < \pi^2$, then the equilibrium is stable. The remarks at the end of section 11.4 apply here too.

11.6 Bifurcation Theory

In Chapter 5 we considered some examples of nonlinear eigenvalue problems, that is, boundary-value problems for differential equations containing a scalar parameter, λ say. We found that for some values of λ the equation has a unique solution (which is often the zero solution), and were led to suspect that for other values of λ there is more than one solution. For the equations commonly arising in practice it is not easy to construct the solutions explicitly. We shall therefore illustrate the phenomenon of multiple solutions with an equation invented so as to be very easy to solve.

Example 11.40 Consider the nonlinear integrodifferential system

$$u'' + [\lambda - 2\textstyle\int_0^1 |u(x)|^2\,\mathrm{d}x]u = 0,\ u(0) = u(1) = 0 \quad . \tag{11.19}$$

Observe that the quantity in square brackets is simply a number, not a function of x. Hence the equation has the form $u'' + \alpha u = 0$, where α is a constant. It has the obvious solution $u \equiv 0$. There is a nonzero solution satisfying the boundary conditions only if $\alpha = n^2\pi^2$ for some positive integer n; the solution is then $u = A\sin(n\pi x)$ for some number A. Thus, $u = A\sin(n\pi x)$ is a solution provided that

$$\alpha = \lambda - 2\textstyle\int_0^1 |A|^2 \sin^2(n\pi x)\,\mathrm{d}x = n^2\pi^2$$

or

$$\lambda - |A|^2 = n^2\pi^2 \quad . \tag{11.20}$$

Now, if $\lambda < \pi^2$ this can never be satisfied, and so (11.19) has the unique solution $u \equiv 0$ if $\lambda \leqslant \pi^2$. If $\pi^2 < \lambda < 4\pi^2$, then (11.20) has a solution with $n = 1$, and (11.19) has exactly three solutions: $u \equiv 0$ and $u = \pm\sqrt{\lambda - \pi^2}\sin(\pi x)$. For $4\pi^2 < \lambda < 9\pi^2$, (11.20) has solutions with $n = 1$ and $n = 2$, and (11.19) has five solutions: the three listed above and the new pair $u = \pm\sqrt{\lambda - 4\pi^2}\sin(2\pi x)$. For $\lambda > 9\pi^2$ a new pair appears, and so on. To illustrate this we draw a diagram with the parameter λ plotted horizontally and the solutions u of (11.19) for each λ plotted vertically. Of course, u is a function of x, and a member of an infinite-dimensional space; to plot it on our diagram we must somehow represent it by a single number. Since all solutions are of the form $u = A\sin(n\pi x)$, we shall represent the solution u by its value of A. Thus $A = 0$ is a solution for all λ; $A = \pm\sqrt{\lambda - \pi^2}$ is a solution for $\lambda > \pi^2$; $A = \pm\sqrt{\lambda - 4\pi^2}$ is a solution for $\lambda > 4\pi^2$; and so on. Thus we obtain Fig. 20, known as the **bifurcation** or **branching diagram** for equation (11.19). It shows clearly how the number of solutions changes as λ increases.

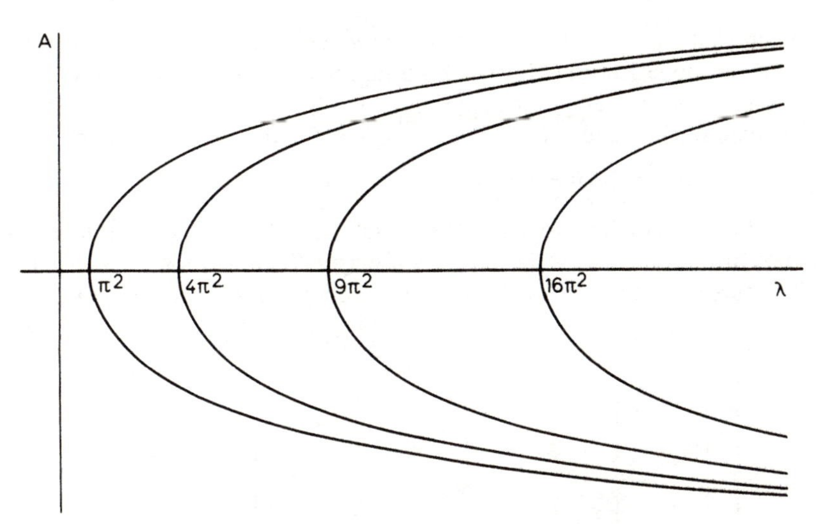

Fig. 20. Branching diagram for equation (11.19).

Although we obtained Fig. 20 for the special equation (11.19), its general shape is typical of a wide class of problems, including the problem of a loaded column considered in 5.19. It is characterised by particular values of λ when new solutions appear. These values of λ are called **bifurcation points.** Thus the bifurcation points of equation (11.19) are $\pi^2, 4\pi^2, \ldots$. More precisely, these points are called the **points of bifurcation from zero**, because the new solutions branch from the zero solution. In this example, all the bifurcations are from zero, but in other cases new solutions may branch from nonzero solutions, and quite complex structures can arise, as illustrated in Fig. 21.

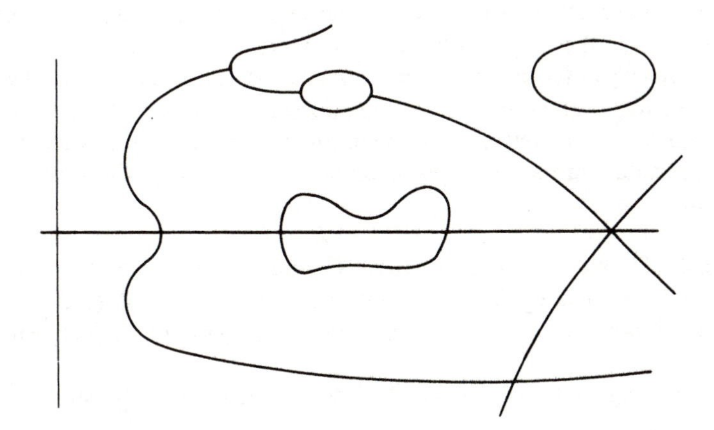

Fig. 21. Hypothetical branching diagram for a non-trivial problem.

In our example, the bifurcation points $n^2\pi^2$ are equal to the eigenvalues of the operator $-D^2$ appearing in (11.19). This is no accident, as we shall now see.

Consider the small-amplitude approximation to (11.19). If u is small, then u^2 will be much smaller, and (11.19) can be replaced by the approximate equation

$$u'' + \lambda u = 0, \quad u(0) = u(1) = 0 \quad .$$

This is a linear eigenvalue problem. Its only solution is $u \equiv 0$ unless $\lambda = n^2\pi^2$ for some positive integer n, in which case $u = A \sin(n\pi x)$ is a solution for all values of A. The bifurcation diagram for this equation is shown in Fig. 22; the vertical

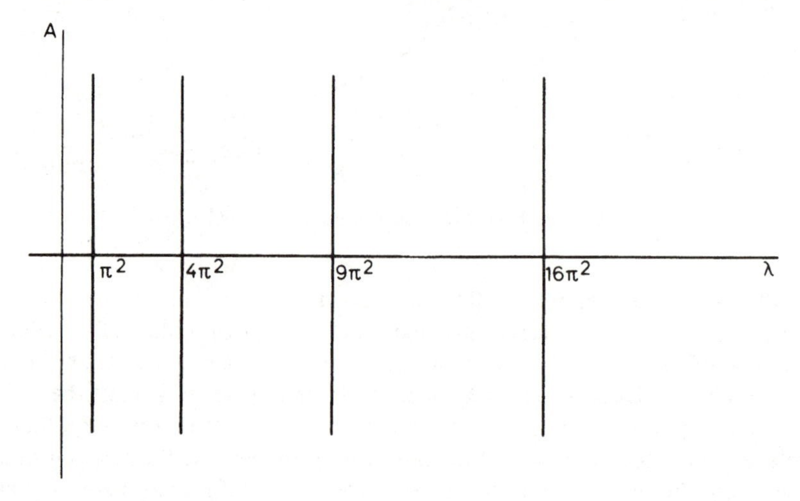

Fig. 22. Branching diagram for the linear equation approximating (11.19) for small u.

lines there indicate that for $\lambda = n^2\pi^2$, any value of A gives a solution. Clearly any linear eigenvalue problem will give a bifurcation diagram looking much like Fig. 22. Notice that in the neighbourhood of the λ axis, Figs. 22 and 20 nearly coincide. This says that the linear approximation is nearly the same as the nonlinear equation for small values of A. Since the bifurcation points of the linear equation are its eigenvalues, we now see why the bifurcation points of the nonlinear equation are the eigenvalues of its linear approximation. The following theorem generalises this result.

Theorem 11.41 (Bifurcation Theorem) Consider the equation $Au = \lambda u$ where A is a compact nonlinear operator, Fréchet-differentiable at $u = 0$, such that $A0 = 0$.

(a) Every point of bifurcation from zero is an eigenvalue of the linear operator $A'(0)$.

(b) Every eigenvalue of $A'(0)$ with odd multiplicity is a bifurcation point. □

The proof of (a) is not very difficult, but the proof of (b) involves topological degree theory, which is beyond the scope of this book; see, for example, Hutson & Pym, or Krasnoselskii (1964). Note that problems of bifurcation from a nonzero solution, w say, can be reduced to a form to which Theorem 11.41 applies, by defining $v = u - w$. We then have $Bv = \lambda v$, where $B : v \mapsto A(v + w) - Aw$, and $v = 0$ corresponds to the solution $u = w$ from which new solutions bifurcate.

The odd-multiplicity condition in 11.41(b) may be unexpected at first sight. But it is easily explained as follows. An operator with a multiple eigenvalue is analogous to a polynomial with coincident roots. Such a polynomial $p(x)$ can be turned into a polynomial with simple roots by adding a small extra term; in other words, $p(x)$ is the limit as $\epsilon \to 0$ of a polynomial $p_\epsilon(x)$ which has simple roots for any $\epsilon > 0$. In the same way, any linear operator with degenerate eigenvalues is the limit of an operator with nondegenerate eigenvalues. Thus, let A be an operator such that the first eigenvalue λ_1 of $A'(0)$ has multiplicity two. Then for any $\epsilon > 0$ there is an operator A_ϵ such that $A_\epsilon \to A$ as $\epsilon \to 0$, $A'_\epsilon(0)$ has simple eigenvalues for all $\epsilon > 0$, and the first two eigenvalues of $A'_\epsilon(0)$ both approach λ_1 as $\epsilon \to 0$. Now, each eigenvalue of $A'_\epsilon(0)$ is a bifurcation point. It is possible that the branch of solutions starting from its first eigenvalue terminates at its second eigenvalue, as shown in Fig. 23a. As $\epsilon \to 0$ and the first two eigenvalues of $A'(0)$ approach each other, the branch of solutions joining them might shrink to zero as suggested in Fig. 23b. This shows how a double eigenvalue might fail to be a bifurcation

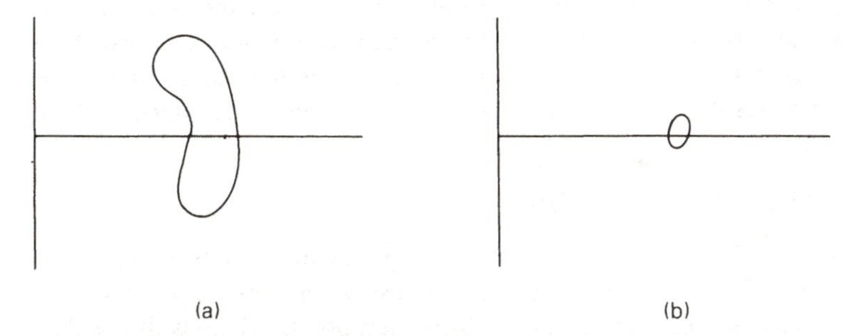

Fig. 23. Hypothetical branching diagrams for an operator A_ϵ such that $A'_\epsilon(0)$ becomes degenerate as $\epsilon \to 0$. (a) large ϵ; (b) small ϵ.

point, even though it is the limit of a pair of simple eigenvalues each of which is a bifurcation point. For a concrete example, consider the operator $A : \mathbb{R}^2 \to \mathbb{R}^2$ defined by $A(x,y) = (x^3 + y, x^2 y - x^3)$, and take $A_\epsilon(x,y) = A(x,y) + (\epsilon x, 0)$. Then $A_\epsilon u = \lambda u$ has bifurcation points at $\lambda = 0$ and $\lambda = \epsilon$, which coalesce in the way described above as $\epsilon \to 0$. The same thing can happen for any even multiplicity, but not for odd multiplicity.

We shall now consider two applications of the Bifurcation Theorem.

Example 11.42 In Example 5.19 we considered the system

$$u'' + \lambda \sin u = 0, \quad u(0) = u(1) = 0 \tag{11.21}$$

governing the angular deflection $u(x)$ at a distance x along a rod in equilibrium under compression λ. We used the contraction mapping theorem to show that if $\lambda < 3\sqrt{10}$ then the only solution is $u \equiv 0$, corresponding to the rod being straight. We shall now prove the existence of other solutions for larger values of λ.

We write (11.21) in the standard form $Au = \lambda u$ where $Au = -(u/\sin u)u'' = -u'' - u^2u''/6 - \ldots$. Then $A'(0)u = -u''$, and this operator, with the given boundary conditions, has non-degenerate eigenvalues $n^2\pi^2$. Hence there are bifurcation points at $\lambda = \pi^2, 4\pi^2, \ldots$, confirming the conjecture made in 5.19 that there are nonzero solutions for $\lambda > 3\sqrt{10}$. Physically they correspond to buckling of the rod.

The reader will have noticed that Theorem 11.41 does not apply to the operator A defined above, because it is not compact. This difficulty is easily removed by using Green's function to replace the differential equation (11.21) by an equivalent integral equation; in this way the result above can be rigorously justified. □

Of course, finding the bifurcation points is only a first step in analysing a nonlinear eigenvalue problem. It does not tell us, for example, whether (11.21) has nonzero solutions for $\lambda < \pi^2$; this depends on whether solutions branch backwards as shown in Fig. 21, or forwards as in Fig. 20. It is difficult to answer global questions about the behaviour of solution branches far from their bifurcation points. But it is relatively easy to find the shape of the branches near the λ-axis; one can use a series-expansion or approximation method, based on the idea that near the λ axis the solutions are small. The technique is illustrated in the next section.

11.7 Bifurcation and Stability

In section 11.5 we saw that the stability of a system depends on whether the eigenvalues of the linearised operator are positive or negative. In section 11.6 we saw that eigenvalues correspond to bifurcation points. There is thus a close connection between bifurcation and stability. In this section we shall return to the reaction-diffusion problem of Example 11.39, and study in more detail the stability of solutions near a bifurcation point.

Consider, then, the system

$$u_t = u_{xx} + h(u,x), \quad u(0,t) = u(1,t) = 0 \quad . \tag{11.22}$$

The zero solution is stable if the linearised operator $\partial^2/\partial x^2 + h_u(0,x)$ has negative eigenvalues only. For the particular case $h(u,x) = au(1 - u/b)$, where a and b are constants, we found in Example 11.39 that the zero solution is stable if $a < \pi^2$ and unstable if $a > \pi^2$. The value π^2 is a simple eigenvalue of the linearised operator, and is therefore a bifurcation point. We shall study the details of the solutions near the bifurcation point, and show that the new solution which appears for $a > \pi^2$ is stable.

We shall use an approximate method to solve the equation, based on the idea that if a is close to π^2, then the nonzero solution will still be close to the zero solution from which it has branched, and will therefore be small. So we write $a = \pi^2 + \epsilon$, and look for a solution of the equilibrium equation which is of the order of magnitude of ϵ. We assume that u can be expanded as a power series in ϵ, and write

$$u = \epsilon u_1(x) + \epsilon^2 u_2(x) + \ldots \quad . \tag{11.23}$$

We cannot be sure that this assumption is correct. Indeed, in many cases u cannot be expressed as a power series in ϵ; in Example 11.40, for instance, u is proportional to $\sqrt{\epsilon}$, which cannot be represented in the form (11.23). We shall show that our equation (11.22) has solutions of the form (11.23) by constructing them.

We substitute the expansion (11.23) into the equilibrium equation

$$u'' + (\pi^2 + \epsilon)u(1-u) = 0, \quad u(0) = u(1) = 0 \quad , \tag{11.24}$$

where we have taken $b = 1$ for simplicity. Equating to zero the coefficients of $\epsilon, \epsilon^2, \epsilon^3, \ldots$ gives

$$u_1'' + \pi^2 u_1 = 0 \quad , \tag{11.25}$$

$$u_2'' + \pi^2 u_2 = \pi^2 u_1^2 - u_1 \quad , \tag{11.26}$$

$$u_3'' + \pi^2 u_3 = 2\pi^2 u_1 u_2 + u_1^2 - u_2 \quad , \tag{11.27}$$

$$\ldots\ldots\ldots\ldots\ldots\ldots\ldots\ldots\ldots$$

and $$u_i(0) = u_i(1) = 0 \quad \text{for all} \quad i \quad . \tag{11.28}$$

We can solve (11.25) for u_1. The right hand side of (11.26) is then known, and we can solve for u_2. The right hand side of (11.27) is then known, and we can solve for u_3, and so on. At first sight it might seem that all the u_i can be determined, and thus a solution of the form (11.23) can always be found. But the story is more subtle than this.

The two-point boundary conditions (11.28) do not determine a unique solution of (11.25). It is satisfied by

$$u_1 = C\sin(\pi x)$$

for any value of the constant C. It is not obvious how to find C. The position is even worse when we consider (11.26). It is an inhomogeneous equation, and the corresponding homogeneous equation has nonzero solutions. The Fredholm Alternative (section 9.4) means that (11.26) will have no solution at all unless its right hand side is orthogonal to all functions satisfying $u_2'' + \pi^2 u_2 = 0$ with the boundary conditions (11.28), and this is not generally true. The way out of this impasse is to observe that the right hand side of (11.26) depends on u_1, which is not completely determined but contains a free constant C. If C can be chosen so that $\pi^2 u_1^2 - u_1$ is orthogonal to solutions of $u_2'' + \pi^2 u_2 = 0, u_2(0) = u_2(1) = 0$, then at one stroke we will have determined u_1 and ensured the existence of u_2.

The solutions of $u_2'' + \pi^2 u_2 = 0$ with $u_2(0) = u_2(1) = 0$ are multiples of $\sin(\pi x)$. The solubility condition for (11.26) is therefore

$$\int_0^1 \sin(\pi x)[\pi^2 u_1^2 - u_1]\,dx = 0 \quad .$$

Putting $u_1 = C\sin(\pi x)$, we find

$$C[(4\pi/3)C - \tfrac{1}{2}] = 0 \quad .$$

This equation has two solutions. $C = 0$ gives the zero solution of (11.24), and $C = 3/8\pi$ gives the nonzero bifurcating solution

$$u = (3/8\pi)\epsilon\sin(\pi x) + \ldots \quad . \tag{11.29}$$

We can now proceed to find higher terms in the same way. Equation (11.26) has, according to the Fredholm Alternative, infinitely many solutions, labelled by a parameter D, say, which is determined by the solubility condition for (11.27), and so on. The calculations become rapidly more laborious for the higher terms, but it is not hard to convince oneself that all the u_i can be determined, and thus a solution in the form of a power series in ϵ can be constructed. We shall not study the convergence of the series, but will use the first term alone as an approximation to u for small ϵ. For a more rigorous discussion see, for example, Iooss & Joseph or Stakgold (1979 or 1967). For a brief account of a problem like equations (11.19) and (5.17), where the solution is not expressible as a power series in ϵ, see section 3 of Drazin & Griffel.

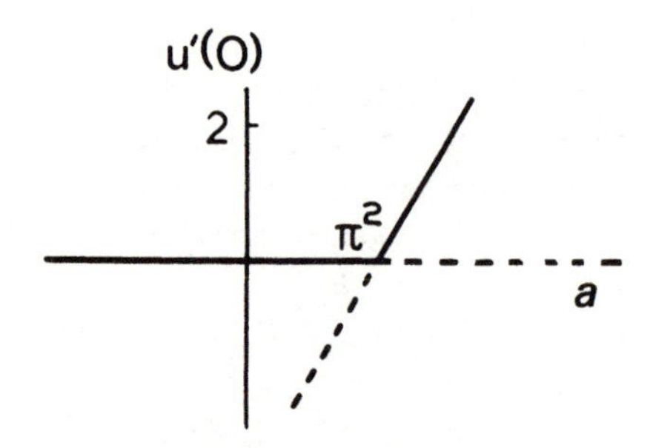

Fig. 24. Branching diagram in the neighbourhood of the first bifurcation point of equation (11.24). Solid lines denote stable solutions, broken lines denote unstable solutions.

We now have an approximate solution $u = (3/8\pi)(a - \pi^2)\sin(\pi x)$ of (11.24) near $a = \pi^2$, as well as the trivial solution. Fig. 24 shows the branching diagram in which u is represented by the number $u'(0)$; this representation is like that in Fig. 20, where $A = u'(0)/n\pi$. Unlike our previous examples, the bifurcating solution exists for $a < \pi^2$ as well as for $a > \pi^2$. This may seem to contradict the uniqueness proved in section 5.4 for $a < 3\sqrt{10}$. We have certainly found two solutions when a is slightly less than π^2. But the nonzero solution is negative-valued when $a < \pi^2$, and our uniqueness proof was for positive solutions, so there is no contradiction. The negative solution for $a < \pi^2$ has no physical meaning in the diffusion context of section 5.4.

We can now proceed to study the stability of the approximate solution (11.29), which we now call U:

$$U(x) = (3/8\pi)\epsilon \sin(\pi x) \quad .$$

If we define an operator P by

$$Pu = u'' + (\pi^2 + \epsilon)u(1 - u) \quad ,$$

then our diffusion equation is $u_t = Pu$, and the equilibrium solution U is stable if the linear operator $-P'(U)$ is strongly positive.

It is easy to calculate $P'(U)$ (or, in other words, to linearise the equation). We have

$$\begin{aligned} P'(U)v &= v'' + (\pi^2 + \epsilon)(1 - 2U)v \\ &= v'' + (\pi^2 + \epsilon)(1 - (3\pi/4)\epsilon \sin(\pi x))v \\ &= v'' + [\pi^2 + \epsilon(1 - (3\pi^3/4)\sin(\pi x))]v + \ldots \quad . \end{aligned} \tag{11.30}$$

We have neglected terms of order ϵ^2, which is reasonable since our approximation U to the equilibrium solution is correct only to order ϵ. We must discover whether the eigenvalues of this operator are all negative. For small ϵ the operator is nearly the same as $D^2 + \pi^2$, so its eigenvalues will be close to $(1 - n^2)\pi^2$, all of which are negative except the first. Hence the nonzero equilibrium solution is stable if the first eigenvalue of the operator in (11.30) is negative.

We estimate the first eigenvalue by means of Rayleigh's Principle. This gives accurate results when the trial function is close to the exact eigenfunction. Now, our operator (11.30) is close to the operator $D^2 + \pi^2$, whose first eigenfunction is $\sin(\pi x)$; hence $\sin(\pi x)$ should be close to the first eigenfunction of (11.30), and should be a good trial function. It gives the Rayleigh quotient

$$\begin{aligned} &2\int_0^1 \sin(\pi x)[-\pi^2 + \pi^2 + \epsilon\{1 - (3\pi^3/4)\sin(\pi x)\}]\sin(\pi x)\,dx \\ &= -(2\pi^2 - 1)\epsilon \quad . \end{aligned}$$

Hence the first eigenvalue is negative if $\epsilon > 0$ and positive if $\epsilon < 0$. This shows that the nonzero solution is stable for $a > \pi^2$ and unstable for $a < \pi^2$, as shown in Fig. 24.

This 'exchange of stabilities' is a common phenomenon in bifurcation problems: when a new solution branches from a stable solution, the original solution branch becomes unstable, and stability is transferred to the new solution. For our diffusion-reaction problem, we can say that if the growth rate is so high that $a > \pi^2$, then although there is an equilibrium with $u = 0$, any small positive u will grow, and the system will approach a stable equilibrium with $u > 0$. This makes good physical sense. It is also satisfying to see that the second solution for $a < \pi^2$ is mathematicall unstable, as well as being physically irrelevant to the diffusion problem.

Finally, we remark that our analysis is valid only near the first bifurcation point. A similar analysis can be carried out in the neighbourhood of the other bifurcation points, but numerical methods must be used to study the bifurcating solutions when they are not small. Fig. 25 shows the results of a numerical computation of the branching diagram. The slope of the branch at the first bifurcation point agrees with our calculated value of 3/8 (equation (11.29)); similar methods give the slopes of the other branches where they cross the axis.

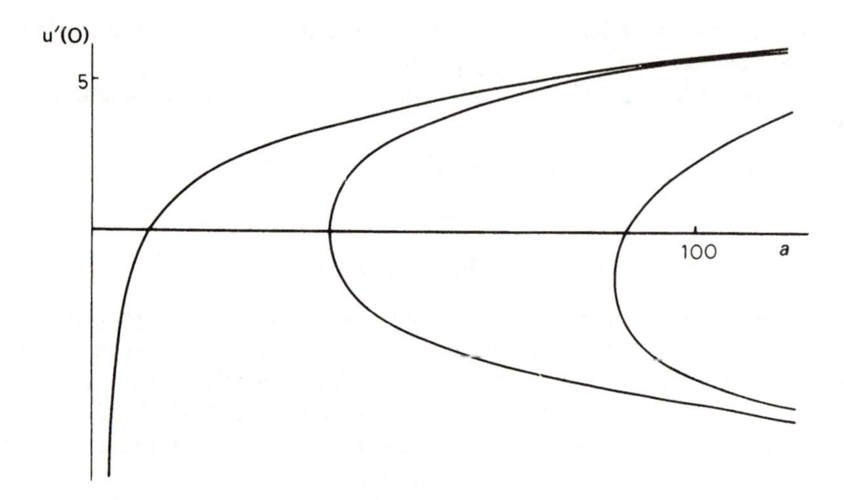

Fig. 25. Branching diagram for equation (11.24).

11.8 Summary and References

In section 11.1 we defined the Fréchet derivative of an operator on a normed space. In section 11.2 we discussed the second derivative, and in section 11.3 we considered maxima and minima of functionals, and the relation of the Fréchet calculus to the classical calculus of variations. The Fréchet derivative is a standard piece of mathematical apparatus, and is discussed in many books; see, for example, Cartan, Dieudonné, Chillingworth, or Moore (1985).

In section 11.4 we showed that the theory of stability of solutions of linear autonomous differential equations is equivalent to an eigenvalue problem. This is a standard piece of applied mathematics; see, for example, Lin & Segel, or books on stability theory. In section 11.5 we discussed the stability of solutions of nonlinear equations; for more extensive accounts see Drazin & Reid, or Joseph.

In section 11.6 we considered some examples of bifurcation; for more examples see Stakgold (1971 or 1979), Keller & Antman, or Rabinowitz. For a brief but accessible account of the theory, with an application to water waves, see Hutson &

Pym. In section 11.7 we used a series expansion technique to study the bifurcation of a nonlinear diffusion equation, and to show how the stability changes at a bifurcation point; see Stakgold, Keller & Antman, Drazin & Reid, or Iooss & Joseph for a discussion of this and other aspects of bifurcation theory.

Chapter 12

Distributional Hilbert Spaces

This last chapter is entirely theoretical. It describes a different approach to the Hilbert spaces L_2, which has the advantage of not needing the theory of Lebesgue integration. This approach elegantly unifies the theory of distributions of Part I with the theory of L_2 spaces in Part II, and it leads very naturally to the introduction of Sobolev spaces, which form the basic framework of a good deal of current research on partial differential equations, from both the pure and the numerical points of view. The disadvantage of this approach to L_2 is that it needs a thorough understanding of distribution theory. That is why we used the conventional theory in Part II, so as to be intelligible to a reader who has not mastered Part I.

12.1 The Space of Square-integrable Distributions

When we discussed $L_2[a,b]$ in 4.60, we were faced with two embarrassing facts. First, in order to obtain a complete space under the L_2 norm, we must include functions which are not Riemann-integrable but are Lebesgue-integrable. It is impossible to expound the theory completely without using the Lebesgue theory. We therefore gave an incomplete discussion, remarking that one can perfectly well use L_2 spaces without understanding all the theoretical details (as is amply illustrated in Parts II and III). The second embarrassing fact is that the vector space of square-integrable functions contains non-zero functions with zero norm, contradicting the axioms of a normed space. We dealt with this problem by redefining L_2 to be the space of equivalence classes of almost-everywhere-equal functions.

Now, the idea of regarding functions as equivalent if they are equal almost everywhere is reminiscent of generalised functions. In 1.18 we saw that piecewise continuous functions which differ only at a single point correspond to the same generalised function. This suggests that we may be able to interpret the equivalence classes of almost-everywhere-equal functions as generalised functions, and thus regard L_2 as a space whose elements are generalised functions. That is the main idea of this section. Starting from the theory of Part I, we shall construct a Hilbert space, and show that it has all the properties of L_2 and is therefore mathematically identical to L_2.

We shall work in $\mathbb{R}^n$, so our starting point is the theory of generalised functions of n variables briefly sketched in section 3.4. We shall now outline a slightly more general version of this theory.

Definition 12.1 The **support** of a function $\phi: \mathbb{R}^n \to \mathbb{C}$ is the closure of the set of points in $\mathbb{R}^n$ at which ϕ is nonzero. □

For the definition of closure see 4.46. We now define test functions; we use a notation more elaborate and explicit than that of Part I.

Definition 12.2 Let Ω be any open set in $\mathbb{R}^n$. Then $C_0^\infty(\Omega)$ is the set of all $\mathbb{R}^n \to \mathbb{C}$ functions with derivatives of all orders in Ω, and with supports which are compact subsets of Ω. □

Thus the space $\mathscr{D}$ of Chapter 1 is $C_0^\infty(\mathbb{R})$. This notation was introduced in 9.23; the subscript 0 is a reminder that functions $\phi \in C_0^\infty(\Omega)$ vanish near the boundary of Ω.

Definition 12.3 We say $\phi_n \to \Phi$ in the sense of test-function convergence if there is a bounded set $S \subset \Omega$ containing the supports of $\Phi, \phi_1, \phi_2, \ldots$, and if all partial derivatives of ϕ_n (including the zeroth, that is, ϕ_n itself) tend to the corresponding derivatives of Φ, uniformly in S. A **distribution on** Ω is a continuous linear functional on $C_0^\infty(\Omega)$, that is, a linear functional f such that $f(\phi_n) \to f(\Phi)$ whenever $\phi_n \to \Phi$ in the sense of test-function convergence. We write $\langle f, \phi \rangle = f(\phi)$ as in Chapter 1. □

Thus a distribution as defined in Chapter 1 is a distribution on $\mathbb{R}$.
Now we make $C_0^\infty(\Omega)$ an inner product space by defining

$$(\phi, \psi) = \int_\Omega \phi \bar{\psi} \, \mathrm{d}^n x \quad \text{for} \quad \phi, \psi \in C_0^\infty(\Omega) \quad . \tag{12.1}$$

It is a subspace of the space of continuous functions, and is incomplete with respect to the norm generated by the inner product (12.1). To obtain a Hilbert space we must add extra elements to $C_0^\infty(\Omega)$ to complete it. We shall see that distributions are just the extra elements that we need.

First we note that for each $\phi \in C_0^\infty(\Omega)$ there is a corresponding regular distribution: the functional $\psi \mapsto \int \phi \psi \, \mathrm{d}^n x = (\phi, \bar{\psi})$ for all $\psi \in C_0^\infty(\Omega)$. Now we consider Cauchy sequences, as in the general method of completion given in Appendix F. (ϕ_n) is a Cauchy sequence if $\| \phi_n - \phi_r \| \to 0$ as $n, r \to \infty$, where $\| \;\|$ denotes the norm generated by the inner product (12.1). This norm carries with it the usual definition of convergence, 4.32, which is different from the test-function convergence defined in 12.3 above. A sequence may converge in norm without converging in the test-function sense. From now on convergence will always mean convergence in norm, unless otherwise stated.

Now, Cauchy sequences in $C_0^\infty(\Omega)$ may not converge. But they always converge weakly.

Lemma 12.4 If (ϕ_n) is a Cauchy sequence in $C_0^\infty(\Omega)$, then for any $\psi \in C_0^\infty(\Omega)$, the sequence (ψ,ϕ_n) converges. □

The proof is not difficult, and will be found in Richtmyer. It follows from 12.4 that for any Cauchy sequence (ϕ_n) we can define a functional $\psi \mapsto \lim(\psi,\phi_n)$. It is easy to show that it is linear and continuous in the sense of Definition 12.3, which gives

Proposition 12.5 For any Cauchy sequence (ϕ_n) in $C_0^\infty(\Omega)$, there is a distribution f defined by $\langle f,\psi\rangle = \lim(\psi,\phi_n)$ for all $\psi \in C_0^\infty(\Omega)$. f is called the **distribution generated by** (ϕ_n). □

If (ϕ_n) converges to Φ in norm, with $\Phi \in C_0^\infty(\Omega)$, then the distribution given by Proposition 12.5 equals the regular distribution generated by Φ in the sense of 1.14. Distributions generated by nonconvergent Cauchy sequences give the extra elements which must be added to $C_0^\infty(\Omega)$ to make it complete.

There is a minor technical problem here. Given $\Phi \in C_0^\infty(\Omega)$, we wish to associate with it a distribution by means of 12.5. But there are many different Cauchy sequences tending to Φ; we must show that they all give the same distribution. Again, two non-convergent Cauchy sequences may be trying, so to speak, to converge to the same generalised function; we must be sure that they generate the same distribution. We express this idea formally by defining two Cauchy sequences (ϕ_n) and (ψ_n) to be **equivalent** if $\|\phi_n - \psi_n\| \to 0$ as $n \to \infty$. This means that the sequences are getting closer to each other, and are trying to converge to the same limit, loosely speaking. One can then prove that two Cauchy sequences generate the same distribution if and only if they are equivalent (see Richtmyer). It follows that for each $\Phi \in C_0^\infty(\Omega)$ there is just one corresponding distribution; $C_0^\infty(\Omega)$ is isomorphic to a subset of the set S of all distributions generated by Cauchy sequences. S is clearly a vector space: if $f,g \in S$ are generated by (ϕ_n) and (ψ_n) respectively, then the distribution $af + bg$ is generated by the Cauchy sequence $(a\phi_n + b\psi_n)$. We can give S an inner product in a natural way; the resulting inner product space is called $L_2(\Omega)$.

Definition 12.6 $L_2(\Omega)$ is the space of all distributions generated by Cauchy sequences in $C_0^\infty(\Omega)$, with inner product defined by

$$(f,g)_{L_2} = \lim_{n\to\infty}(\phi_n,\psi_n) \quad , \tag{12.2}$$

where (ϕ_n), (ψ_n) are Cauchy sequences generating the distributions f and g. □

The subscript in $(f,g)_{L_2}$ is to distinguish the inner product in the space of distributions $L_2(\Omega)$ from the inner product in the function space $C_0^\infty(\Omega)$. To justify using (12.2) to define $(f,g)_{L_2}$, one should show that it is independent of which of the many equivalent Cauchy sequences generating f or g is used; that is easy. To justify calling this space $L_2(\Omega)$ one must show that it is a Hilbert space, and that it contains the functions whose squares are integrable over Ω. It is not too difficult to show that $L_2(\Omega)$ is complete in the norm corresponding to the inner product (12.2). To show that it contains the square-integrable functions, we proceed as follows.

Lemma 9.23 shows that for any square-integrable function f and any $\epsilon > 0$ there is a $\phi \in C_0^\infty(\Omega)$ such that $\| f - \phi \| < \epsilon$, where $\| \; \|$ denotes the square-integral norm. Although 9.23 was stated in terms of three-dimensional space, it applies essentially unchanged in the n-dimensional case. Thus for each integer n there is a $\phi_n \in C_0^\infty(\Omega)$ with $\| f - \phi_n \| < 1/n$. It is easy to show that (ϕ_n) is a Cauchy sequence, and therefore generates an element of $L_2(\Omega)$ as defined in 12.6. This element of $L_2(\Omega)$ is the regular distribution generated by the square-integrable function f. In this sense, L_2 contains the square-integrable functions. It also contains distributions corresponding to functions which are pathological from the point of view of Riemann integration, but these do not normally appear in practice.

We have now shown that the space of distributions $L_2(\Omega)$ contains elements corresponding to the elements of the function space L_2 used in Part II. Distributions can be added, multiplied by constants, differentiated, and integrated, so in many respects they can be treated like ordinary functions. But we saw in Chapter 1 that distributions cannot in general be multiplied, and therefore cannot be used to solve nonlinear equations. Yet in Part II we used fixed-point theorems in L_2 to obtain solutions of nonlinear integral equations. How can we deal with these problems from the distributional point of view?

The answer is that those distributions which belong to L_2 are sufficiently well-behaved for multiplication to be possible. The difficulties arise when one multiplies by singular distributions such as $\delta(x)$. But $\delta \notin L_2(\Omega)$ – there is no Cauchy sequence (ϕ_n) of test functions such that $(\phi_n, \psi) \to \psi(0)$ for all test functions ψ. There is therefore no difficulty in multiplying elements of L_2 together.

However, the product of two ordinary square-integrable functions f,g need not be square-integrable – consider $(x^{-\frac{1}{3}})^2$, for example. Thus the distribution fg may or may not belong to L_2. It is defined in terms of Cauchy sequences in a natural way, as follows.

Definition 12.7 (Multiplication) If $f,g \in L_2(\Omega)$ are generated by Cauchy sequences (ϕ_n), (ψ_n) respectively, then their product is the distribution fg: $\theta \mapsto \lim(\theta, \phi_n \psi_n)$ for all $\theta \in C_0^\infty(\Omega)$. □

Of course, this needs justification; one can prove that the limit exists, defines a continuous linear functional on $C_0^\infty(\Omega)$, and is independent of the choice of Cauchy sequences representing f and g. Note that the sequence $(\phi_n \psi_n)$ need not be a Cauchy sequence; thus fg need not belong to $L_2(\Omega)$. However, in the special case where g is (generated by) a bounded continuous function, it can be shown that $fg \in L_2(\Omega)$. Details can be found in Richtmyer.

Finally, we consider briefly the operation of integration in $L_2(\Omega)$. In Chapter 2 we showed that a distribution f in one dimension always has an indefinite integral F such that $F' = f$. We cannot define definite integrals in general, because $\int_a^b f(x)\,dx = F(b) - F(a)$ is meaningless: a generalised function does not have a 'value' at $x = a$. However, if $f \in L_2(\mathbb{R})$, it turns out that its indefinite integral is a continuous function (more precisely, a regular distribution generated by a continuous function), and we can then define $\int_a^b f(x)\,dx = F(b) - F(a)$. Furthermore, if $f,g \in L_2(\mathbb{R})$, then again the integral of the distribution fg is a continuous function, and therefore definite integrals of products can be defined. This applies also in n dimensions. In this way one can define integrals such as $\int_\Omega fg\,d^n\boldsymbol{x}$, where $f,g \in L_2(\Omega)$, and prove that $\int f\bar{g}\,d^n\boldsymbol{x} = (f,g)_{L_2}$ as defined by equation (12.2) above. $L_2(\Omega)$ can now be described as the space of distributions f over Ω such that $\int |f|^2\,d^n\boldsymbol{x}$ exists, and all the theory and applications of L_2 spaces described in Parts II and III work in our distribution-theoretical version.

12.2 Sobolev Spaces

In Part II of this book we applied functional-analytic techniques to differential equations. A fundamental difficulty in this work is the fact that differential operators are unbounded, and are not defined on the whole of the Banach spaces $C[a,b]$ or $L_2[a,b]$ (which contain non-differentiable functions). We noted that by using a different norm, differential operators can be made bounded and continuous (see Problem 6.3). But with respect to that norm our spaces are incomplete, so we are no better off. We avoided these difficulties by a circuitous route: we used Green's functions to replace differential equations by integral equations, which involve continuous and often compact operators on Banach spaces, and are therefore easier to deal with.

We shall now see how the distributional approach to L_2 spaces leads to a theory into which differential equations fit naturally and can be treated directly without reduction to integral equations. Roughly speaking, the idea is that while spaces of differentiable ordinary functions are incomplete, the greater richness of distribution theory means that spaces of differentiable generalised functions are complete, and the powerful theories of Banach and Hilbert spaces are therefore available. These spaces of differentiable generalised functions are called Sobolev spaces. We shall describe only the simplest such spaces, and make no attempt to describe the full structure of the Sobolev theory; see the references mentioned in section 12.4.

In Problem 6.3 we saw that differentiation is a continuous operator on the space $C_1^1[a,b]$ of continuously differentiable functions with norm

$$[\int_a^b(|u|^2 + |u'|^2)\,dx]^{\frac{1}{2}} \quad . \tag{12.3}$$

This space is incomplete, but we shall now construct its distributional analogue H^1, which is complete. We shall generalise from $[a,b]$ to an n-dimensional region Ω, which may be $\mathbb{R}^n$ itself, or may be the region inside some surface. Then $H^1(\Omega)$ is a space of distributions on Ω with a norm which is an n-dimensional analogue of (12.3):

$$\| u \|_1 = \left[\int_\Omega \left(|u|^2 + \sum_1^n \left|\frac{\partial u}{\partial x_i}\right|^2\right) d^n x\right]^{\frac{1}{2}} \quad . \tag{12.4}$$

This norm can only be applied to those distributions in $L_2(\Omega)$ whose derivatives are square-integrable. A function with a step discontinuity, for example, gives a member of L_2, but its derivative involves the delta function, which is not in L_2. Hence the set of distributions to which the above norm can be applied is a proper subspace of L_2; it is the simplest example of a Sobolev space.

Definition 12.8 The **Sobolev space** $H^1(\Omega)$ is the set of all $f \in L_2(\Omega)$ such that all the first partial derivatives $\partial f/\partial x_i$ belong to $L_2(\Omega)$. The inner product in $H^1(\Omega)$ is

$$(f,g)_1 = \int (f\bar{g} + \nabla f \cdot \nabla \bar{g})\, d^n x \tag{12.5}$$

where $\nabla f \cdot \nabla \bar{g}$ denotes $\Sigma_1^n \partial f/\partial x_i\, \partial \bar{g}/\partial x_i$. □

This inner product clearly gives the norm (12.4). If we denote the L_2 inner product by a subscript zero:

$$(f,g)_0 = \int f\bar{g}\, d^n x \quad , \tag{12.6}$$

then (12.5) reads

$$(f,g)_1 = (f,g)_0 + (\nabla f, \nabla g)_0$$

where $(\nabla f, \nabla g)$ is an abbreviation for $\Sigma_i(\partial f/\partial x_i, \overline{\partial g/\partial x_i})$. In particular,

$$\| f \|_1^2 = \| f \|_0^2 + \| \nabla f \|_0^2 \quad .$$

It is not difficult to show from this that Cauchy sequences in $H^1(\Omega)$ converge to elements of $H^1(\Omega)$. In other words, $H^1(\Omega)$ is a Hilbert space. It is in fact the Hilbert space obtained by completing the set of smooth functions with respect to the norm $\| \ \|_1$, in the same way that $L_2(\Omega)$ is the Hilbert space obtained by completing the set of smooth functions with respect to the L_2-norm. Another way of saying this is to say that $C^\infty(\Omega)$ is dense in $L_2(\Omega)$ with the 0-norm, and $C^\infty(\Omega)$ is dense in $H^1(\Omega)$ with the 1-norm.

Now, when we constructed $L_2(\Omega)$, we started not from $C^\infty(\Omega)$ but from the space $C_0^\infty(\Omega)$ of smooth functions which vanish near the boundary of Ω. $C_0^\infty(\Omega)$ and $C^\infty(\Omega)$ are both dense subspaces of $L_2(\Omega)$, which may be regarded as the completion of either of them. It is natural to ask if the same is true of $H^1(\Omega)$. The answer is no. $C^\infty(\Omega)$ is dense in $H^1(\Omega)$, which may be regarded as the completion of $C^\infty(\Omega)$ with respect to the 1-norm. But $C_0^\infty(\Omega)$ is not dense in $H^1(\Omega)$. The point is that $H^1(\Omega)$ consists of differentiable functions (roughly speaking), and its norm involves the derivative; two functions which are close together according to the 1-norm must have nearly the same derivatives as well as nearly the same values. Since functions in $C_0^\infty(\Omega)$ are zero near the boundary of Ω, it is plausible that any limit, in the 1-norm, of members of $C_0^\infty(\Omega)$ must also vanish on the boundary. This means that elements of $H^1(\Omega)$ which do not vanish on the boundary cannot be approximated by elements of $C_0^\infty(\Omega)$; hence $C_0^\infty(\Omega)$ is not dense in $H^1(\Omega)$, and its closure is a proper subspace of $H^1(\Omega)$. This subspace is very useful in the theory of partial differential equations, because it consists of distributions which, roughly speaking, vanish on the boundary of Ω, and thus satisfy boundary conditions of the Dirichlet type often associated with partial differential equations. It is called the Sobolev space $H_0^1(\Omega)$. The discussion above is, of course, imprecise; we cannot strictly speak of the value of a distribution on the boundary of Ω. The purpose of the discussion is to motivate the following definition.

Definition 12.9 $H_0^1(\Omega)$ is the closure of $C_0^\infty(\Omega)$ in the space $H^1(\Omega)$. □

Since $H_0^1(\Omega)$ is by definition a closed subspace of a Hilbert space, it is itself a Hilbert space. In general it is difficult to give a clearer description of $H_0^1(\Omega)$ than the definition above. But in the one-dimensional case, when Ω is an interval of the real line, it can be shown that $H_0^1(\Omega)$ consists of those continuous functions f vanishing at the end-points of the interval such that $f' \in L_2(\Omega)$. This is a special case of Sobolev's Embedding Theorem; see, for example, Hutson & Pym. In higher dimensions, the members of $H_0^1(\Omega)$ need not all be continuous functions, but it can be proved that all continuous functions in $H_0^1(\Omega)$ must vanish on the boundary, and conversely all continuously differentiable functions which vanish on the boundary belong to $H_0^1(\Omega)$. It is thus reasonable to think of $H_0^1(\Omega)$ as consisting of (generalised) functions satisfying Dirichlet conditions on the boundary of Ω.

The above remarks suggest that the integration-by-parts formula should apply to elements of $H_0^1(\Omega)$, with zero contributions from the boundary terms. We now prove this result, which we shall use below.

Proposition 12.10 (Integration by Parts) If $u \in H_0^1(\Omega)$ and $\partial\phi/\partial x_i \in H^1(\Omega)$ for $i = 1, \ldots, n$, then

$$\int_\Omega u \nabla^2 \phi \, \mathrm{d}^n x = -\int_\Omega \nabla u \cdot \nabla \phi \, \mathrm{d}^n x \quad ,$$

using the ∇ notation of 12.8.

Proof Since $C_0^\infty(\Omega)$ is dense in $H_0^1(\Omega)$, there is a sequence (ψ_r) in $C_0^\infty(\Omega)$ with $\psi_r \to u$ in the 1-norm as $r \to \infty$. Hence

$$\int u\nabla^2\phi\, d^n x = \lim \int \psi_r \nabla^2 \psi\, d^n x$$
$$= \lim[-\int \nabla\psi_r \cdot \nabla\phi\, d^n x]$$

(since ψ_r vanishes on the boundary)

$$= -\int \nabla u \cdot \nabla\phi\, d^n x \quad ,$$

because $\psi_r \to u$ in the 1-norm implies $\nabla\psi_r \to \nabla u$ in the 0-norm. □

12.3 Application to Partial Differential Equations

We shall now consider the Sobolev-space approach to partial differential equations. We illustrate the theory by considering the Dirichlet problem for Poisson's equation. Essentially the same ideas and methods apply to elliptic equations in general, though there are some extra technical complications. The main aim of this theory is to prove existence and uniqueness results of the type of Theorem 3.32.

In Chapter 2 we discussed differential equations in the context of distribution theory, and distinguished between classical and generalised solutions. A generalised solution of Laplace's equation $\nabla^2 u = 0$ is a distribution u such that $\langle \nabla^2 u, \phi \rangle = 0$ for all test functions ϕ. If u is a classical solution, that is, a twice-differentiable function such that $\nabla^2 u = 0$, then obviously $\langle \nabla^2 u, \phi \rangle = 0$ for all test functions ϕ, so u is a generalised solution. However, the converse is not obvious; indeed, in Chapter 2 there are examples of generalised solutions of differential equations which are not classical solutions. Thus there are more generalised solutions than classical solutions; at any rate, generalised solutions satisfy a weaker condition than classical solutions, and should be easier to find. It is therefore sensible to look first for generalised solutions. Having shown the existence and uniqueness of a generalised solution, one can then sometimes prove that every generalised solution is a classical solution; in that case, the result is a classical existence and uniqueness theorem.

We shall illustrate this strategy by considering the Dirichlet problem for Poisson's equation, that is

$$\left.\begin{aligned} &\nabla^2 u = f \quad \text{in} \quad \Omega \\ \text{with} \quad &u = 0 \quad \text{on} \quad S \quad , \end{aligned}\right\} \tag{12.7}$$

where S is the boundary of the region Ω. A generalised solution of $\nabla^2 u = f$ is a distribution u such that $\langle u, \nabla^2 \phi \rangle = \langle f, \phi \rangle$ for all $\phi \in C_0^\infty(\Omega)$. The boundary condition $u = 0$ on S cannot be applied directly to a distribution u. But in section 12.2 we saw that there is a distributional version of the condition '$u = 0$ on S'; it is '$u \in H_0^1(\Omega)$'. Hence the generalised version of the problem (12.7) is to find a $u \in H_0^1(\Omega)$ such that

$$\langle u, \nabla^2 \phi \rangle = \langle f, \phi \rangle \quad \text{for all} \quad \phi \in C_0^\infty(\Omega) \quad ; \tag{12.8}$$

here f is a given distribution in $L_2(\Omega)$. We shall now reformulate this problem in such a way that its solution becomes obvious.

First, we note that $C_0^\infty(\Omega)$ can be regarded as a subspace of $H_0^1(\Omega)$, and that for $u \in H_0^1(\Omega)$ and $\phi \in C_0^\infty(\Omega)$,

$$\langle u,\phi\rangle = \int u\phi \, d^n x = (u,\bar{\phi})_0$$

where $(\,,\,)_0$ denotes the L_2 inner product as in (12.6). Hence the generalised Dirichlet problem (12.8) is equivalent to finding $u \in H_0^1(\Omega)$ such that

$$(u, \nabla^2\phi)_0 = (f,\phi)_0 \quad \text{for all} \quad \phi \in C_0^\infty(\Omega) \ .$$

Now we integrate by parts, using Proposition 12.10, which gives

$$(\nabla u, \nabla\phi)_0 = -(f,\phi)_0 \quad \text{for all} \quad \phi \in C_0^\infty(\Omega) \ . \tag{12.9}$$

But both sides of this equation make sense when ϕ is any member of H_0^1. Since C_0^∞ is a dense subset of H_0^1, it easily follows that (12.9) is equivalent to

$$(\nabla u, \nabla\phi)_0 = -(f,\phi)_0 \quad \text{for all} \quad \phi \in H_0^1(\Omega) \ . \tag{12.10}$$

We now introduce yet another inner product: we write

$$(f,g)_E = \int_\Omega \nabla f \cdot \nabla \bar{g} \, d^n x \tag{12.11}$$

for any $f,g \in H_0^1(\Omega)$. This is called the **energy inner product**, because the corresponding norm can often be interpreted as an energy. In fluid mechanics, for example, the 'velocity potential' ϕ satisfies Poisson's equation under certain conditions; $\nabla\phi$ is the velocity, and $\|\phi\|_E^2 = (\phi,\phi)_E$ is proportional to the kinetic energy. Similar interpretations apply in elasticity, electromagnetism, and other fields.

We remark in passing that the energy norm is relevant to problems of numerical approximation. If ϕ is a velocity potential and (ϕ_n) is a sequence of approximations to ϕ in the L_2-norm, so that $\|\phi_n - \phi\|_0 \to 0$ as $n \to \infty$, the velocities $\nabla\phi_n$ may not approach the exact velocity $\nabla\phi$ as $n \to \infty$. If, for example, a numerical approximation-procedure gives an approximation ϕ_n differing from ϕ by a small-amplitude small-wavelength ripple $n^{-1}\sin(nx)$, then $\|\phi_n - \phi\|_0 \to 0$ as $n \to \infty$, but the error in $\nabla\phi$ remains large as $n \to \infty$. Thus an approximate velocity potential which is accurate in the L_2 norm may give a very bad approximation to the physically relevant quantity, the velocity. But if $\|\phi_n - \phi\|_E \to 0$, we can be sure that the velocities given by ϕ_n approach the true velocity as $n \to \infty$.

We now apply the energy norm to the Dirichlet problem. Note first that $(f,g)_1 = (f,g)_0 + (f,g)_E$ for all $f,g \in H_0^1(\Omega)$. Combining this relation with the fact that $-\nabla^2$ is a strongly positive operator, it is not hard to show that the norms $\|\ \|_E$ and $\|\ \|_1$ are equivalent, and hence that the space of distributions in H_0^1 with inner product $(\,,\,)_E$ is a Hilbert space, which we shall call H_E. We now recast our Dirichlet problem as a problem in the Hilbert space H_E. Write (12.10) as

$$(u,\phi)_E = -(f,\phi)_0 \quad \text{for all} \quad \phi \in H_E \ . \tag{12.12}$$

It is easy to show that the functional $\phi \mapsto (\phi,f)_0$ is a continuous linear functional on H_E. By the Riesz representation theorem it follows that there is a unique $g \in H_E$ such that $(\phi,f)_0 = (\phi,g)_E$. Then (12.12) reads

$$(u,\phi)_E = -(g,\phi)_E \quad \text{for all} \quad \phi \in H_E \quad . \tag{12.13}$$

But this, as an equation for the unknown u, is completely trivial. It has exactly one solution, namely $u = -g$ (*proof:* take $\phi = u + g$ in (12.13), then $\| u + g \| = 0$, so $u = -g$). We have now proved the following.

Theorem 12.11 The generalised Dirichlet problem $\nabla^2 u = f$ has exactly one solution $u \in H_0^1(\Omega)$ for any $f \in L_2(\Omega)$. □

We now ask whether the generalised solution of $\nabla^2 u = f$ is a classical solution. If f is discontinuous, then $\nabla^2 u$ need not (classically) exist everywhere, and u will not be a classical twice-differentiable solution. But if f is smooth, then the generalised solution given by Theorem 12.11 is in fact a classical solution. The proof of this is quite difficult, and we shall just quote some results.

If f is smooth (infinitely differentiable), then u is smooth in Ω; if the boundary of Ω is also smooth, then u is a smooth function on $\overline{\Omega}$, that is, smooth up to and including the boundary, on which u vanishes. If f is not smooth but $f \in H^1$ (which means that f' is square-integrable), then u is s times continuously differentiable in Ω for any $s < 3 - \frac{1}{2}n$, where n is the dimension of space. Thus if Ω is a domain in $\mathbb{R}^3$ and $f \in H^1$, then $u \in C^1(\Omega)$. More generally, if $f^{(k)} \in L_2(\Omega)$, then u is s times continuously differentiable for any $s < k + 2 - \frac{1}{2}n$, and if the boundary of Ω is also well-behaved, then $u \in C^s(\overline{\Omega})$. Thus the degree of smoothness of the solution depends on that of the right hand side of the equation, and also on the dimension of the space.

The last fact is familiar from Part I. We found that Green's function for ordinary differential equations is continuous; for Laplace's equation in two dimensions it has a logarithmic singularity (and is therefore locally integrable and belongs to L_2), and for Laplace's equation in three dimensions it has a more severe singularity, r^{-1}. Thus as n increases, the solutions behave worse, in general agreement with the results quoted above.

12.4 Summary and References

In section 12.1 we reconstructed the space L_2 as a space of distributions, thus avoiding the need for any appeal to Lebesgue integration theory. This construction is given in more detail in Richtmyer. In section 12.2 we introduced two Sobolev spaces, one corresponding roughly to the space of differentiable functions, the other to the space of differentiable functions vanishing on the boundary of a given domain. In section 12.3 we used these spaces to show that the Dirichlet problem for Poisson's equation has exactly one generalised solution, and gave some results about the degree of smoothness of the solution. Results of this kind apply to a wide class of elliptic partial differential equations. A fairly simple introduction to the theory is given by Hutson & Pym, who also discuss its applications to the theory of the finite-element method of numerical approximation. For more detailed accounts of the theory, see Showalter, Friedman, Schechter, or Gilbarg & Trudinger.

Appendices

Appendix A Sets and Mappings

This Appendix outlines our basic terminology and conventions. We assume that the reader knows what is meant by a set of objects; we do not try to analyse this idea, since such analysis leads to great logical difficulties. We write $a \in A$ to mean a belongs to the set A; we specify sets either (i) by listing their members thus: $\{1,2,3,\ldots\}$ is the set of positive integers; or (ii) by specifying conditions for membership thus: $B = \{x \in A : x > 0\}$ means that B is the set of all elements x of A such that $x > 0$. The list of elements of a set may have repetitions, but no significance is attached to the repetition; that is, $\{1,2,2\}$ is the same set as $\{1,2\}$. A set with no members is called **empty**. The **union** $A \cup B$ of two sets A and B is the set of all elements in A or B or both; thus $\{1,2\} \cup \{2,3\} = \{1,2,3\}$. The **intersection** $A \cap B$ of two sets A and B is the set of elements common to A and B; thus $\{1,2\} \cap \{2,3\} = \{2\}$. If all elements of A also belong to another set B, we say A is **contained in** B, and write $A \subset B$ or $B \supset A$; A is said to be a **subset** of B. For any set A, $A \subset A$; if $A \subset B$ and $A \neq B$, then A is said to be a **proper subset** of B. If $A \subset B$, the set $B - A = \{x \in B : x \notin A\}$ is called the **complement** of A with respect to B; it is non-empty when A is a proper subset of B.

A **function** or **mapping** from a set X to a set Y is a rule which, given any $x \in X$, associates with it a unique $y \in Y$. We say 'f maps X into Y', written $f: X \to Y$. According to context, '$f: X \to Y$' may also be read as 'f mapping X into Y', as in the phrase 'we define $f: X \to Y$ by ...'. If f associates the element $y \in Y$ with $x \in X$, we say f maps x to y, and write $f: x \mapsto y$; the arrow $\mapsto$ denotes the action of a function on the elements of X. We also use the conventional notation $y = f(x)$. Instead of giving a function a single letter as a name, we sometimes refer to it by specifying its rule of action; thus, the function $x \mapsto x^2$ (read 'x maps to x^2' or 'x goes to x^2') is the function which squares any given number.

$\mathbb{R}$ denotes the set of real numbers, and $\mathbb{C}$ the set of complex numbers; thus $f: \mathbb{R} \to \mathbb{R}$ means that f is a real-valued function of a real variable. Strictly speaking, such a function should be defined for all real numbers, but we sometimes write $f: \mathbb{R} \to \mathbb{R}$ even when f is defined only on a subset of $\mathbb{R}$ (for example, the function $x \mapsto \sqrt{x}$).

We use the conventional notation for intervals in $\mathbb{R}$: $(a,b) = \{x \in \mathbb{R}: a < x < b\}$, $[a,b] = \{x \in \mathbb{R}: a \leqslant x \leqslant b\}$, $[a,b) = \{x \in \mathbb{R}: a \leqslant x < b\}$, $[a,\infty) = \{x \in \mathbb{R}: x \geqslant a\}$, etc. (a,b) is called an open interval and $[a,b]$ a closed interval. There is a clash of notation between open intervals and two-component vectors, but the context should always make it clear which is meant.

Appendix B Sequences, Series, and Uniform Convergence

This Appendix outlines briefly some of the main ideas and results of the theory of uniform convergence. For a fuller account, see Kreider *et al.*, Titchmarsh (1939), or other textbooks of analysis. We begin by sketching the elementary theory of convergence, with which the reader should already be familiar.

A **sequence** of numbers is an infinitely long list of numbers, written $(a_1, a_2, \ldots)$ or (a_n). Thus $(1,4,9,\ldots)$ or (n^2) denotes the sequence of perfect squares. A sequence (u_n) differs from a set $\{u_n\}$ in the following two respects. Firstly, the elements of a sequence, unlike the elements of a set, are arranged in a definite order (two sets are equal if they consist of the same elements, regardless of order). Secondly, a sequence may contain repetitions: the three sequences $(1,1,2,3,4,\ldots)$, $(1,2,3,4,\ldots)$, $(1,2,1,3,\ldots)$ are regarded as all different, whereas $\{1,1,2,3,\ldots\}$ is merely a redundant way of writing the set $\{1,2,3,4,\ldots\}$. One can construct sequences of elements of any set, though for the moment we consider only numbers.

Another way of looking at a sequence is to regard it as a mapping from the positive integers to the real or complex numbers: for each integer n, a number u_n is specified. This point of view may be helpful when considering doubly infinite sequences of the form $(\ldots, u_{-2}, u_{-1}, u_0, u_1, u_2, \ldots)$. The idea of a list which has no beginning may give rise to feelings of unease. But regarded as a mapping from the set of all integers (positive and negative) to the numbers, it should cause no difficulty.

A **subsequence** of a sequence (u_n) is a sequence obtained from (u_n) by choosing certain members and deleting others. Thus $(u_1, u_3, u_5, \ldots)$ is a subsequence of (u_n). The idea is very simple, but a consistent and explicit notation can be quite clumsy. A subsequence of (u_n) can be defined by a sequence of integers $(s_1, s_2, \ldots)$ specifying which members of (u_n) are included in the subsequence; thus for the above example, $(s_n) = (1,3,5,\ldots)$. The subsequence is then often written (u_{s_n}). I have tried to avoid such double subscripts when working with subsequences in Parts II and III.

A sequence (u_n) of real or complex numbers is said to **converge** to a number U if for each $\epsilon > 0$ there is a number N, depending on ϵ, such that $|u_n - U| < \epsilon$ for all $n > N$. We write $u_n \to U$. It is easy to prove that if $u_n \to U$ and $v_n \to V$, then $au_n + bv_n \to aU + bV$ for any numbers a and b, and $u_n v_n \to UV$. If for each $A > 0$ there is an N such that $u_n > A$ for all $n > N$, we say $u_n \to \infty$ as $n \to \infty$; if for each $B < 0$ there is an N such that $u_n < B$ for all $n > N$, we say $u_n \to -\infty$ as $n \to \infty$; these two definitions apply to real sequences only. Let L denote either a number or ∞ or $-\infty$; if $u_n \to L$, then any subsequence of (u_n) also tends to L.

The two main results about sequences in $\mathbb{R}$ and $\mathbb{C}$ are the following.

Theorem B1 (Cauchy's Criterion) A sequence of numbers (u_n) converges if and only if for each $\epsilon > 0$ there is an N such that $|u_n - u_m| < \epsilon$ for all $n,m > N$.

BORDERS®

BORDERS
BOOKS AND MUSIC
335 HOWE AVE
CUYAHOGA FALLS
(330) 945-7683

STORE: 0226 REG: 05/03 TRAN#: 8041
SALE 08/13/2009 EMP: 00433

STORY OF MATHEMATICS
9780681211940 BI T 2.39
3.99 40% COUPON
COUPON 15904333000000000

Subtotal 2.39
BR: 8272953095 S

Subtotal 2.39
OHIO 6.5% .16
1 Item Total 2.55
CASH 5.00
Cash Change Due 2.45

You Saved $1.60

08/13/2009 02:52PM

permitted only if presented in saleable condition accompanied by the original sales receipt or Borders gift receipt within the time periods specified below. Returns accompanied by the original sales receipt must be made within 30 days of purchase and the purchase price will be refunded in the same form as the original purchase. Returns accompanied by the original Borders gift receipt must be made within 60 days of purchase and the purchase price will be refunded in the form of a return gift card.

Exchanges of opened audio books, music, videos, video games, software and electronics will be permitted subject to the same time periods and receipt requirements as above and can be made for the same item only.

Periodicals, newspapers, comic books, food and drink, digital downloads, gift cards, return gift cards, items marked "non-returnable," "final sale" or the like and out-of-print, collectible or pre-owned items cannot be returned or exchanged.

Returns and exchanges to a Borders, Borders Express or Waldenbooks retail store of merchandise purchased from Borders.com may be permitted in certain circumstances. See Borders.com for details.

BORDERS.

Returns

Returns of merchandise purchased from a Borders, Borders Express or Waldenbooks retail store will be permitted only if presented in saleable condition accompanied by the original sales receipt or Borders gift receipt within the time periods specified below. Returns accompanied by the original sales receipt must be made within 30 days of purchase and the purchase price will be refunded in the same form as the original purchase. Returns accompanied by the original Borders gift receipt must be made within 60 days of purchase and the purchase price will be refunded in the form of a return gift card

Theorem B2 An increasing bounded real sequence converges. That is, if the real numbers (u_n) satisfy $u_1 \leqslant u_2 \leqslant u_3 \leqslant \ldots$, and there is an M such that $u_r \leqslant M$ for all r, then (u_n) converges. □

From the point of view of this book, the main thing about these theorems is that they are true. From the point of view of pure mathematics, their logical status varies from one exposition of the subject to another (see, for example, Kreider *et al.*, Marsden, and Rudin). Sometimes B1 is regarded as a basic property of the number system and B2 is deduced from it; sometimes B2 is regarded as the basic property and B1 is deduced from it; sometimes both B1 and B2 are deduced from some other basic property. However, the resulting structure is the same no matter in what order it is built up.

An infinite **series** is a list of numbers to be added together (whereas a sequence is a list of numbers to be considered in turn, with no addition). The series $\Sigma_1^\infty u_n$ is defined in terms of the sequence (s_n) of partial sums: $s_n = u_1 + \ldots + u_n$. We say $\Sigma_1^\infty u_n$ converges to the sum U, or diverges to ∞ or $-\infty$, whenever the sequence (s_n) tends to U or to ∞ or $-\infty$. The basic theorems for sequences can thus be translated into theorems for series.

A series Σu_n is said to **converge absolutely** if $\Sigma |u_n|$ converges. It is easy to prove that if $\Sigma |u_n|$ converges then Σu_n converges; in other words, an absolutely convergent series is convergent in the usual sense. The terms of an absolutely convergent series can be rearranged without affecting its sum, and absolutely convergent series can be multiplied together. That is, if Σa_n and Σb_n converge absolutely to A and B respectively, then $\Sigma a_r b_s$, summed over all possible combinations of r and s, in any order, converges to AB. These properties do not always hold for non-absolutely convergent series.

We now turn to the theory of convergence of sequences and series of functions. Unless otherwise stated, we shall consider $\mathbb{R} \to \mathbb{C}$ functions, though everything is very easily generalised to $\mathbb{R}^n \to \mathbb{C}$, or indeed $\mathbb{R}^n \to B$ functions for any Banach space B. We shall use I to denote an interval of $\mathbb{R}$, which may be an open interval (a,b), a closed interval $[a,b]$, a semi-infinite interval such as $[a,\infty) = \{x : a \leqslant x\}$, or $\mathbb{R}$ itself.

A sequence (f_n) of functions is said to **converge pointwise** to F on an interval I if $f_n(x) \to F(x)$ as $n \to \infty$ for all $x \in I$. This is the obvious and natural way to define convergence for functions, but it has some awkward consequences.

Example B3 Consider the sequence $f_n(x) = x^n$ on the interval $[0,1]$. It converges to the function F defined by $F(x) = 0$ for $0 \leqslant x < 1$, $F(1) = 1$. We thus have a sequence of continuous functions converging to a discontinuous limit (which greatly confused eighteenth-century mathematicians, to whom it seemed obvious that a limit of continuous functions must be continuous in the same way that a limit of non-negative functions must be non-negative – see Kline).

Example B4 The sequence $f_n(x) = n^2 x e^{-nx}$ displays more surprising behaviour; $f_n(x) \to 0$ for $0 \leqslant x \leqslant 1$, as $n \to \infty$, yet $\int_0^1 f_n(x)\,dx \to 1$ as $n \to \infty$. This has the air of a conjuring trick; the function vanishes as $n \to \infty$, for all x, yet contrives to leave a non-zero integral behind, as it were.† If the reader will sketch the graphs of $f_1, f_2, f_3, \ldots$ he will see how the trick is worked. As n increases, the region in which $f_n(x)$ takes large values shrinks, so that for any fixed x, there is an N such that $f_n(x)$ is small for all $n > N$, and $f_n(x) \to 0$; at the same time, the values of f_n, in the small region where it takes appreciable values, grow as $n \to \infty$, so that for large n the graph of f_n has a very tall and narrow spike, the area under which remains non-zero as $n \to \infty$. □

The above examples show that sequences which converge pointwise can behave quite oddly. We therefore introduce a stronger definition of convergence, called uniform convergence, designed to exclude such odd behaviour.

Definition B5(a) A sequence of functions (f_n) is said to **converge uniformly** to F on an interval I if for each $\epsilon > 0$ there is a number N, depending on ϵ but not on x, such that $|f_n(x) - F(x)| < \epsilon$ for all $n > N$ and all $x \in I$. □

This definition can easily be shown to be equivalent to the following.

Definition B5(b) $f_n \to F$ uniformly on I if $\sup\{|f_n(x) - F(x)| : x \in I\} \to 0$ as $n \to \infty$.

We extend the definition from sequences to series in the usual way: a series of functions converges uniformly if the sequence of its partial sums converges uniformly. The following then gives a useful way of identifying uniformly convergent series.

Theorem B6 (Weierstrass *M*-Test) If ΣM_n is a convergent series of numbers and $|f_n(x)| \leqslant M_n$ for all n and $x \in I$, then Σf_n converges absolutely and uniformly for $x \in I$. □

Examples B3 and B4 converge non-uniformly to zero. For other examples, and for proof of the results quoted here, see, for example, Kreider *et al.* or Titchmarsh (1939). The general theory can be broadly summed up by saying that uniform convergence guarantees that things work out as they ought to – that is, continuous functions tend to continuous limits, series can be differentiated term by term, etc. More precisely, we can say that uniform convergence justifies the

† cf. Carroll, Chapter 6, penultimate paragraph.

interchange of limiting processes. In Example B4, for instance, we applied two limit processes to $f_n(x)$: the limit as $n \to \infty$ and the limit process implied in the definition of integration. We found that the limit as $n \to \infty$ of the integral is not the same as the integral of the limit. Thus the value of a double limit depends on which order they are taken in. Most of the theorems below can be interpreted as saying that if the convergence is uniform, then the limits can be taken in either order, giving the same result.

Theorem B7 (Continuity) If f_n is continuous on I for all n, and $f_n \to F$ uniformly on I, then F is continuous on I.

Theorem B8 (Differentiation) If f_n has a continuous derivative on $[a,b]$, if $f_n \to F$ pointwise on $[a,b]$, and if the sequence (f_n') converges uniformly on $[a,b]$, then F is differentiable and $f_n' \to F'$ on $[a,b]$.

Theorem B9 (Integration) If f_n is continuous for each n, and $f_n \to F$ uniformly on $[a,b]$, then $\int_a^x f_n(t)\,dt \to \int_a^x F(t)\,dt$ uniformly for $x \in [a,b]$. □

These theorems apply to infinite series as well as sequences, the uniform convergence of a series being identified with that of the sequence of partial sums. They apply also when the integer index n is replaced by a continuous variable y, so that $f_n(x)$ becomes $f(x,y)$. We then have the definition: $f(x,y) \to F(x)$ as $y \to c$, uniformly for $x \in I$, if for each $\epsilon > 0$ there is a $\delta > 0$, independent of x, such that $|f(x,y) - F(x)| < \epsilon$ for all y such that $|y - c| < \delta$ and all $x \in I$. In this definition c can be replaced by ∞, with obvious modifications elsewhere. Theorems B7, B8 and B9 have straightforward analogues; for example, the analogue of B7 is

Theorem B10 (Continuity) If $f(x,y) \to F(x)$ uniformly for $x \in I$ as $y \to c$, and f is a continuous function of x for all y, then F is continuous. □

We now consider the uniform convergence of integrals. An integral over an infinite range is defined by a limit process: if $\int_a^Y f(y)\,dy \to I$ as $Y \to \infty$, then we say the improper integral $\int_a^\infty f(y)\,dy$ exists or converges, and equals I. Similarly, an integral over a finite region may have convergence problems if the integrand has a singularity. Thus if f is singular at $y = b$, then we consider $\int_a^{b-\epsilon} f(y)\,dy$ for $\epsilon > 0$; if it tends to a limit I as $\epsilon \downarrow 0$, then we say the improper integral $\int_a^b f(y)\,dy$ exists and equals I. If f has a singularity inside the region of integration, at $y = c$, say, then we write $\int_a^b f(y)\,dy = \int_a^c f(y)\,dy + \int_c^b f(y)\,dy$, and interpret these two improper integrals by the above definition and its obvious analogue for singularity at the lower limit. We used this procedure in 1.39.

Considerations of uniformity arise when the integrand is a function of two variables. Given a function $f(x,y)$, we define

$$g(x) = \int_a^b f(x,y)\,dy \quad , \tag{B1}$$

assuming that the integral exists. We shall suppose that $f(x,y)$ is singular at $y = b$, for all x in some interval. Similar results hold when the position of the singularity depends on x, and for infinite integrals.

Definition B11 Suppose that $f(x,y)$ is continuous for $a \leqslant y < b, \alpha \leqslant x \leqslant \beta$. The integral $\int_a^b f(x,y)\,dy$ is said to **converge uniformly** to $g(x)$ on the interval $\alpha \leqslant x \leqslant \beta$ if $g_\epsilon(x) \to g(x)$ uniformly for $x \in [\alpha,\beta]$ as $\epsilon \downarrow 0$, where $g_\epsilon(x) = \int_a^{b-\epsilon} f(x,y)\,dy$. □

The analogue of the Weierstrass M-test is the following:

Theorem B12 (Test for Uniformity) Suppose $\int_a^b f(x,y)\,dy$ converges for $\alpha \leqslant x \leqslant \beta$, and $|f(x,y)| \leqslant M(y)$ for $\alpha \leqslant x \leqslant \beta$, where M is a function such that $\int_a^b M(y)\,dy$ converges. Then $\int_a^b f(x,y)\,dy$ converges uniformly for $x \in [\alpha,\beta]$. □

The analogues of Theorems B7 and B8 are as follows.

Theorem B13 (Continuity) If $f(x,y)$ is continuous for $\alpha \leqslant x \leqslant \beta$ and $a \leqslant y < b$, and if $\int_a^b f(x,y)\,dy$ converges uniformly for $x \in [\alpha,\beta]$, then the function $g(x) = \int_a^b f(x,y)\,dy$ is continuous for $x \in [\alpha,\beta]$.

Theorem B14 (Differentiation of Integrals) Suppose f and $f_x \equiv \partial f/\partial x$ are continuous for $\alpha \leqslant x \leqslant \beta$ and $a \leqslant y < b$. If $\int_a^b f(x,y)\,dy$ converges and $\int_a^b f_x(x,y)\,dy$ converges uniformly for $x \in [\alpha,\beta]$, then the function $g(x) = \int_a^b f(x,y)\,dy$ is differentiable, and $g'(x) = \int_a^b f_x(x,y)\,dy$. □

Theorem B9 also has a straightforward analogue, expressing $\int_\alpha^\beta g(x)\,dx$ in terms of a double integral of f. The following theorem deals with a situation one degree more complicated: both the x and the y integrals are improper. It justifies changing the order of integration in such a double integral. We state the theorem in terms of infinite integrals; similar results hold for improper integrals over finite regions.

Theorem B15 Suppose that $\int_a^\infty f(x,y)\,dx$ converges, and $\int_b^\infty f(x,y)\,dy$ converges uniformly for $x \geqslant a$. Then if

$$\int_a^\infty \{ \int_b^\infty f(x,y)\,dy \}\,dx$$

converges, it equals

$$\int_b^\infty \{ \int_a^\infty f(x,y)\,dx \}\,dy \quad .$$ □

Appendix C Sup and Inf

Roughly speaking, the supremum of a set of numbers means its maximum value, and the infimum means its minimum value. This explanation should make sup and inf intelligible where they appear in the text, but it glosses over subtleties which are explained in this Appendix.

A set of real numbers may or may not have a largest member. The closed interval $[0,1]$ has a largest member, 1, but the open interval $(0,1)$ has no largest member: for any $x \in (0,1)$ there is a larger member of $(0,1)$, for example, $\frac{1}{2}(x+1)$. Yet even though $1 \notin (0,1)$, it has some of the properties characterising the largest member; it forms an upper boundary, so to speak, for $(0,1)$. The following theory makes this vague idea precise.

Definition C1 Let S be a set of real numbers. x is said to be an **upper bound** for S if $x \geqslant s$ for all $s \in S$. y is said to be a **lower bound** for S if $y \leqslant s$ for all $s \in S$.

Definition C2 If a set S of real numbers has an upper bound x, S is said to be **bounded above by** x, or simply **bounded above.** If S has a lower bound y, S is said to be **bounded below by** y, or simply **bounded below**.

Remark C3 If x is an upper bound for S, so is $x + a$ for any $a \geqslant 0$. If y is a lower bound for S, so is $y - a$ for any $a \geqslant 0$.

Examples C4 For $S = (0,1)$, any number $\geqslant 1$ is an upper bound; any number $\leqslant 0$ is a lower bound.

The set of all positive numbers is bounded below, by 0, but is not bounded above. □

If the set S is bounded above, it follows from C3 that the set U of all upper bounds of S is a semi-infinite interval. U is bounded below, since any element of S is a lower bound for U. A set which is bounded below need not in general have a smallest member, but the set U has.

Proposition C5 If a set S of real numbers is bounded above, the set of all its upper bounds has a least member. If S is bounded below, the set of all its lower bounds has a greatest member. □

The logical status of this Proposition varies from one exposition of analysis to another. Sometimes, as in Kreider *et al.*, it is taken as a basic property of the real number system, and everything else is deduced from it. Sometimes, as in Marsden, Proposition C5 is deduced from the fact that a bounded increasing sequence converges (Appendix B Theorem B2); other logical arrangements are possible. From our point of view, the fact that C5 is true is more important than the details of its proof, for which systematic textbooks of analysis should be consulted.

Definition C6 If a set S is bounded above, the least upper bound is called the **supremum** of S, written $\sup S$. If S is bounded below, the greatest lower bound is called its **infimum**, written $\inf S$.

Proposition C7 u is the supremum of a set S if and only if u is an upper bound for S and for each $\epsilon > 0$ there is an $x \in S$ with $x > u - \epsilon$. Similarly l is the infimum of S if and only if l is a lower bound for S and for each $\epsilon > 0$ there is an $x \in S$ with $x < l + \epsilon$. □

The proof of C7 is easy, and will not be given here.

We now return to the examples considered above. The closed interval $[0,1]$ has a smallest member 0 and greatest member 1, and we have $\sup[0,1] = 1$, $\inf[0,1] = 0$. In this case the sup and inf belong to the set. For the open interval $(0,1)$, every number $\geqslant 1$ is an upper bound, and $\sup(0,1) = 1$; similarly $\inf(0,1) = 0$. In this case the sup and inf do not belong to the set.

As another example consider the function

$$f(x) = 1 - e^{-x^2} \quad .$$

It is bounded above – that is, the set of its values $\{f(x): x \in \mathbb{R}\}$ is bounded above. Hence $\{f(x)\}$ has a supremum, and clearly $\sup\{f(x): x \in \mathbb{R}\} = 1$. But $f(x)$ has no maximum value; it takes values arbitrarily close to 1, but never equals 1. This is an example where $\max\{f(x)\}$ does not exist but $\sup\{f(x)\}$ exists. That is why we use sup rather than max in 4.30.

Appendix D Countability

It is clear that sets can be classified as either finite or infinite. This Appendix is devoted to the perhaps less obvious fact that there are two kinds of infinite set, one of which is more infinite than the other, so to speak. This distinction is important in the theory of Hilbert space; we use it in section 7.3, in proving that if a space has a basis, then every subspace has a basis. To explain the ideas involved, we begin by considering more carefully what it means to say that a set is finite.

A set has n elements if its elements can be counted off against the numbers from 1 to n, that is, if there is a correspondence between the elements of the set and the integers $1, \ldots, n$, so that each element is assigned to exactly one number and each number is assigned to exactly one element. Such a correspondence is called a **one-to-one correspondence**, and two sets are called **similar** if there is a one-to-one correspondence between them. We then say that a set has n elements if it is similar to $\{1,2, \ldots, n\}$. A set with n elements is called **finite**; an **infinite** set is one which is not similar to $\{1,2, \ldots, n\}$ for any n. For finite sets, similarity means having the same number of elements.

There is an obvious next step to this argument. If an infinite set is so large that it is not similar to any finite subset of the integers, perhaps it is similar to the set of all integers. A set is said to be **countably infinite** if it can be put into one-to-one correspondence with the set of all positive integers. A set is said to be **countable** if it is either finite or countably infinite, that is, if it can be put into one-to-one correspondence with either a finite set of integers or all the integers. If this is possible then we can label each element with the corresponding integer thus: a_n. The elements of a countable set can then be written as an ordered list $\{a_1, a_2, \ldots\}$, and in this sense they can be counted. The counting process may never end, but at least one can be sure that any given element of the set will appear in the list if one counts far enough. In this sense a countable set is more manageable than an uncountable set, which has so many elements that it is impossible to arrange them in a single list. Indeed, if the elements of a set can be arranged in a list, then it is countable, for each element can be uniquely assigned to a positive integer, namely, its position in the list. It is perhaps not obvious that there exist sets which are not countable; we shall prove it later in this Appendix.

We now consider some examples of countable sets. The set of positive integers is the classic example, of course. The even integers can be listed thus: 2,4,6, . . . , and they also form a countable set. We thus have a one-to-one correspondence between the set of all positive integers and the subset of even positive integers. No finite set can be similar to a proper subset of itself, because they have different numbers of elements. But it is a characteristic feature of infinite sets that they can be put into one-to-one correspondence with proper subsets of themselves.

Consider now the set of all rational numbers, that is, all numbers of the form p/q where p and q are integers. They are densely distributed along the real line, in the sense that given any two rational numbers $r \neq s$, no matter how close together they are there is always another rational lying between them. It is therefore impossible to arrange them in an ordered sequence $r_1 < r_2 < \ldots$; infinitely many rationals would always be left out of such a list. It might seem then that the rationals are uncountable. But this is not so. We shall now show how the rationals can be arranged as a sequence.

Consider first the positive rationals only. Each one can be uniquely expressed as a fraction p/q in its lowest terms (that is, with p and q having no common divisor). There is one rational with $p + q = 2$, namely $1 = 1/1$. There are two with $p + q = 3$, two with $p + q = 4$, and in general for any n, a finite number with $p + q = n$. We can therefore list them in ascending order of $(p + q)$: $\frac{1}{1}$; $\frac{1}{2}, \frac{2}{1}$; $\frac{1}{3}, \frac{3}{1}$; $\frac{1}{4}, \frac{2}{3}, \frac{3}{2}, \frac{4}{1}$; $\frac{1}{5}, \frac{5}{1}$; $\frac{1}{6}, \frac{2}{5}, \frac{3}{4}, \frac{4}{3}, \frac{5}{2}, \frac{6}{1}$; $\frac{1}{7}, \frac{3}{5}, \ldots$

Each positive rational appears exactly once in this sequence, which shows that they are countable. It is now clear that the set of all rationals is countable: if our list of positive rationals is $\{r_1, r_2, r_3, \ldots\}$, then the set of all rationals can be listed as $\{0, r_1, -r_1, r_2, -r_2, \ldots\}$.

We can extend the above argument as follows. The rationals $\{p/q\}$ form a set whose elements can be labelled or specified by a pair of integers p,q. We can consider more general sets labelled by several integers. The following theorem shows that such sets are countable, a fact which is used in section 7.3.

Theorem If for some integer N the elements of a set can be labelled by means of N indices, each of which takes positive integral values, then the set is countable.

Proof For any integer $r \geqslant N$ there are only finitely many elements whose indices add up to r. We can therefore write down the element whose indices add up to N, then those whose indices add up to $N + 1$, and so on. Each element appears somewhere in this list; we have shown that the set is countable. □

Finally we shall show that the set of all real numbers is not countable. This fact is not strictly needed for our purposes; but to put the idea of countability into proper perspective one should realise that uncountable sets are extremely common in mathematics.

Suppose, then, that the real numbers are countable; we shall obtain a contradiction. Express the real numbers in decimal form $N{\cdot}abcd\ldots$, where N is an integer and $a,b\ldots$ are digits from 0 to 9. We suppose that they can all be listed thus:

$$\left.\begin{array}{l} N_1{\cdot}a_1b_1c_1\ldots \\ N_2{\cdot}a_2b_2c_2\ldots \\ N_3{\cdot}a_3b_3c_3\ldots \\ \ldots\ldots\ldots\ldots \end{array}\right\} \tag{D1}$$

To avoid ambiguity, we do not allow a number to be represented by a decimal ending in repeated 9s. Now consider the number

$$0{\cdot}\alpha\beta\gamma\ldots \tag{D2}$$

where $\alpha \neq a_1, \beta \neq b_2, \gamma \neq c_3, \ldots$, and none of $\alpha,\beta,\gamma,\ldots$ equals 9. There are at least eight choices for each digit in (D2), so there are very many such numbers. Each number (D2) differs from the first number in the list (D1) because the first digit after the decimal point is different, and differs from the second number in (D1) because the second digit is different, and so on. Since no digit in (D2) is 9, there is no ambiguity in the decimal representation (D2), and we conclude that the number (D2) does not appear in the list (D1). But (D1) is by definition a complete list of all real numbers, under the assumption of countability. We therefore have a contradiction, and the assumption must be false. This proves the uncountability of the real numbers.

The ideas and proofs given here may be found hard to digest at first. When they were first put forward, by Georg Cantor in the eighteen-seventies, they provoked considerable discussion, and some hostility. Cantor's intellectual career was not an orthodox one; it included research and lecturing on the theory that Shakespeare's plays were written by Francis Bacon, and a deep preoccupation with the theological aspects of infinite set theory. A colourful account of his life and work is given by Bell; for a more rigorous and careful treatment, see Dauben. Cantor's proof of the uncountability of the reals is discussed in a very broad philosophical context by Hofstadter.

Appendix E Equivalence Relations

Let $*$ denote any relation between pairs of members of a set. For example, if a and b are real numbers, $a * b$ could stand for the statement $a < b$, or for $a = 2b$; if f and g are $\mathbb{R} \to \mathbb{R}$ functions, $f * g$ could stand for $f(t) > g(t)$ for all t, or for $\int_0^1 f(t)\,dt = \int_0^1 g(t)\,dt$. Generally, if S is any set, a relation is defined by specifying certain ordered pairs of elements of S, the first of which is related to the second.

A relation is said to be **reflexive** if $a * a$ holds for all a, to be **symmetric** if $a * b$ implies $b * a$, and **transitive** if $a * b$ and $b * c$ together imply $a * c$. For example, for real numbers the relation $a \leqslant b$ is transitive and reflexive but not symmetric, the relation $a \neq b$ is symmetric but not reflexive or transitive, and the relation $a = b$ has all three properties.

A relation which is reflexive, symmetric and transitive is called an **equivalence relation.** Equality is the most familiar example of an equivalence relation. Another example is the relation between $\mathbb{R} \to \mathbb{R}$ functions given by $\int_0^1 f(x)\,dx = \int_0^1 g(x)\,dx$.

If $\sim$ denotes an equivalence relation on a set S, we say a is **equivalent** to b if $a \sim b$. The **equivalence class determined by** a is $\{b \in S: b \sim a\}$. It is easy to show that all elements of an equivalence class are equivalent to one another, and that any element of S equivalent to any member of a class belongs to that class. We shall prove that S can be divided into equivalence classes, such that each element of S belongs to exactly one class.

Firstly, each $a \in S$ belongs to at least one equivalence class, namely, the class determined by a (it may happen that a is the only member of this class). To prove that no element belongs to more than one class, suppose that $a \in K$ and $a \in L$ where K and L are equivalence classes; we shall prove that $K = L$. For any element k of K, $k \sim a$ because $a \in K$, hence $k \in L$ because k is equivalent to the element a of L. Thus all elements of K belong to L. A similar argument shows that all elements of L belong to K, hence $K = L$. This completes the proof that S can be divided into mutually exclusive equivalence classes.

Appendix F Completion

This Appendix proves that every incomplete normed space can be completed, that is, turned into a Banach space by adding extra elements. This proof is not necessary for understanding the rest of the book; it is given here because it may be interesting to theoretically-inclined readers, and because the method is closely related to the theory of section 12.1.

It is an oversimplification to say that an incomplete space N is turned into a complete space by adding more elements. What one does is construct a completely new space Ω, and show that Ω is complete and contains a subspace whose elements correspond exactly to those of N. Then N is regarded as embedded in Ω, in the sense that Ω contains a copy of N.

The idea behind the construction of Ω is to associate with each element x of N a Cauchy sequence in N which converges to x. If we consider the set K of all Cauchy sequences in N, and regard it as a vector space in its own right, it will contain convergent Cauchy sequences corresponding to elements of N as above, and also nonconvergent Cauchy sequences, which correspond to the extra elements needed to complete the space. Thus K is a possible candidate for being the completion of N. However, the construction is complicated by the fact that there are many different Cauchy sequences in N converging to the same $x \in N$. Thus each $x \in N$ is associated not with a single Cauchy sequence but with a whole class of equivalent sequences in the following sense.

Definition F1 Two Cauchy sequences (f_n), (g_n) in N are **equivalent** if $\|f_n - g_n\| \to 0$ as $n \to \infty$. □

It is easy to show that for convergent sequences, equivalence means that they converge to the same limit.

Lemma F2 If $f_n \to F$ in N, and (g_n) is a Cauchy sequence equivalent to (f_n), then $g_n \to F$.

Proof $\|g_n - F\| \leqslant \|g_n - f_n\| + \|f_n - F\| \to 0$ as $n \to \infty$. □

Definition F1 is framed so as to apply also to non-convergent Cauchy sequences. In this case equivalence means that the sequences are coming closer and closer together, and are trying, so to speak, to converge to the same (nonexistent) element of N.

It is easy to show that Definition F1 gives an equivalence relation in the sense of Appendix E. It follows that the set of all Cauchy sequences can be divided into equivalence classes. These classes of equivalent Cauchy sequences are the elements of the completion of N.

Definition F3 Given any normed space N, let Ω be the set of all equivalence classes of Cauchy sequences in N. Define the sum of two elements ϕ and ψ of Ω to be the class of all Cauchy sequences equivalent to $(f_n + g_n)$, where (f_n) is a sequence in the class ϕ and (g_n) is a sequence in the class ψ. For any scalar a, define $a\phi$ to be the class of all Cauchy sequences equivalent to (af_n). For any $\phi \in \Omega$, define $\|\phi\|_\Omega = \lim_{n\to\infty} \|f_n\|$, where (f_n) is a Cauchy sequence in the class ϕ. □

This definition should be justified by showing that the operations do not depend on the choice of Cauchy sequence (f_n) representing the class ϕ, and by showing that the sequence of numbers $(\|f_n\|)$ converges for any Cauchy sequence (f_n). This is quite straightforward. It is also easy to show that Ω is a normed space with these definitions. Note that the space K of Cauchy sequences in N is not a normed space; there are many Cauchy sequences which converge to zero, therefore many elements of K have norm zero (norm being defined as in F3), which is not allowed. That is why we must consider the space Ω of equivalence classes.

We shall now show that Ω is the completion of N. The argument proceeds in two stages: first we show that Ω contains an isomorphic copy M of N, then we show that Ω is complete. The argument is not difficult, but it needs clear thinking to deal with Cauchy sequences of elements of Ω which are themselves equivalence classes of Cauchy sequences of elements of N. We shall denote elements of Ω by Greek letters; thus if (x_n) is a Cauchy sequence of elements of N, then $(x_n) \in \xi$ and $\xi \in \Omega$ where ξ is the class of Cauchy sequences in N equivalent to (x_n).

Theorem F4 Let M be the subspace of Ω consisting of all equivalence classes of convergent Cauchy sequences in N. Then M is a dense subspace of Ω, and M is isometrically isomorphic to N – that is, there is one-to-one correspondence between M and N such that if $\mu_1, \mu_2 \in M$ correspond to $u_1, u_2 \in N$, then $a\mu_1 + b\mu_2$ corresponds to $au_1 + bu_2$ for any scalars a and b, and $\| \mu_1 \|_\Omega = \| u_1 \|_N$.

Proof For each $u \in N$, the sequence $(u, u, u, \ldots)$ is convergent; let μ be the class of Cauchy sequences equivalent to $(u, u, u, \ldots)$. Conversely, given any $\mu \in M$, let $u \in N$ be the limit of the convergent sequences of which μ is composed. It is easy to show that this gives a one-to-one correspondence between M and N, in which $a\mu_1 + b\mu_2$ corresponds to $au_1 + bu_2$. Since $(u, u, \ldots) \in \mu$, $\| \mu \|_\Omega$ is the limit of the sequence $(\| u \|_N, \| u \|_N, \| u \|_N, \ldots)$, therefore $\| \mu \|_\Omega = \| u \|_N$. This shows that M and N are isometrically isomorphic ('isometric' means having the same lengths, that is, the same norms for corresponding elements).

It is easy to show that M is a vector subspace of Ω. To show that it is dense in Ω, take any $\alpha \in \Omega$, and let (x_n) be a Cauchy sequence in N which belongs to the class α. For each n, define $\xi_n \in M$ to be the class of sequences equivalent to $(x_n, x_n, x_n, \ldots)$. We calculate $\| \xi_n - \alpha \|_\Omega$ using $(x_n, x_n, x_n, \ldots)$ as a representative of the class ξ_n and $(x_1, x_2, x_3, \ldots)$ as a representative of the class α. Thus

$$\| \xi_n - \alpha \|_\Omega = \lim_{r \to \infty} \| x_n - x_r \|_N$$

$$\to 0 \quad \text{as} \quad n \to \infty$$

because (x_n) is a Cauchy sequence in N. Thus we can make $\xi_n \in M$ as close as we please to $\alpha \in \Omega$ by taking n sufficiently large, which shows that M is dense in Ω. □

Theorem F5 Ω is complete.

Proof Let (α_n) be a Cauchy sequence in Ω; we shall show that it converges. By Theorem F4, for any n there is a $\mu_n \in M$ such that

$$\| \alpha_n - \mu_n \| < \tfrac{1}{n} \quad . \tag{F1}$$

An easy application of the triangle inequality shows that (μ_n) is a Cauchy sequence in Ω. Now for each n, μ_n is an equivalence class of convergent Cauchy sequences in N; let $x_n \in N$ be the limit of those sequences. Then $(x_n, x_n, x_n, \ldots) \in \mu_n$. We shall show that $(x_1, x_2, x_3, \ldots)$ is a Cauchy sequence, and hence determines an element of Ω which is the limit of (α_n).

We calculate $\| \mu_n - \mu_r \|_\Omega$, using $(x_n, x_n, x_n, \ldots)$ as a representative of the class μ_n. Then $\| \mu_n - \mu_r \|_\Omega = \| x_n - x_r \|_N$. Since (μ_n) is a Cauchy sequence in Ω, it follows that (x_n) is a Cauchy sequence in N. Let $\xi \in \Omega$ be the class of sequences equivalent to (x_n). Using (F1) we have

$$\begin{aligned} \| \alpha_n - \xi \|_\Omega &\leqslant \| \alpha_n - \mu_n \|_\Omega + \| \mu_n - \xi \|_\Omega \\ &\leqslant \frac{1}{n} + \lim_{r \to \infty} \| x_n - x_r \|_N \\ &\to 0 \quad \text{as} \quad n \to \infty \end{aligned}$$

because (x_n) is Cauchy in N. Thus $\alpha_n \to \xi$, and we have shown that the Cauchy sequence (α_n) converges in Ω, and thus that Ω is complete. □

Appendix G Sturm-Liouville Systems

We shall illustrate the idea of a Sturm-Liouville system with a simple example before giving the general definition. Consider the differential equation

$$u'' + \lambda u = 0 \quad \text{for} \quad 0 < x < 1 \tag{G1}$$

with boundary conditions

$$u(0) = u(1) = 0 \quad . \tag{G2}$$

In (G1) λ is regarded as a parameter, that is, it is a constant, and we consider various possible values for λ. For any λ, (G1) and (G2) are satisfied by $u(x) \equiv 0$. For most values of λ this is the only solution, as is easily verified by solving (G1) explicitly and applying (G2). But if $\lambda = \pi^2$, then there is a nonzero solution, in fact $u(x) = A \sin(\pi x)$ satisfies (G1) and (G2) for any A if $\lambda = \pi^2$. Similarly, if $\lambda = n^2 \pi^2$ where n is an integer, then there is a nonzero solution $A \sin(n\pi x)$. These exceptional values of λ are called **eigenvalues**, and the corresponding solutions $A \sin(n\pi x)$ are called **eigenfunctions.** The eigenvalues and eigenfunctions play a leading role in the solution of partial differential equations by separation of variables.

Note that it is essential that the boundary conditions involve two different values of x, 0 and 1. If instead of (G2) we apply two conditions at the same value of x, such as $u(0) = u'(0) = 0$, then for all values of λ there is exactly one solution $(u \equiv 0)$, and there are no eigenvalues.

We now generalise. A **Sturm-Liouville equation** is an equation of the form

$$\frac{\mathrm{d}}{\mathrm{d}x}\left[p(x)\frac{\mathrm{d}u}{\mathrm{d}x}\right] + [q(x) + \lambda r(x)]u = 0 \quad \text{for} \quad a < x < b \quad , \tag{G3}$$

where λ is a parameter and p,q,r are given real-valued functions with p,p',q continuous for $a \leqslant x \leqslant b$ and $r(x) > 0$, $p(x) \neq 0$ for $a < x < b$. These conditions are imposed to make the properties of (G3) similar to those of (G1). If $p(a) = 0$ or $p(b) = 0$ or both, then the Sturm-Liouville equation is said to be **singular**; otherwise it is said to be **regular**.

There are two common types of boundary condition for (G3). The first type is a generalisation of the conditions (G2):

$$\left.\begin{aligned} \alpha u(a) + \beta u'(a) = 0 \quad , \\ \gamma u(b) + \delta u'(b) = 0 \quad , \end{aligned}\right\} \tag{G4}$$

where $\alpha,\beta,\gamma,\delta$ are given real numbers with α and β not both zero and γ and δ not both zero. These are called **separated end-point conditions.** Another type of boundary condition arising in practice is the following, called **periodic boundary conditions**:

$$u(a) = u(b), \;\; u'(a) = u'(b) \quad . \tag{G5}$$

(G4) and (G5) are special cases of the following. A pair of equations of the form

$$\left.\begin{aligned} \alpha_1 u(a) + \alpha_2 u'(a) + \alpha_3 u(b) + \alpha_4 u'(b) = 0 \quad , \\ \beta_1 u(a) + \beta_2 u'(a) + \beta_3 u(b) + \beta_4 u'(b) = 0 \quad , \end{aligned}\right\} \tag{G6}$$

where α_i, β_i are real numbers, is said to be a set of **Sturm-Liouville boundary conditions** for (G3) if, for any two functions u and v satisfying (G6), we have

$$[p(u\bar{v}' - \bar{v}u')]_a^b = 0 \quad , \tag{G7}$$

where $[f]_a^b$ denotes $f(b) - f(a)$. A **Sturm-Liouville system** is a Sturm-Liouville equation with a set of Sturm-Liouville boundary conditions. The reason for imposing (G7) is given in section 9.2: equation (9.10) shows that if (G7) holds then the system is self-adjoint.

Note that if $p(a) = p(b) = 0$, then (G7) holds for any u,v. In this case no boundary conditions are needed, which means that all the coefficients in (G6) are zero so that it is satisfied for any u. If p vanishes at only one endpoint, then one nontrivial boundary condition is needed. Such singular systems are not considered further in this book (except for Problem 9.8).

In the same way as for the system (G1) and (G2), there is a set of values of λ, called eigenvalues, for which the system has nonzero solutions, called eigenfunctions. In many applications, the eigenvalues are interpreted in terms of vibration frequencies. A vibrating string is the classic example. A nonuniform string is stretched between the points a and b on the x-axis; its mass per unit length is $r(x)$, $p(x)$ is the tension, and the medium in which it is embedded exerts a force qu proportional to its transverse displacement u. The tension p is often constant, but if there are body forces, such as gravity, then p can vary with x. Simple mechanics gives the equation of motion

$$r\frac{\partial^2 u}{\partial t^2} = \frac{\partial}{\partial x}\left[p\frac{\partial u}{\partial x}\right] + qu \quad .$$

If we look for solutions in which the string vibrates sinusoidally with angular frequency ω, $u(x,t) = U(x)\sin\omega t$, then we have

$$\frac{\mathrm{d}}{\mathrm{d}x}\left[p\frac{\mathrm{d}U}{\mathrm{d}x}\right] + [q + \omega^2 r]U = 0 \quad ,$$

which is the standard Sturm-Liouville equation with eigenvalue ω^2. The natural boundary conditions in this problem are $U(a) = U(b) = 0$, corresponding to a string with fixed ends, but other conditions are possible.

In this interpretation of the Sturm-Liouville equation, the eigenvalues are the squares of the possible vibration frequencies, and the eigenfunctions give the shape of the string in the various modes of vibration. This suggests that the eigenfunctions may be qualitatively similar to those for the simple uniform string, corresponding to equation (G1). We thus expect the first eigenfunction, corresponding to the lowest eigenvalue, to resemble $\sin(\pi x)$ in its general shape, and in particular to be nonzero in (a,b); we expect the n-th eigenfunction to vanish $(n - 1)$ times in (a,b). This and other properties of Sturm-Liouville systems are proved in Coddington & Levinson, for example.

Appendix H Fourier's Theorem

We shall prove that the Fourier series of a function f converges uniformly to f if f is smooth and vanishes at the endpoints. The proof can be extended to a much wider class of functions, but the theorem below is sufficient for our purposes (see 7.34).

Lemma If g is smooth in $[\alpha,\beta]$, then

$$\lim_{R\to\infty}\int_\alpha^\beta g(y)\frac{\sin Ry}{y}\,\mathrm{d}y = \begin{cases} \pi g(0) & \text{if} \quad 0\in(\alpha,\beta) \\ 0 & \text{if} \quad 0\notin[\alpha,\beta] \end{cases} \quad .$$

Proof Suppose first that $0\notin[\alpha,\beta]$. Then all factors in the integrand are differentiable in $[\alpha,\beta]$, and integration by parts gives

$$\begin{aligned}\int_\alpha^\beta g(y)\frac{\sin Ry}{y}\,\mathrm{d}y &= \frac{1}{R}\left\{-\left[\frac{\cos Ry\, g(y)}{y}\right]_\alpha^\beta + \int_\alpha^\beta \cos Ry\left[\frac{1}{y}g(y)\right]'\mathrm{d}y\right\} \\ &= \frac{1}{R}\times(\text{bounded function of } R) \\ &\to 0 \quad \text{as} \quad R\to\infty \quad .\end{aligned}$$

Now suppose that $\alpha < 0 < \beta$. Write $g(y) = g(0) + yG(y)$, where $G(y) = [g(y) - g(0)]/y$ is smooth because g is smooth. Then

$$\int_\alpha^\beta g(y)\frac{\sin Ry}{y}\,\mathrm{d}y = \int_\alpha^\beta \left[g(0) + yG(y)\right]\frac{\sin Ry}{y}\,\mathrm{d}y$$

$$= \int_{\alpha R}^{\beta R} g(0)\frac{\sin u}{u}\,\mathrm{d}u + \int_\alpha^\beta G(y)\sin Ry\,\mathrm{d}y$$

$$= g(0)\int_{\alpha R}^{\beta R}\frac{\sin u}{u}\,\mathrm{d}u - \frac{1}{R}\left[\cos Ry\; G(y)\right]_\alpha^\beta + \frac{1}{R}\int_\alpha^\beta \cos Ry\; G'(y)\,\mathrm{d}y \quad .$$

The second and third terms are of the form R^{-1} x (bounded function of R), hence tend to zero as $R \to \infty$. Since $\alpha < 0 < \beta$, the first term tends to $g(0)\int_{-\infty}^{\infty}\mathrm{d}u(\sin u)/u = \pi g(0)$ (see Problem 1.19). □

Theorem If f is smooth on $[a,b]$ and $f(a) = f(b) = 0$, then the Fourier sine series of f converges to f uniformly on $[a,b]$.

Proof The transformation $x \mapsto \pi(x - a)/(b - a)$ maps $[a,b]$ to $[0,\pi]$. The sine series for f is then $\Sigma c_n \sin(nx)$, where

$$c_n = \tfrac{2}{\pi}\int_0^\pi f(u)\sin nu\,\mathrm{d}u \quad .$$

Let $$S_N(x) = \tfrac{2}{\pi}\int_0^\pi \sum_1^N \sin nx \sin nu\, f(u)\,\mathrm{d}u \quad .$$

By some straightforward manipulation (write $\sin\theta = (\mathrm{e}^{i\theta} - \mathrm{e}^{-i\theta})/2i$ and sum the series), we have

$$S_N(x) = \tfrac{1}{\pi}\int_0^\pi \left\{\frac{\sin(N+\frac{1}{2})(u-x)}{2\sin\frac{1}{2}(u-x)} - \frac{\sin(N+\frac{1}{2})(u+x)}{2\sin\frac{1}{2}(u+x)}\right\} f(u)\,\mathrm{d}u$$

$$= \tfrac{1}{\pi}\left\{\int_{-x}^{\pi-x} f(x+v)\frac{\sin(N+\frac{1}{2})v}{2\sin\frac{1}{2}v}\,\mathrm{d}v - \int_x^{\pi+x} f(v-x)\frac{\sin(N+\frac{1}{2})v}{2\sin\frac{1}{2}v}\,\mathrm{d}v\right\}$$

Applying the above lemma to the first term, with

$$g(y) = \frac{y}{2\sin\frac{1}{2}y}f(x+y) \quad ,$$

shows that it tends to $f(x)$ if $0 < x < \pi$. Similarly the second term tends to zero for $x > 0$, so we have

$$S_N(x) \to f(x) \quad \text{as} \quad N \to \infty \quad \text{for} \quad 0 < x < \pi \quad .$$

But $S_N(0) = 0 = f(0)$, and $S_N(\pi) = 0 = f(\pi)$, so S_N converges pointwise to f for $0 \leqslant x \leqslant \pi$.

To prove that the convergence is uniform, integrate the formula for c_n twice by parts, which gives

$$c_n = -\frac{2}{n^2\pi}\int_0^\pi f''(u)\sin nu \, \mathrm{d}u \quad .$$

$$\therefore |c_n| \leqslant A/n^2$$

where $A = 2\int_0^\pi |f''(u)|\,\mathrm{d}u/\pi$. The M-test (Appendix B) now shows that the series converges uniformly. □

Appendix I Proofs of 9.24 and 9.25

We shall omit some of the technical details because they are given in detail in Smirnov, section 200. We prove a slightly stronger version of 9.24, the extra clause in (b) being needed in the proof of 9.25.

Lemma 9.24′ Let g be the Dirichlet Green's function for the region $V \subset \mathbb{R}^3$ inside a well-behaved closed surface.

(a) If $u \in L_2[V]$, then the function

$$v(x) = \int_V g(x;y)u(y)\,\mathrm{d}^3y \tag{I1}$$

is continuous on V.

(b) If u is continuous on V, then v is continuously differentiable on V, and its derivatives are given by differentiating under the integral sign.

Proof · In Theorem 3.33 it is proved that

$$g(x;y) = (4\pi|x-y|)^{-1} + \phi(x;y) \tag{I2}$$

where ϕ satisfies $\nabla^2\phi = 0$. The function g is singular at $x = y$, and the convergence of the integral in (I1) is a little delicate. We construct a smoothed version g_ϵ of g as follows. Replace $(4\pi|x-y|)^{-1}$ by a function f which is continuously differentiable everywhere, and satisfies $0 < f(x;y) \leqslant (4\pi|x-y|)^{-1}$ everywhere, and $f(x;y) = (4\pi|x-y|)^{-1}$ for $|x-y| \geqslant \epsilon$. For example, we could take $f = 3/8\pi\epsilon - |x-y|^2/8\pi\epsilon^3$ for $|x-y| \leqslant \epsilon$, but there are many other possibilities. Now set $g_\epsilon = \phi + f$. One can think of g_ϵ as being obtained from g by chopping off the spike, and rounding off the stump.

Now set

$$v_\epsilon(x) = \int g_\epsilon(x;y)u(y)\,\mathrm{d}^3y \quad .$$

$g_\epsilon \in C^1[V]$ (the set of continuously differentiable functions), and so $v_\epsilon \in C^1[V]$ because the integral converges uniformly with respect to x (Appendix B, Theorems B13, B14). We shall show that $v_\epsilon \to v$.

(a) Suppose $u \in L_2[V]$. We have

$$|v_\epsilon(x) - v(x)| = |\int [g_\epsilon(x;y) - g(x;y)]u(y)\, d^3y$$
$$\leqslant (\int |g_\epsilon - g|^2 d^3y \int |u|^2 d^3y)^{\frac{1}{2}} \quad .$$

Now, $g = g_\epsilon$ everywhere except inside the sphere

$$\Sigma_\epsilon = \{y : |y - x| \leqslant \epsilon\} \quad , \tag{I3}$$

where $|g_\epsilon - g| \leqslant (4\pi |x - y|)^{-1}$. Hence

$$|v_\epsilon(x) - v(x)| \leqslant (1/4\pi)(\int_{\Sigma_\epsilon} |x - y|^{-2} d^3y)^{\frac{1}{2}} \|u\|$$
$$= \frac{\|u\|}{4\pi}\sqrt{4\pi\epsilon} \quad .$$

Since this is independent of x, it follows that $v_\epsilon \to v$ uniformly as $\epsilon \to 0$, and the continuity of v therefore follows from that of v_ϵ.

(b) Now suppose that u is continuous, and let q be the function obtained by differentiating under the integral in (I1): $q(x) = \int \partial g/\partial x_1 (x;y)u(y)\, d^3y$. In the same way as above one shows that $|\partial v_\epsilon/\partial x_1 - q| \to 0$ uniformly in x as $\epsilon \to 0$. It follows that $v \in C^1[V]$ and $\partial v/\partial x_1 = q$, as required. □

We now prove 9.25.

Proposition 9.25 If $u \in C^1[V]$ and v is defined by (I1), then $v \in C^2[V]$, $\nabla^2 v = -u$, and v vanishes on the boundary S of V.

Proof Note first that $v(x) = 0$ for $x \in S$ because $g(x;y) = 0$ for $x \in S$. Now, the function ϕ in (I2) above satisfies $\nabla_x^2 \phi = 0$, and therefore $\nabla_x^2 \int \phi(x;y)u(y)\, d^3y = 0$; differentiating under the integral sign here is easily justified because ϕ, unlike g, is well behaved. It only remains to show that $\nabla^2 \psi = -4\pi u$, where

$$\psi(x) = \int u(y)|x - y|^{-1} d^3y \quad .$$

Let $V_\epsilon = V - \Sigma_\epsilon$ where Σ_ϵ is the sphere given by (I3). Writing $Q \equiv |x - y|^{-1}$, we have $\partial Q/\partial x_i = -\partial Q/\partial y_i$, and

$$\frac{\partial \psi}{\partial x_i} = \int_{V_\epsilon} \frac{\partial Q}{\partial x_i} u(y)\, d^3y - \int_{\Sigma_\epsilon} \frac{\partial Q}{\partial y_i} u(y)\, d^3y$$
$$= \int_{V_\epsilon} \frac{\partial Q}{\partial x_i} u(y)\, d^3y + \int_{\Sigma_\epsilon} Q \frac{\partial u}{\partial y_i} d^3y - \int_{S_\epsilon} Q u e_i \cdot d\mathbf{S} \quad ,$$

where S_ϵ is the surface of the sphere Σ_ϵ, and $\boldsymbol{e}_i$ is a unit vector in the i direction. In the first and third integrals here, the integrand is well behaved throughout the domain of integration, and therefore we can differentiate under the integral. The lemma above justifies differentiating the second term under the integral too. Since $\nabla^2 Q = 0$ for $\boldsymbol{x} \neq \boldsymbol{y}$, we have

$$\nabla^2 \psi = \int_{\Sigma_\epsilon} \nabla Q \cdot \nabla u \, \mathrm{d}^3 y - \int \mathrm{d}\boldsymbol{S} \cdot \nabla Q u(y) \quad . \tag{I4}$$

We now let $\epsilon \to 0$. A simple estimate of the first term in (I4) shows that it vanishes as $\epsilon \to 0$, and a calculation very similar to that in the proof of 3.31 shows that the second term tends to $4\pi u(\boldsymbol{x})$, which completes the proof. □

Notes on the Problems

These notes give hints, answers (which are not guaranteed correct), and references for some of the problems. Complete solutions can be obtained from **http://www.maths.bris.ac.uk/~madhg/afa/**

Chapter 1

Problem 1.1 $\langle x^n\delta^{(m)},\phi\rangle = \langle\delta^{(m)},x^n\phi\rangle = (-1)^m(x^n\phi)^{(m)}(x=0)$; differentiate by Leibniz' rule. **Problem 1.2** f is even if $f(-x)=f(x)$, in the notation of 1.23; f is odd if $f(-x)=-f(x)$. **Problem 1.3** Note that for any $\phi\in\mathscr{D}$ the series terminates, so no convergence problems. The distribution is $\Sigma_0^\infty\delta^{(n)}(x-n)$, and integrating this any number of times gives a singular distribution. **Problem 1.4** See Korevaar, section 8.3. **Problem 1.5** If an analytic function vanishes along any curve, then it is identically zero. **Problem 1.6** Integrate $\int\sin kx\,\{[\phi(x)-\phi(0)]/x\}\,dx$ by parts, and show it $\to 0$. **Problem 1.7** $f_t\to F$ as $t\to a$ if, for all $\phi\in\mathscr{D}$, $\langle f_t,\phi\rangle\to\langle F,\phi\rangle$ as $t\to a$. **Problem 1.9** (a) Integrate $\int\phi(x)x/(x^2+a^2)\,dx$ by parts, and use equation (1.12). (b) (i) A convergent integral equals its principal value; (ii) show $P\int_{-\infty}^{\infty}\cos Rx\,\phi(x)/x\,dx = \int_0^\infty\cos Rx\,\psi(x)\,dx\to 0$ as $R\to\infty$, where $\psi(x)=[\phi(x)-\phi(-x)]/x$. **Problem 1.10** $1/(x\pm i\alpha)=(x\mp i\alpha)/(x^2+\alpha^2)$; use (1.7). **Problem 1.11** Use 3.6(d). **Problem 1.12** (c) $-1/2a^4+(\pi^4/b^3)(\sinh b+\sin b)/(\cosh b-\cos b)$ where $b=\sqrt{2}\pi a$. **Problem 1.14** See Melzak.

Chapter 2

Problem 2.1 Show that f is the zero functional on the subset $Z=\{\phi\in\mathscr{D}:\phi(0)=0\}$, then split any $\phi\in\mathscr{D}$ into a member of Z plus a remainder, as in the proof of 2.2. **Problem 2.2** $f=c+d_0H(x)+d_1H'(x)+\ldots+d_{k-1}H^{(k-1)}(x)$ is the general solution. Prove it by induction. **Problem 2.3** Use 1.42. **Problem 2.4** (a) Must show $h^{-1}\int_{a-h/2}^{a+h/2}\phi(x)\,dx\to\phi(a)$ as $h\to 0$, easy. (b) The sum gives a "staircase function" which equals $f(x)$ at all points $x=nh$, and approximates f everywhere when h is small. **Problem 2.5** $g(x;y)=\cos kx\cos k(y-1)/k\sin k$ if $x<y$; $g(y;x)=g(x;y)$. **Problem 2.7** $u(x)=-[a\sin(x-1)-b\sin x]/\sin 1+\int_0^1 g(x;y)f(y)\,dy$. **Problem 2.8** $g(x;y)=0$ if $x<y$, $=(x-y)^{n-1}/(n-1)!$ if $x>y$. **Problem 2.10** $g(x;y)=-x^2(y-a)^2[x(1+2y/a)-3y]/6a^2$ for $x<y$, and g is symmetric. **Problem 2.11** $p\in K^1$ (cf. Problem 1.4), q locally integrable.

Chapter 3

Problem 3.1 $\widetilde{f}(k) = 2\sin k/k$, not absolutely integrable because $\int_{-\infty}^{\infty}|\widetilde{f}(k)|\,dk = \sum_{-\infty}^{\infty} I_n$ where $I_n = \int_{n\pi}^{(n+1)\pi} \cong (2/n\pi)\int_{n\pi}^{(n+1)\pi} |\sin k|\,dk = 0(n^{-1})$ as $n \to \infty$. Hence $\widetilde{f}$ is not Fourier-transformable. 3.5 does not apply because f is not continuous. **Problem 3.2** Integrand in $\int \widetilde{f} e^{ikx}\,dk$ is well-behaved at $k = 0$. To show the integral converges, use the fact that $\int_c^\infty [\sin ak/k]\,dk$ and $\int_c^\infty [\cos ak/k]\,dk$ converge, easily shown by integrating by parts. For $x = \pm 1$, the integral fails to converge: $\int_K^L \widetilde{f}(k)e^{ikx}\,dk$ does not tend to a definite limit as K and L independently tend to infinity. But taking the Cauchy Principal Value $\lim_{K\to\infty}(1/2\pi)\int_{-K}^{K} \widetilde{f} e^{-ikx}\,dk$ (cf. 1.39) gives $\frac{1}{2}[f(x+) + f(x-)]$. **Problem 3.4** $-i\pi[\delta(k+a) - \delta(k-a)]$; $\pi[\delta(k+a) + \delta(k-a)]$; $\pi(-i)^{n+1}[\delta^{(n)}(k+a) - \delta^{(n)}(k-a)]$; $\pi(-i)^n[\delta^{(n)}(k+a) + \delta^{(n)}(k-a)]$; $[(-ik)^{n-1}/(n-1)!]\,\mathrm{sgn}(k)$. **Problem 3.5** (a) Must show $\int_0^\infty(e^{-\epsilon x} - 1)\phi(x)\,dx \to 0$ as $\epsilon \to 0$ for all $\phi \in \mathscr{D}$, easy using the Mean Value Theorem. **Problem 3.6** Coloured glass changes the phase as well as the intensity of the light. Thus the light reaching X's eyes is $(1/\pi)\int_{\omega_1}^{\omega_2} \cos[\omega(t-6) - \alpha]\,d\omega$, where α depends on ω in such a way that the integral is zero for $t < 6$. See Pippard p. 107, or books on electromagnetic theory, such as Jackson, under "Kramers-Kronig Relations". **Problem 3.7** See Zemanian (1954). **Problem 3.11** In three dimensions, $W'' + (2/r)W' + k^2 W = 0$; $W = v/r$ gives $v'' + k^2 v = 0$. In two dimensions, one gets Bessel's equation. **Problem 3.12** (a) Apply Green's Theorem to ϕ and $\nabla^2\psi$, then to ψ and $\nabla^2\phi$, then add. **Problem 3.13** Apply the divergence theorem to $\nabla\phi$. **Problem 3.16** These fundamental solutions are complex, and not physically interesting. **Problem 3.17** One dimension: the ω-integral is the same as equation (3.40); do it first, and you get a combination of integrals of the form $P\int_{-\infty}^{\infty}(e^{ik\alpha}/k)\,dk$; evaluate using a contour with a small semicircle around the origin. Answer is a step function spreading out sideways. Two dimensions: do the ω-integral first, then either do the θ-integral, giving a Bessel function, or do the k-integral, which is the Fourier transform of sin. Answer: $1/(2\pi\sqrt{\tau^2 - \xi^2})$ for $\tau > \xi$ (τ, ξ defined as in section 3.6). **Problem 3.19** See Wallace, for example.

Chapter 4

Problem 4.1 mn. (a) no, (b) yes, (c) no. **Problem 4.2** $\{1, i\}$ is a basis. **Problem 4.3** Continuous on $[a,b]$ implies continuous on $[c,d]$. **Problem 4.5** One must be contained in the other.† **Problem 4.6** Suppose $\|x\| \geqslant \|y\|$, and remove the modulus sign. **Problem 4.7** and **Problem 4.9** Proof is nearly the same as for Example 4.25. **Problem 4.10** cf. section 8.2. For the 2×2 case, take $x = (\cos\theta, \sin\theta)$, to ensure $\|x\| = 1$, and maximise with respect to θ. **Problem 4.11** Must show that if $f_n \to F$ and $f_n \leqslant p$, then $F \leqslant p$, which is obvious. **Problem 4.12** If $g_n \to G \in C[-1,1]$, the

† or vice versa.

continuity of G means that it cannot keep pace with the rapid change in g_n near $x = 0$ for large n, so the area under the graph of $|g_n - G|^2$ near $x = 0$ cannot tend to 0 as $n \to \infty$, contradiction. This argument can be made precise, using the fact that for any $\epsilon > 0$ there is a $\delta > 0$ such that $|G(x) - G(0)| < \epsilon$ whenever $|x| < \delta$. **Problem 4.13** Use components (Problem 4.4) with respect to a basis for the subspace. **Problem 4.14** Show that $\int_0^\infty u^{-2/3} \dots$ converges. The moral is that a sequence of continuous functions can converge in the mean to a function F with a singularity, hence such functions must be included in the completion of $C[a,b]$. **Problem 4.15** For any $\epsilon > 0$ and $c \in C$ there is a $b \in B$ with $\|b - c\| < \epsilon/2$ and an $a \in A$ with $\|a - b\| < \epsilon/2$, $\therefore \|a - c\| < \epsilon$. **Problem 4.16** Use 4.50 and Problem 4.15. **Problem 4.18** False, for example $A = [a,b)$, $B = [a,b]$, $C = \{b\}$. **Problem 4.19** (i) $f_n(x) = 1$ for $x = r_{n1}, r_{n2}, \dots, r_{nK}$, $f_n(x) = 0$ otherwise, r_{ni} as in (ii). **Problem 4.20** If $(z_n) = (x_n + iy_n)$ is a Cauchy sequence in $\mathbb{C}$, show that (x_n) and (y_n) are Cauchy in $\mathbb{R}$. **Problem 4.21** Consider the sequence (x_n) of sequences $x_n = (1, 1/2, 1/3, \dots, 1/n, 0, 0, \dots)$. **Problem 4.22** See, for example, Kolmogorov & Fomin or Kreyszig. **Problem 4.23** To prove the triangle inequality is straightforward but not trivial. **Problem 4.24** This space is called $L_1[a,b]$ in the standard textbooks. **Problem 4.25** Bounded and nonzero almost everywhere. Banach space if r is bounded away from zero ($r(x) \geqslant m > 0$ for all x), not otherwise. **Problem 4.26** See Theorem 8.18. Counterexample: $f_n(t) = \tanh(n+1)t - \tanh nt$; Σf_n is not convergent in $C[-1,1]$, but $\|f_n\|^2 \leqslant (2/n^3) \int_0^\infty u^2 \operatorname{sech}^4 u \, du$, so $\Sigma \|f_n\|$ converges. **Problem 4.27** $\|x_n - x_m\| \leqslant \|x_n - y_n\| + \|y_n - y_m\| + \|y_m - x_m\|$; each term can be made $< \epsilon/3$ for $m,n > N$ by choosing (y_n) and N suitably, hence (x_n) is Cauchy. **Problem 4.28** Use 4.67 and 4.68. **Problem 4.29** (a) and (b). **Problem 4.31** $1, \sqrt{n}$. **Problem 4.32** Consider $\sin(nx)$. **Problem 4.33** No. **Problem 4.34** r is positive and continuous, so $m \leqslant r(x) \leqslant M$ for some $m,M > 0$, $\therefore m\|\cdot\|_{60} \leqslant \|\cdot\|_{61} \leqslant M\|\cdot\|_{60}$.

Chapter 5

Problem 5.1 $\int_0^1 x^a u \, dx$ should converge, a more stringent condition in $L_2[0,1]$ than in $C[0,1]$, because $u \in L_2$ may be singular; $a > -1$ on $C[0,1]$, $a > -\frac{1}{2}$ on $L_2[0,1]$. **Problem 5.2** Counterexample: $a_{nm} = 1/m$. **Problem 5.3** Needs 6.37. **Problem 5.4** $x = (1+x)^{\frac{1}{3}}$. **Problem 5.5** In first quadrant, iteration converges to (1,1), in third quadrant iteration converges to (−1,−1). In second and fourth quadrants iteration does not converge, even though T is a contraction on most of each quadrant; it does not map these quadrants into themselves. **Problem 5.7** Let x be the fixed point of T^n, and consider Tx. **Problem 5.8** Use the space of Problem 4.7 but with the norm $\|x\| = \Sigma 2^{-n}|x_n|$. Take $T(x_1, x_2, \dots) = (1, x_1, x_2, \dots)$. **Problem 5.10** Start with $n = 2$ and proceed by induction; be careful with the limits when reversing the order of integration. **Problem 5.12** $|\lambda| < \{\int\int y^4 |g|^2 dx dy\}^{-\frac{1}{2}}$. **Problem 5.13** (a) To find a counterexample, take the simplest possible case, for example $K \equiv 1$, $f(s,x) = x$. **Problem**

5.14 Simple inequalities give $\| Tu - Tv \| \leqslant |\lambda| \sup\{\int N \, dy\} \| u - v \|$. **Problem 5.15** Use the contraction mapping theorem in $C[a,b]$. It is easy to find Green's function g, and show that $\int |g(x;y)| \, dy \leqslant (b-a)^2/8$. For the last part, see Problem 2.7. **Problem 5.17** $g(x;y) = y - 1$ if $y > x$. **Problem 5.20** The maximum value of dh/du is $4ae^{-2}$, by calculus. Since g and h are positive, T clearly maps G into G. Easy to show that G is closed.

Chapter 6

Problem 6.1 Apply the Cauchy-Schwarz inequality to $\|A(x-y)\|$. **Problem 6.2** K continuous in x,y. **Problem 6.4** If $x,y \in S$ and $z = \alpha x + (1-\alpha)y$, then $z - a = \alpha(x-a) + (1-\alpha)(y-a)$. **Problem 6.5** (a) For any convex $T \supset \hat{S}$, show that if $x,y \in S$ then $ax + (1-a)y \in T$. (b) is obvious. (d) Use induction. (e) Use (b) and (d). **Problem 6.6** See Collatz. **Problem 6.7** Manipulate $B(x+h) - B(x)$ into the form $\{Ax(\|Ax\| - \|Ax + Ah\|) - Ah\|Ax\|\}/(\|Ax\| \, \|Ax + Ah\|)$. **Problem 6.8** Need only $A_{ij} \geqslant 0$ provided that A is invertible; see Gantmacher. **Problem 6.11** Any infinite sequence must contain an element which occurs infinitely often, giving a convergent subsequence. **Problem 6.12** Use Arzelà's theorem. Easy to show the set is bounded; equicontinuity follows from $|F(x) - F(y)| \leqslant A|x-y|$ where $A = \sup\{\|f\| : f \in M\}$. **Problem 6.13** Consider $f: Y \to \mathbb{R}$; show that it is continuous, deduce from 6.25 that it has a minimum value for some $\bar{x} \in Y$, then show $f(\bar{x}) = 0$. Uniqueness proof similar to 5.15. This result generalises the Contraction Mapping Theorem to operators which are not strict contractions, like that of 5.14; it does not apply to 5.14 because that operator does not map $\mathbb{R}$ into a compact subset of itself. **Problem 6.14** $\{(x,y) : x > 0, 0 < y < e^{-x}\}$. **Problem 6.15** K continuous on the closed rectangle, therefore equicontinuous by 6.37, therefore $\delta > 0$ can be chosen independent of y. **Problem 6.16** (a) $M = \max \| e_n \|$ where $\{e_n\}$ is the basis for V. (b) Use Problem 4.6. (c) Use 6.25. **Problem 6.17** Not as easy as it might look. Let N be a finite ϵ-net for S. Use Problem 6.5(e) to show that $\hat{N}$ is finite-dimensional. Show that $\hat{N}$ is relatively compact. Let M be a finite ϵ-net for $\hat{N}$, and show that M is a 2ϵ-net for $\hat{S}$. See Kantorovich & Akilov, p. 644. **Problem 6.18** An easy extension of the proof of 6.45. Cannot deduce existence of a solution of the given equation, because there is no value of A for which $R \subset S$. **Problem 6.20** Very similar to proof of 6.45: T maps the sphere of radius M into the sphere of radius BC, so need $BC < M$ for Schauder's theorem to apply. **Problem 6.21** Reduce to an integral equation as in section 6.6; $C = \pi^2/4\beta^2$ and $B = M + aM^3 + 2b/\pi$. If $\beta > \pi/2$, find a limiting amplitude b_0 of the external force, in terms of α and β, such that there is a periodic solution if $b < b_0$. Note that $b_0 \to \infty$ as $a \to 0$, as one would expect. **Problem 6.22** See Hochstadt, under the heading of Jentzsch's Theorem.

Chapter 7

Problem 7.1 (a) Square and use the Schwarz inequality. (b) is easily deduced from (a). It says that the length of one side of a triangle equals the sum of the lengths of the other two when one vertex lies on the line joining the other two. **Problem 7.2** (a) Counterexample: in $\mathbb{R}^2$ with $\|(x_1,x_2)\| = \max\{|x_1|, |x_2|\}$, take $x = (1,2)$, $y = (2,1)$. (b) $4(x,y) = \|x+y\|^2 - \|x-y\|^2 + i\|x+iy\|^2 - i\|x-iy\|^2$. **Problem 7.3** V is a vector subspace of l_2 (Example 7.4). $(x^{(1)},x^{(2)},\ldots)$ is a Cauchy sequence, converging to $(1,\frac{1}{2},\frac{1}{3},\ldots)\in l_2$, and thus does not converge to an element of V. **Problem 7.4** Let (u_n) be a Cauchy sequence in the closed subspace S of H. Then $u_n \to U\in H$, and U is a limit point of S, $\therefore U\in S$. **Problem 7.5** Immediate from the triangle inequality. **Problem 7.6** $C[a,b]$ is not complete in the L_2 norm. **Problem 7.10** $f_n(x) = 2^{(n+1)/2}$ if $2^{-n-1} < x < 2^{-n}$, $f_n(x) = 0$ otherwise. **Problem 7.11** Write $y = \Sigma d_n e_n$ and use 7.13. **Problem 7.12** Show that for any x, $x - \Sigma(x,\phi_n)\|\phi_n\|^{-2}\phi_n = 0$, $\therefore x = \Sigma\ldots$. **Problem 7.13** No, unless S is a Hilbert subspace. **Problem 7.14** Use Problem 7.12. **Problem 7.15** Consider the sequence defined by $c_n = n^{-1/3}$ when n is a perfect square, $c_n = 0$ otherwise. **Problem 7.16** Take a Cauchy sequence (x_r), where $x_r = \Sigma_n c_{rn} e_n$. Set $C_r = (c_{r1}, c_{r2},\ldots)$, and show that (C_r) is a Cauchy sequence in l_2 (Problem 4.22). **Problem 7.18** Use the Gram-Schmidt process to get the orthogonal set $1, x-\frac{1}{2}, x^2 - x + \frac{1}{6}$; then evaluate a lot of integrals and get $Q(x) = \mathrm{e} - 1 + 6(3-\mathrm{e})(x-\frac{1}{2}) + 30(7\mathrm{e}-19)(x^2 - x + \frac{1}{6})$. Q is more accurate than a Taylor polynomial beyond about 0.3 from the point about which the Taylor expansion is taken. Q gives better overall accuracy, Taylor polynomial gives better local accuracy. **Problem 7.19** Odd functions. **Problem 7.20** Easy to show $N^\perp \subset M^\perp$, and $M \subset M^{\perp\perp}$. To show that $M^{\perp\perp}\subset M$ (and hence that $M^{\perp\perp} = M$), take any $x\in M^{\perp\perp}$, write it as the sum of elements of M and $M^\perp$ by 7.47, and show that the component in $M^\perp$ is zero, hence $x\in M$. **Problem 7.21** For $M^{\perp\perp} \neq M$, take $M = C[a,b]\subset L_2[a,b]$, and show that $M^\perp = \{0\}$, so $M^{\perp\perp} = L_2[a,b]$. **Problem 7.22** $f(u) = \int_{-1}^{1} u(x)H(x)\,\mathrm{d}x$ where H is Heaviside's function. Apply the Riesz representation theorem in $L_2[-1,1]$ to deduce that there is no such $\phi\in C[-1,1]$. The theorem does not apply in $C[a,b]$ because it is incomplete under this norm. **Problem 7.23** Bounded, not linear. **Problem 7.24** f is linear and continuous, g is continuous but not linear. **Problem 7.25** See almost any functional analysis book. **Problem 7.26** If $x_n \to x$ weakly and $\|x_n\| \to \|x\|$, then $\|x_n - x\|^2 = \|x_n\|^2 - (x,x_n) - (x_n,x) + \|x\|^2 \to 0$. The converse is easy too.

Chapter 8

Problem 8.1 $\|f\| = a$, $\|g\| = \Sigma_i|a_i|$. **Problem 8.2** Easy to show that $\|A\| \leqslant \max\Sigma|a_{kj}|$. To prove equality, must find an x such that $\|Ax\| = (\max\Sigma|a_{kj}|)\|x\|$. Take $x_j = \mathrm{sgn}(a_{\alpha j})$ where α is the value of k which maximises $\Sigma_j|a_{kj}|$. **Problem 8.4** (b) Maximised for $f(x) = \mathrm{sgn}(\sin(2\pi x))$, giving $2/\pi$. Can find $f\in C[0,1]$ as close as we please to $\mathrm{sgn}(\sin(2\pi x))$, hence norm $= \sup\{\ldots\} = 2/\pi$. (c) $\int_0^1|\phi|\,\mathrm{d}x$. **Problem 8.5** Last part: consider $f_n(x) = n^{-1}\sin(nx)$, and show $\|\mathrm{D}f_n\|/\|f_n\| \to 1$ as $n\to\infty$.

Problem 8.6 P_n is $n + 1$ dimensional, and all finite dimensional spaces are complete: see section 4.6. **Problem 8.7** may look long, but is easy. **Problem 8.8** $A_nB_n - AB = A_n(B_n - B) + (A_n - A)B$. **Problem 8.9** See books on differential equations or linear systems. **Problem 8.11** $\| x_a \| \leqslant \alpha \| b \|$ shows $\alpha^2 \delta \| b \|/(1 - \alpha\delta) \geqslant \alpha\delta \| x_a \|/(1 - \alpha\delta) \geqslant \| x - x_a \|$, hence this formula is further from the exact solution than (8.12). **Problem 8.13** Take $u_n = \log\| A^n \|$, deduce that $n^{-1}\log\| A^n \| \to L$ where L is a number or $-\infty$, hence $\| A^n \|^{1/n}$ tends to a positive number or zero. (8.9) shows that $r(A) \leqslant \| A \|$. For the Volterra operator L, $\| L^n \| \leqslant M^n/(n-1)!$. **Problem 8.15** $(x,B^*A^*y) = (Bx,A^*y) = (ABx,y)$ for all x,y, $\therefore B^*A^* = (AB)^*$. $(x,\bar{c}A^*y) = c(x,A^*y) = (cAx,y)$, $\therefore \bar{c}A^* = (cA)^*$. $AB - (AB)^* = AB - B^*A^* = AB - BA$ if A,B are self-adjoint. **Problem 8.16** $\| T \| \leqslant 1$ and $T^* = T$. **Problem 8.17** To show that $N(A)$ is complete, take a Cauchy sequence (x_n) in $N(A)$, then $x_n \to \xi \in H$, $\therefore A\xi = \lim(Ax_n) = 0$, $\therefore \xi \in N(A)$. Finally, $x \in [R(A)]^\perp \Leftrightarrow (Ay,x) = 0$ for all $y \in H \Leftrightarrow (y,A^*x) = 0$ for all $y \in H \Leftrightarrow A^*x = 0$. **Problem 8.18** To show that $R(P)$ is complete, take a Cauchy sequence $(x_n) = (Py_n)$ in $R(P)$. Then $x_n \to \xi \in H$, and $P\xi = \lim(Px_n) = \lim(P^2 y_n) = \lim(x_n) = \xi$, $\therefore \xi \in R(P)$. See section 9.5. **Problem 8.20** $T = \frac{1}{2}(T + T^*) + i\frac{1}{2}(T - T^*)/i$. **Problem 8.21** If $R(A) = \mathbb{R}^n$, $Ax = 0$ implies $x = 0$. *Proof:* if $x \neq 0$, can take x to be the first member of a basis (e_r), then $R(A) = \{A\Sigma_1^n c_r e_r\} = \{\Sigma_2^n c_r A e_r\}$ is $(n-1)$-dimensional, contradiction. Rest is as in proof of 8.35. Converse is easy. **Problem 8.22** If (e_n) is a basis, let $A : x \mapsto \Sigma_1^\infty c_{n+1} e_n$, where $c_n = (x,e_n)$. Then $A(ae_1) = 0$ for any scalar a, so A is not invertible. But for any c_i, $A(\Sigma_2^\infty c_{n-1} e_n) = \Sigma_1^\infty c_n e_n$, so $R(A) = H$. **Problem 8.24** To show orthogonality, let $Ax = \lambda x$, $Ay = \mu y$, and consider (Ax,Ay); remember that $\bar{\lambda} = \lambda^{-1}$ since $|\lambda| = 1$. Use the power series for e^{iB}, and the result of Problem 8.15, to show $(e^{iB})^* = e^{-iB}$. **Problem 8.26** Follows from $A^n u = A^{n-1}Au = A^{n-1}\lambda u = \ldots = \lambda^n u$, where u is an eigenvector of A. **Problem 8.27** Follows from the fact that eigenvectors u_r belonging to different eigenvalues λ_r are linearly independent. *Proof:* for $n = 2$, if $c_1 u_1 + c_2 u_2 = 0$, then applying $A - \lambda_1 I$ (where I is the identity) shows that $c_2 = 0$; similarly $c_1 = 0$; now use induction on n. **Problem 8.28** (b) First show $\int \phi(x,y) p_i(x)\, dx = 0$ for all i, then $\phi = 0$ (= means almost-everywhere equality). (c) Use Problem 7.12.

Chapter 9

Problem 9.1 (a) Show that A^* is the operator $u \mapsto du - i\int Ku$, then some calculation shows $AA^*u = A^*Au$ for all u. (b) is easy. (c) Write $A = \frac{1}{2}(A + A^*) + i(A - A^*)/2i$. **Problem 9.2** A common eigenvector of B and C is clearly an eigenvector of A. **Problem 9.3** Define a truncated operator K_n of rank n by $(K_n x)_i = \Sigma_{j=1}^n k_{ij} x_j$ if $i \leqslant n$, $(K_n x)_i = 0$ otherwise. K_n is compact and self-adjoint, and some calculation shows $\| K_n - K \|^2 \leqslant 2\Sigma_{j=n+1}^\infty \Sigma_{i=1}^\infty | k_{ij} |^2 \to 0$ as $n \to \infty$; now use 8.51. **Problem 9.4** Use Problem 8.28 with $p_i = \phi_i$, $q_i = \bar{\phi}_i$. The expression for $\Sigma | \lambda_i |^2$ comes from

Parseval's relation in the space H of Problem 8.28. **Problem 9.5** (a) We have $|c_n\phi_n(x)| = |c_n/\lambda_n|\ |\lambda_n\phi_n(x)|$. Use the results of Problem 9.4 to show that $\Sigma|\lambda_i\phi_i(x)|^2 \leqslant A^2 = (b-a)\max|K|^2$, and deduce that for each n, $|\lambda_n\phi_n(x)| \leqslant A$. Now use the Weierstrass M-test. (b) $(f,\phi_n) = (Ag,\phi_n) = \lambda_n(g,\phi_n)$. Parseval's relation for g now shows that $c_n = (f,\phi_n)$ satisfies the condition in (a). If $f \notin R(A)$, the result is not true – a discontinuous function has a nonuniformly convergent expansion in terms of the sine functions. **Problem 9.6** In a Hilbert space with orthonormal basis (ϕ_n), consider the operator $u \mapsto \Sigma_r r^{-\frac{1}{2}}(u,\phi_r)\phi_r$. Show that its eigenvalues are $\lambda_n = n^{-\frac{1}{2}}$, and use 8.51 to show that it is compact (take the n-th partial sum). **Problem 9.7** The eigenfunctions are $\sin(2n\pi x/a)$, $\cos(2n\pi x/a)$. 9.12 shows that f is in the range of Γ, so Problem 9.5(b) applies. **Problem 9.8** $g(r;s) = r^p(s^p - s^{-p})/2p$ for $r < s$, if $p > 0$. If $p = 0$, $g(r;s) = \log(s)$ if $r < s$. $g(r;s) = g(s;r)$. In both cases, $rs|g(r,s)|^2$ is continuous and bounded, hence $\iint |g|^2 r\,dr\ s\,ds$ converges. **Problem 9.11** The proof is the same as that of 9.29, using 9.16 in place of 9.4. **Problem 9.14** $L^* = \mathrm{D}^2 - 2k\mathrm{D} + \omega^2$. Solubility condition: $\int_0^\pi e^{2x}\sin(x)f(x)\,dx = 0$. **Problem 9.15** Use 9.38, Problem 8.15, and easy calculations. $R(P_1P_2) = R(P_1) \cap R(P_2)$, $R(P_1 + P_2 - P_1P_2) = \{x_1 + x_2 : x_1 \in R(P_1),\ x_2 \in R(P_2)\}$. **Problem 9.16** $\begin{pmatrix}\alpha^2 & \alpha\beta \\ \alpha\beta & \beta^2\end{pmatrix}$. **Problem 9.17** $25P_+ - 25P_-$, where $P_+ = \frac{1}{25}\begin{pmatrix}9 & 12 \\ 12 & 16\end{pmatrix}$, $P_- = \frac{1}{25}\begin{pmatrix}16 & -12 \\ -12 & 9\end{pmatrix}$. **Problem 9.20** Show that if e_r is an eigenvector of A with eigenvalue λ_r, then e_r is an eigenvector of $p(A)$ with eigenvalue $p(\lambda_r)$. Then prove that $p(A)$ is compact, and use the spectral theorem. **Problem 9.22** (a) follows at once from $A^n = \Sigma_r \lambda_r^n P_r$. (b) Express it in the form $\lambda_1(1 + \ldots)/(1 + \ldots)$. (c) Show that the series converge uniformly with respect to n. (d) Show that $x_n - P_1x/\|P_1x\| \to 0$. (e) Not necessarily. But to make the method work, just replace A by $-A$. For further details see Fröberg.

Chapter 10

Problem 10.1 No, for example $T = \begin{pmatrix}-1 & -1 \\ 0 & 0\end{pmatrix}$. But $(x,T^*Tx) = (Tx,Tx) \geqslant 0$, so T^*T is positive. **Problem 10.2** If $p(x) < 0$, $q(x) \geqslant 0$ for $x \in [a,b]$. **Problem 10.3** $\{\phi \in C^2[V] : \phi = \boldsymbol{n}\cdot\nabla\phi = 0$ on the boundary$\}$. **Problem 10.4** Use Problem 9.4. Converse is false: consider $K(x,y) = \sin(x)\sin(y) - a\sin(2x)\sin(2y)$ for suitable a. **Problem 10.5** (a) Compact self-adjoint operator, therefore the eigenvalues are bounded. 10.12 gives $\sup\{\lambda_n\} \geqslant \int_0^1 x^{-1}(e^x - 1)\,dx = \int_0^1 [1 + x/2! + x^2/3! + \ldots]\,dx$. (b) $(f,Af) = 1.3179 + 0.1187a - 0.3179a^2$. Rayleigh quotient is maximised at $a = -1.05$, giving $\sup\{\lambda_n\} \geqslant 2.654$. **Problem 10.6** $\iint \{y[(1 - y/\sqrt{3})^2 - x^2]\}^2\,dx\,dy$, easy but tedious to integrate. **Problem 10.7** Eigenfunctions of $-\nabla^2$ on the rectangle $0 \leqslant x \leqslant a$, $0 \leqslant y \leqslant b$ are $\sin(n\pi x/a)\sin(m\pi y/b)$, giving eigenvalues $\pi^2(n^2/a^2 + m^2/b^2)$, for $n,m = 1,2,\ldots$. Lowest eigenvalue for $m = n = 1$; if ab is fixed, this is least when $a = b$: a square. **Problem 10.8** $1 - x^2/a^2 - y^2/b^2$ is the obvious trial function. It gives a Rayleigh quotient $3(a^{-2} + b^{-2})$. For fixed area (that is, ab fixed), this is least when $a = b$. **Problem 10.10** $\{\alpha^2[v(1)]^2 + \int_0^1 v'^2\,dx\}/\int_0^1 v^2\,dx$.

Problem 10.11 The matrix is close to the diagonal matrix with diagonal 1,2,3,1, so expect the first two eigenvalues to be close to 3 and 2, and the eigenvectors to be close to (0,0,1,0) and (0,1,0,0). Hence use trial vector $c(0,0,1,0) + d(0,1,0,0)$, giving $R = (2a^2 + 0.4a + 3)/(1 + a^2)$, where $a = c/d$. $dR/da = 0$ when $a = 0.1926$ or -5.1926, that is, $R = 3.04$, 1.96. $\therefore \lambda_1 \geqslant 3.04$. $(0,1,-0.1926,0) \perp f_1$ (notation of 10.13), giving $R = 1.96$, so $\lambda_2 \geqslant 1.96$. **Problem 10.13** Use 10.19. **Problem 10.14** (a) Note that any function defined on V_1 can be extended to V by making it zero on $V - V_1$. (b) Consider the eigenvalue problem for L on V with boundary condition of vanishing on all boundaries of $V, V_1, V_2, \ldots$. See Courant-Hilbert vol. 1, pp. 407–411. **Problem 10.15** $A - C = (A - B) + (B - C)$, the sum of two positive operators is easily seen to be positive, $\therefore A > C$, etc. Counterexample for non-self-adjoint operators: let A be the operator of rotation through 90° in $\mathbb{R}^2$, $B = 0$. Then $A \geqslant 0$ and $A \leqslant 0$ but $A \neq 0$. $\lambda_n^A \geqslant \lambda_n^B$ follows at once from 10.18(b). Set $L = -D^2 - r(x)$. If $r(x) \geqslant 0$ then $L \leqslant -D^2$, which has eigenvalues $n^2\pi^2$. If $r(x) \leqslant R$ for $0 \leqslant x \leqslant 1$, then $L \geqslant -D^2 - R$, which has eigenvalues $n^2\pi^2 - R$. Hence $n^2\pi^2 - R \leqslant \lambda_n \leqslant n^2\pi^2$. **Problem 10.16** Eigenvalues $-\alpha, \pi^2 - \alpha, 4\pi^2 - \alpha, \ldots$, so need $\alpha < 0$ for the operator to be positive. Take trial function $v \equiv c$, get $(u,f) \geqslant -\frac{1}{\alpha}[\int_0^1 f dx]^2$. **Problem 10.19** Plugging into (10.46) gives $\alpha = (5e^2 + 1)\delta/(1 - e^2)$, $\beta = 2e^2\delta/(e^2 - 1)$, $\gamma = 2e\delta/(e^2 - 1)$, where $\delta = [e(e^2 - 1)/(4 - 2e + 24e^2 - 4e^3)]^{\frac{1}{2}}$.

Supplementary Problems

These problems have been added for the second edition.

1.17 The distribution $\boldsymbol{P/x}$ can be approached from a different point of view as follows. Given numbers $a,b > 0$, define a functional by

$$\phi \mapsto \int_{-\infty}^{-a} x^{-1}\phi(x)\mathrm{d}x + \int_{-a}^{b} x^{-1}\,[\phi(x) - \phi(0)]\,\mathrm{d}x + \int_{b}^{\infty} x^{-1}\phi(x)\mathrm{d}x \ .$$

Show that it is a distribution; call it $\boldsymbol{R}_{ab}$. Show that $x\boldsymbol{R}_{ab} = 1$. Deduce that $\boldsymbol{R}_{ab} = \boldsymbol{P/x} + \boldsymbol{C}$ where $\boldsymbol{C}$ is a constant which you should find in terms of a and b. Show that $\boldsymbol{R}_{aa} = \boldsymbol{P/x}$ for any $a > 0$.

1.18 Given any numbers $a < b$ and any $c > 0$, show how to find a test function ϕ such that $\phi(x) = 1$ for $a < x < b$, $\phi(x) = 0$ for $x < a - c$ and $x > b + c$, and $0 \leqslant \phi(x) \leqslant 1$ for all x. (Hint: use integrals of the function of Example 1.4.)

1.19 Use Problem 1.6 to show that $\int_{-\infty}^{\infty} x^{-1}\sin x\,\mathrm{d}x = \pi$. (Hint: use the function of Problem 1.18, with $-a = b = 1$, and note that for every x, $\phi(x/k) = 1$ for large k.)

1.20 In section 1.5 we defined the distribution $\boldsymbol{x}^{-n}$ for integer n. Here is an example of a fractional power.

Show that the functional $\phi \mapsto 2\int_0^{\infty} \phi'(x)x^{-1/2}\mathrm{d}x$ defines a generalised function, f, say. Show that $xf = \mathbf{g}$, where $\mathbf{g}$ is the regular generalised function corresponding to the locally integrable function $H(x)x^{-1/2}$. Why can we not define a distribution corresponding simply to $x^{-3/2}$? What about $|x|^{-3/2}$? Generalise to other powers.

2.12 Let f be a continuous function, and set $u(x) = \int_0^1 g(x;y)f(y)\mathrm{d}y$ where g is given by equation (2.27). Calculate the first two derivatives of u, and hence show that it is a classical solution of $u'' + k^2u = f(x)$, $u(0) = u(1) = 0$. (This result can be extended to the general Sturm-Liouville system: see Proposition 9.12.)

2.13 The stress at a distance r from the centre of a circular elastic sheet satisfies $u'' + (3/r)u' + (a + u)^{-2} = 0$ for $0 < r < 1$, with $u(1) = 0$ and u well-behaved as $r \to 0$. Use Green's function for the operator $\mathrm{D}^2 + (3/r)\mathrm{D}$ to convert this differential equation into the integral equation $u(r) = \int_0^1 g(r;s)\,[a + u(s)]^{-2}\,\mathrm{d}s$ where $g(r;s) = (s - s^3)/2$ for $r < s$, $g(r;s) = s^3(r^{-2} - 1)/2$ for $r > s$. (The solution of this equation is considered in Problem 6.25.)

5.21 Prove that if $0 \leqslant \lambda \leqslant 1$, then the equation

$$u(x) = \lambda \int_0^1 ds/[1 + x + u(s)] \quad \text{for } 0 \leqslant x \leqslant 1$$

has exactly one continuous non-negative solution.

5.22 Let P be the set of all ordered pairs $f = (f_1, f_2)$ of real-valued continuous functions on $[0,1]$. Show that P is a Banach space if we define addition and scalar multiplication in the obvious way, and define $\|f\|_P = \max\{\|f_1\|_C, \|f_2\|_C\}$, where $\|\cdot\|_C$ denotes the usual sup norm in $C[0,1]$.

Show that the coupled integral equations

$$u(x) = \lambda \int_0^1 e^{xy} u(y) dy/[1 + u^2(y) + v^2(y)]$$

$$v(x) = \mu \int_0^1 e^{xy} u(y) v(y) dy/[1 + u^2(y) + v^2(y)]$$

have no nontrivial continuous solutions if $|\lambda| < 1/2e$ and $|\mu| < 1/e$.

5.23 This is a generalisation of Problem 5.15. Consider the equation $u''(x) = f(x, u(x), u'(x))$ for $a < x < b$, with $u(a) = p, u(b) = q$. Suppose the function $f(x,y,z)$ is continuous, and $|\partial f/\partial y| \leqslant m$, $|\partial f/\partial z| \leqslant n$ for all x,y,z.

Write the equation in the form $u = Tu$ where T is an integral operator. Let B be the space of all continuously differentiable functions on $[a,b]$, with norm $\|u\| = m \sup|u| + n \sup|u'|$. Show that B is a Banach space. Use the contraction mapping theorem to show that the above boundary-value problem has a unique solution provided that the length of the interval $[a,b]$ is less than δ, where δ is the positive root of the equation $(m/8)\delta^2 + (n/2)\delta = 1$.

6.25 Return to the elastic stress equation considered in Problem 2.13. Set $P(r) = a^{-2} \int_0^1 g(r;s) ds$. Use Schauder's theorem to show that the equation has a solution u satisfying $0 \leqslant u(r) \leqslant P(r)$ for $0 \leqslant r \leqslant 1$.

6.26 Consider the equation

$$3u(x) = x + [u(x)]^2 + \int_0^1 |x - u(s)|^{1/2} ds \ .$$

Show that it has a continuous solution u satisfying $0 \leqslant u(x) \leqslant 1$ for $0 \leqslant x \leqslant 1$ by means of the following, which is a combination of the contraction mapping and Schauder theorems.

Krasnoselskii's fixed point theorem: let S be a closed bounded convex set in a Banach space N; then if an operator $P: S \to S$ is the sum of a contraction $S \to N$ and a continuous map $S \to T$ where T is relatively compact, then P has a fixed point in S. (For a proof, see Smart.)

6.27 Prove that the following is (like Example 6.19) a continuous operator on a closed bounded convex set which has no fixed point. The set S is $\{f \in C[0,1] : \|f\| \leqslant 1, f(0) = 0, f(1) = 1\}$; the operator $T: S \to S$ is defined by $(Tf)(x) = f(x^2)$.

7.28 In the vector space $\mathbb{R}^n$ use the norm $\|u\| = \Sigma\ |u_r|$. Let $x = (1,-1,0,0,\ldots,0)$ and let E be the subspace $\{(t,t,0,0,\ldots,0): t \in \mathbb{R}\}$. Writing $y_t = (t,t,0,\ldots,0)$ for elements of E, show that all y_t with $|t| \leqslant 1$ are the same distance from x, and are closer to x than any y_t with $|t| > 1$. This shows that best approximations in a subspace can be nonunique in normed spaces, though in Hilbert space they are unique (Corollary 7.51). Deduce that the norm $\Sigma\ |u_r|$ cannot be obtained from any inner product.

Give an example illustrating nonunique best approximations in the space of continuous functions on $[0,1]$ with norm $\int_0^1 |f(t)|\,\mathrm{d}t$.

8.29 Prove that a bounded operator maps a weakly convergent sequence into a weakly convergent sequence. (Compare this with the statement 8.55 for compact operators.)

8.30 Prove that the adjoint of a compact operator is compact.

8.31 A bounded linear operator $A: H \to H$ is called a Hilbert-Schmidt operator if the series $\sum_{ij} |(Ae_i,f_j)|^2$ converges whenever (e_i) and (f_j) are orthonormal bases for the Hilbert space H. Show that this sum equals $\Sigma\ |Ae_i|^2$, and deduce that it is independent of the choice of bases (e_i) and (f_j).

Show that the set of all Hilbert-Schmidt operators on a given Hilbert space H is a vector space, and that $N(A) = \{\Sigma\ \|Ae_i\|^2\}^{1/2}$ is a norm on that space. Show that $N(A) \geqslant \|A\|$ where $\|A\|$ is the usual operator norm. Give an example in which $N(A) > \|A\|$.

If A and B are Hilbert-Schmidt operators, show that $\Sigma(Ae_i,Be_i)$ converges absolutely for every orthonormal basis (e_i), and is independent of the choice of (e_i). Show that one can define an inner product $[A,B]$ on the space of Hilbert-Schmidt operators on H by $[A,B] = \Sigma(Ae_i,Be_i)$.

If A and B are linear integral operators on $L_2[a,b]$ with continuous kernels K and L respectively, show that they are Hilbert-Schmidt operators, and $[A,B] = \int\int K(s,t)\overline{L(s,t)}\,\mathrm{d}s\mathrm{d}t$.

(See Balakrishnan for further developments and applications.)

8.32 Let $H = \{f \in L_2[0,1] : f' \in L_2[0,1]\}$, and for $f,g \in H$ define $(f,g) = f(0)g(0) + \int_0^1 f'(s)g'(s)\,\mathrm{d}s$. Take L_2 here to the space of real functions. Show that H is a Hilbert space. For each $t \in [0,1]$ define a function $R_t \in H$ by $R_t(s) = 1 + \min(s,t)$, where $\min(s,t)$ denotes the smaller of s and t. Show that $(f,R_t) = f(t)$ for all $f \in H$.

Now consider the following problem in approximation theory. The interval $[0,1]$ is divided into subintervals by given numbers $0 = t_1 < t_2 < \ldots < t_n = 1$. Given a function f, we wish to approximate it by a piecewise linear function F which is linear in each subinterval. Show that the set of all such functions F is the subspace of H spanned by $\{R_{t_i}: i = 1,\ldots,n\}$. Show that the best piecewise linear approximation to f (in the sense of the norm corresponding to the above inner product in H) is the piecewise linear function F which equals f at the points t_i.

(R_t is called a 'reproducing kernel' for H, because its inner product with any function in H reproduces the values of that function at t. It is a Hilbert-space relative of the delta function. For more details and applications in numerical analysis, see Moore (1985), Chapter 9.)

9.23 Discuss the following paradox. Let A be a compact self-adjoint operator on a Hilbert space H, and suppose that all the eigenvalues λ_n of A are non-zero. If e_n are the corresponding orthonormal eigenfunctions, then an easy calculation shows that the solution of the equation $Au = f$ is $u = \Sigma\, \lambda_n^{-1}(f,e_n)e_n$.

But this is nonsense. Suppose $f = \Sigma\, n^{-1/2}\lambda_n e_n$ (it is easy to find an A such that $\lambda_n \to 0$ fast enough to make this series converge). The above formula then gives $u = \Sigma\, n^{-1/2} e_n$, which contradicts Bessel's inequality. Since the formula $u = \Sigma\, \lambda_n^{-1}(f,e_n)e_n$ can give meaningless results, it should be discarded.

10.22 Define an operator P on $L_2[0,1]$ by $(Pu)(x) = u(x) - \int_0^1 e^{-xy}u(y)\mathrm{d}y$. Show that P is a positive operator. Obtain an upper bound for $\int_0^1 xv(x)\mathrm{d}x$ where v is a solution of $v(x) = x + \int_0^1 e^{-xy}v(y)\,\mathrm{d}y$.

10.23 Let A be a strongly positive symmetric operator on a real inner product space V, with inner product denoted by $(,)$. Show that a new inner product space W can be defined with the same elements as V but with inner product $[u,v] = (u,Av)$.

Let (ϕ_n) be a basis for V, and let (ψ_n) be the vectors obtained by applying the Gram-Schmidt orthonormalisation process in W to the (ϕ_n). Show that the n-th order Rayleigh-Ritz approximation to the solution of the equation $Au = f$ can be written in the form $u^{(n)} = \sum_{r=1}^{n} c_r \psi_r$ where $c_r = (\psi_r, f)$.

Show that $(\psi_r,f) = [\psi_r,u]$, and deduce that $u^{(n)} \to u$ in W. Finally, use the fact that A is strongly positive to show that convergence in W implies convergence with respect to the original norm in V. This proves the convergence of the Rayleigh-Ritz method.

Index of Symbols

References and Name Index

The numbers in square brackets following each entry give the pages of this book on which reference to the entry is made.

ARSAC, J. (1966) *Fourier Transforms and the Theory of Distributions.* Prentice-Hall. [63, 78]
ARTHURS, A. M. (1980) *Complementary Variational Principles*, 2nd edition. Oxford U.P. [301]
BALAKRISHNAN, A. V. (1976) *Applied Functional Analysis.* Springer. [206, 209]
BANKS, W. H. H. & M. B. ZATURSKA (1981) The Unsteady Boundary-Layer Development on a Rotating Disc in Counter-Rotating Flow. *Acta Mechanica*, **38**, 143–155. [169]
BARNSLEY, M. F. & P. D. ROBINSON (1977) Bivariational Bounds for Non-Linear Problems. *J. Inst. Maths. Applics.*, **20**, 485–504. [301]
BATCHELOR, G. K. (1951) Note on a Class of Solutions of the Navier-Stokes Equations Representing Steady Rotationally Symmetric Flow. *Q. J. Mech. Appl. Math.*, **4**, 29–41. [168]
BATCHELOR, G. K. (1967) *An Introduction to Fluid Mechanics.* Cambridge U. P. [165, 169, 171]
BELL, E. T. (1937) *Men of Mathematics.* Gollancz. [356]
BELLMAN, R. (1960) *Introduction to Matrix Analysis.* McGraw-Hill. [147, 236]
BELLMAN, R. (1970/73) *Methods of Nonlinear Analysis*, 2 vols., Academic Press. [301, 319]
BERNKOPF, M. (1966) The Development of Function Spaces with Particular Reference to their Origins in Integral Equation Theory. *Arch. Hist. Exact Sciences*, **3**, 1–136. [112]
BODONYI, R. J. (1978) On the Unsteady Similarity Equations for the Flow above a Rotating Disc in a Rotating Fluid. *Q. J. Mech. Appl. Math.*, **31**, 461–472. [168]
BODONYI, R. J. & K. STEWARTSON (1977) The unsteady laminar boundary layer on a rotating disc in a counter-rotating fluid. *J. Fluid Mech.*, **79**, 669–688. [169]
BRACEWELL, R. (1965) *The Fourier Transform and its Applications.* McGraw-Hill. [63, 78]
BRAUN, M. (1975) *Differential Equations and their Applications.* Springer. [320]
CARROLL, L. (1865) *Alice's Adventures in Wonderland.* Macmillan. [350]
CARTAN, H. (1971) *Differential Calculus.* Herrman/Kershaw. [314, 317, 334]
CHESTER, W. (1975) The Forced Oscillations of a Simple Pendulum. *J. Inst. Maths. Applics.*, **15**, 289–306. [171]
CHILLINGWORTH, D. (1976) *Differential Topology with a View to Applications.* Pitman. [334]
CODDINGTON, E. A. & N. LEVINSON (1955) *Theory of Ordinary Differential Equations.* McGraw-Hill, [51, 325, 362]
COLE, R. J. & D. C. PACK (1975) Some Complementary Bivariational Principles for Linear Integral Equations of Fredholm Type. *Proc. Roy. Soc.*, **A 347**, 239–252. [301, 305]
COLLATZ, L. (1966) *Functional Analysis and Numerical Mathematics.* Academic Press. [134, 171, 172, 370]
COLOMBEAU, J. F. (1984) *New Generalised Functions and Multiplication of Distributions.* North-Holland (Mathematics Studies No. 84). [33]
COURANT, R. & D. HILBERT (1953/62) *Methods of Mathematical Physics*, 2 vols., Interscience. [48, 51, 71, 73, 78, 158, 171, 301, 318, 374]
COURANT, R. & H. ROBBINS (1941) *What is Mathematics?* Oxford U.P. [170]
CRONIN, J. (1964) *Fixed Points and Topological Degree in Nonlinear Analysis.* Amer. Math. Soc. [171]

CURTAIN, R. F. & A. J. PRITCHARD (1977) *Functional Analysis in Modern Applied Mathematics.* Academic Press. [236]
DAUBEN, J. W. (1979) *Georg Cantor: His Mathematics and Philosophy of the Infinite.* Harvard U.P. [356]
DENNERY, P. & A. KRZYWICKI (1967) *Mathematics for Physicists.* Harper & Row. [78]
DETTMAN, J. W. (1969) *Mathematical Methods in Physics and Engineering*, 2nd edition. McGraw-Hill. [53, 78, 301]
DIEUDONNÉ, J. (1960) *Foundations of Modern Analysis.* Academic Press. [314, 334]
DIJKSTRA, D. (1980) On the Relation Between Adjacent Inviscid Cell Type Solutions to the Rotating-Disk Equations. *J. Eng. Math.*, **14**, 133–154. [168]
DIRAC, P. A. M. (1930) *The Principles of Quantum Mechanics.* Oxford U.P. [12, 33]
DRAZIN, P. G. & D. H. GRIFFEL (1977) On the Branching Structure of Diffusive Climatological Models. *J. Atmos. Sci.*, **34**, 1690–1706. [332]
DRAZIN, P. G. & W. H. REID (1980) *Hydrodynamic Stability.* Cambridge U.P. [334]
DUNFORD, N. & J. T. SCHWARTZ (1958) *Linear Operators, Part I, General Theory.* Interscience. [171]
ELCRAT, A. R. (1980) On the Flow Between a Rotating Disk and a Porous Disk. *Arch. Rat. Mech. Anal.*, **73**, 63–68. [168, 170]
ERDÉLYI, A. (1962) *Operational Calculus and Generalised Functions.* Holt, Rinehart & Winston. [33]
EVANS, D. V. & C. A. N. MORRIS (1972) The Effect of a Fixed Vertical Barrier on Obliquely Incident Waves in Deep Water. *J. Inst. Maths. Applics.*, **9**, 198–204. [293]
EVANS, J. W. (1971) Nerve Axon Equations: 1 Linear Approximations. Indiana Univ. *Math. J.*, **21**, 877–885. [325]
FIFE, P. C. (1979) Mathematical Aspects of Reacting and Diffusing Systems. Springer; *Lecture Notes in Biomathematics,* **28**, [134]
FRIEDMAN, A. (1969) *Partial Differential Equations.* Holt, Rinehart & Winston. [346]
FRÖBERG, C.-E. (1965) *Introduction to Numerical Analysis.* Addison-Wesley. [373]
GANTMACHER, F. R. (1959) *Applications of the Theory of Matrices.* Interscience. [148, 370]
GEL'FAND, I. M. & G. E. SHILOV (1964) *Generalised Functions*, vol. 1. Academic Press. [33, 62, 78]
GILBARG, D. & N. S. TRUDINGER (1977) *Elliptic Partial Differential Equations of Second Order.* Springer. [346]
GOULD, S. H. (1957) *Variational Methods for Eigenvalue Problems.* Toronto U.P. [301]
GREENSPAN, D. (1972) Numerical Studies of Flow Between Rotating Coaxial Discs. *J. Inst. Maths. Applics.*, **9**, 370–377. [168]
HADAMARD, J. (1923) *Lectures on Cauchy's Problem.* Yale U.P. [29]
HALE, J. (1977) *Theory of Functional Differential Equations.* Springer. [134]
HALMOS, P. R. (1958) *Finite-Dimensional Vector Spaces.* Van Nostrand. [112]
HARDY, G. H. (1949) *Divergent Series.* Oxford U.P. [33]
HASTINGS, S. P. (1970) An Existence Theorem for some Problems from Boundary Layer Theory. *Arch. Rat. Mech. Anal.*, **38**, 308–316. [170]
HELMBERG, G. (1969) *Introduction to Spectral Theory in Hilbert Space.* North-Holland. [205, 269]
HIGGINS, J. R. (1977) *Completeness and Basis Properties of Sets of Special Functions.* Cambridge U.P. [190, 206]
HOCHSTADT, H. (1973) *Integral Equations.* Wiley. [134, 145, 158, 171, 301, 370]
HOFSTADTER, R. (1979) *Gödel, Escher, Bach: an Eternal Golden Braid.* Basic Books. [192, 356]
HOSKINS, R. F. (1979) *Generalised Functions.* Horwood. [33, 108]
HOWARD, L. N. (1961) A Note on the Existence of Certain Viscous Flows. *J. Math. & Phys.*, **40**, 172–176. [169]
HUTSON, V. & J. S. PYM (1980) *Applications of Functional Analysis and Operator Theory.* Academic Press. [112, 171, 206, 269, 329, 342, 346]
IOOSS, G. & D. D. JOSEPH (1980) *Elementary Stability and Bifurcation Theory.* Springer. [332]
JACKSON, J. D. (1975) *Classical Electrodynamics.* Second Edition. Wiley. [368]

JONES, D. S. (1966) *Generalised Functions.* McGraw-Hill. [33, 78]
JORDAN, D. W. & P. SMITH (1977) *Nonlinear Ordinary Differential Equations.* Oxford U.P. [165]
JOSEPH, D. D. (1976) *Stability of Fluid Motions.* Springer. [334]
KANTOROVICH, L. V. & G. P. AKILOV (1964) *Functional Analysis in Normed Spaces.* Pergamon. [112, 171, 370]
KEADY, G. & J. NORBURY (1975) The Jet from a Horizontal Slot Under Gravity. *Proc. Roy. Soc.*, **A 344**, 471–487. [134]
KIELHOFER, H. (1976) On the Liapounov Stability of Stationary Solutions of Semilinear Parabolic Differential Equations. *J. Diff. Eqs.*, **22**, 193–208. [325]
KLINE, M. (1972) *Mathematical Thought from Ancient to Modern Times.* Oxford U.P. [112, 350]
KOLMOGOROV, A. N. & S. V. FOMIN (1957) *Elements of the Theory of Functions and Functional Analysis*, vol. 1. Graylock Press. [101, 112, 134, 171, 260]
KOLMOGOROV, A. N. & S. V. FOMIN (1970) *Introductory Real Analysis.* Prentice-Hall. [101, 112, 134, 171, 206, 268]
KORDYLEWSKI, W. (1979) Critical Parameters of Thermal Explosion. *Combustion and Flame*, **34**, 109–117. [134]
KOREVAAR, J. (1968) *Mathematical Methods*, vol. 1. Academic Press. [33, 34, 180, 367]
KRASNOSELSKII, M. A. (1964) *Topological Methods in the Theory of Nonlinear Integral Equations.* Pergamon. [171, 329]
KREIDER, D. L., R. G. KULLER, D. R. OSTBERG & F. W. PERKINS (1966) *An Introduction to Linear Analysis.* Addison-Wesley. [51, 84, 112, 257, 348]
KREISS, H.-O. & S. V. PARTER (1983) On the Swirling Flow Between Rotating Coaxial Disks: Existence and Nonuniqueness. *Comm. Pure Appl. Math.*, **36**, 55–84. [168]
KREYSZIG, E. (1978) *Introductory Functional Analysis with Applications.* Wiley. [112, 134, 136, 205, 269, 277]
LANCZOS, C. (1961) *Linear Differential Operators.* Van Nostrand. [51]
LANCZOS, C. (1966) *Discourse on Fourier Series.* Oliver & Boyd. [35, 78]
LIN, C. C. & L. A. SEGEL (1974) *Mathematics Applied to Deterministic Problems in the Natural Sciences.* MacMillan. [334]
LLOYD, N. G. (1978) *Degree Theory.* Cambridge U.P. [171]
LUENBERGER, D. G. (1969) *Optimization by Vector Space Methods.* Wiley. [112, 206]
LIGHTHILL, M. J. (1959) *Introduction to Fourier Analysis and Generalised Functions.* Cambridge U.P. [78]
LUSTERNIK, L. A. & V. J. SOBOLEV (1974) *Elements of Functional Analysis.* Third English Edition, Hindustan Publishing Co./Wiley. [206]
McLEOD, J. B. (1969) Von Kármán's Swirling Flow Problem. *Arch. Rat. Mech. Anal.*, **33**, 91–102. [170]
McLEOD, J. B. (1970) A Note on Rotationally Symmetric Flow Above an Infinite Rotating Disc. *Mathematika*, **17**, 243–249. [168]
McLEOD, J. B. & S. V. PARTER (1974) On the Flow Between Two Counter-Rotating Infinite Plane Discs. *Arch. Rat. Mech. Anal.*, **54**, 301–327. [168]
MARSDEN, J. E. (1974) *Elementary Classical Analysis.* W. H. Freeman. [349]
MASSEY, W. S. (1961) *Algebraic Topology: an Introduction.* Harcourt, Brace & World. [170]
MATKOWSKY, B. J. & W. L. SIEGMANN (1976) The Flow Between Counter-Rotating Discs at High Reynolds Number. *S.I.A.M.J. Appl. Math.*, **30**, 720–727. [168]
MAYNARD SMITH, J. (1974) *Models in Ecology.* Cambridge U.P. [320]
MELZAK, Z. A. (1973/76) *Companion to Concrete Mathematics.* 2 vols., Wiley. [92, 367]
MIKHLIN, S. G. (1964) *Variational Methods in Mathematical Physics.* Pergamon. [301]
MILNE, R. D. (1980) *Applied Functional Analysis.* Pitman. [112, 206, 301]
MOORE, F. K. (1956) Three-Dimensional Boundary Layer Theory. in: *Advances in Applied Mechanics*, vol. 4. Academic Press. [167]
MOORE, R. E. (1985) *Computational Functional Analysis.* Ellis Horwood. [172, 235, 334]
MORSE, P. M. & H. FESHBACH (1953) *Methods of Theoretical Physics.* 2 vols., McGraw-Hill. [78, 301]
MURRAY, J. D. (1977) *Lectures on Nonlinear Differential-Equation Models in Biology.*

Oxford U.P. [134]
NAYLOR, A. W. & G. R. SELL (1971) *Linear Operator Theory in Engineering and Science.* Holt, Rinehart & Winston. [206, 269]
NOWINSKI, J. L. (1981) *Applications of Functional Analysis in Engineering.* Plenum Press. [301]
OLDHAM, K. B. & J. SPANIER (1974) *The Fractional Calculus.* Academic Press. [52]
PIPPARD, A. B. (1978) *The Physics of Vibration*, vol. 1. Cambridge U.P. [368]
POLYA, G. (1954) *Induction and Analogy in Mathematics.* Princeton U.P. [302]
POLYA, G. & G. SZEGÖ (1951) *Isoperimetric Inequalities in Mathematical Physics.* Princeton U.P. [302]
POLYA, G. & G. SZEGÖ (1972) *Problems and Theorems in Analysis*, vol. 1. Springer. [237]
PRYCE, J. D. (1973) *Basic Methods of Linear Functional Analysis.* Hutchinson. [206, 269]
RALL, L. B. (Ed.) (1971) *Nonlinear Functional Analysis and Applications.* Academic Press. [319]
REED, M. & B. SIMON (1972) *Methods of Modern Mathematical Physics, Volume 1: Functional Analysis.* Academic Press. [171]
RICHTMYER, R. D. (1979) *Principles of Advanced Mathematical Physics*, vol. 1. Springer. [338, 340, 345]
RIESZ, F. & B. SZ-NAGY (1955) *Functional Analysis*, 2nd Edition, Ungar. [206, 208, 277]
ROGERS, M. H. & G. N. LANCE (1960) The Rotationally Symmetric Flow of a Viscous Fluid in the Presence of an Infinite Rotating Disc. *J. Fluid Mech.*, 7, 617–631. [167]
ROSS, B. (1977) Fractional Calculus. *Math. Magazine*, **50**, 115–122. [52]
RUDIN, W. (1976) *Principles of Mathematical Analysis*, 3rd Edition, McGraw-Hill. [108, 349]
SATTINGER, D. H. (1973) Topics in Stability and Bifurcation Theory. Springer, *Lecture Notes in Mathematics*, vol. **309**. [325]
SAWYER, W. W. (1978) *A First Look at Numerical Functional Analysis.* Oxford U.P. [134, 221]
SCHECHTER, M. (1977) *Modern Methods in Partial Differential Equations.* McGraw-Hill. [346]
SCHWARTZ, L. (1950/51) *Théorie des Distributions.* Herrmann. [33]
SCHWARZENBERGER, R. L. E. (1969) *Elementary Differential Equations.* Chapman & Hall. [40]
SHILOV, G. E. (1968) *Generalised Functions and Partial Differential Equations.* Gordon & Breach. [33]
SHOWALTER, R. E. (1977) *Hilbert Space Methods for Partial Differential Equations.* Pitman. [236, 346]
SMART, D. R. (1974) *Fixed Point Theorems.* Cambridge U.P. [134, 147, 148, 171]
SMIRNOV, V. I. (1964) *A Course of Higher Mathematics*, vol. II. Pergamon. [364]
SNEDDON, I. N. (1972) *The Use of Integral Transforms.* McGraw-Hill. [53, 78]
SOBOLEV, S. L. (1964) *Partial Differential Equations of Mathematical Physics.* Pergamon. [78]
STAKGOLD, I. (1967/68) *Boundary-Value Problems of Mathematical Physics.* 2 vols., Macmillan. [33, 51, 52, 65, 73, 78, 112, 269, 301]
STAKGOLD, I. (1971) Branching of Solutions of Nonlinear Equations. *S.I.A.M. Rev.*, **13**, 289–332. [334]
STAKGOLD, I. (1979) *Green's Functions and Boundary Value Problems.* Wiley. [33, 51. 52, 73, 78, 112, 269, 301, 332, 334]
STEWARTSON, K. (1953) On the Flow Between Two Rotating Coaxial Disks. *Proc. Camb. Phil. Soc.*, **49**, 333–341. [167]
STOKER, J. J. (1950) *Nonlinear Vibrations.* Interscience. [165]
TITCHMARSH, E. C. (1937) *Introduction to the Theory of Fourier Integrals.* Oxford U.P. [33, 36, 54, 78]
TITCHMARSH, E. C. (1939) *The Theory of Functions.* Second Edition; Oxford U.P. [31, 108, 348]
TRICOMI, F. G. (1957) *Integral Equations.* Interscience. [171]
VAN DER POL, B. & H. BREMMER (1950) *Operational Calculus Based on the Two-Sided Laplace Integral.* Cambridge U.P. [33]

VLADIMIROV, V. S. (1971) *Equations of Mathematical Physics.* Dekker. [33, 78]
VULIKH, B. Z. (1963) *Introduction to Functional Analysis for Scientists and Technologists.* Pergamon. [112, 206]
WALLACE, P. R. (1972) *Mathematical Analysis of Physical Problems.* Holt, Rinehart & Winston. [78, 368]
WEINBERGER, H. F. (1974) *Variational Methods for Eigenvalue Approximation.* S.I.A.M. [301]
WHEELER, J. A. & R. P. FEYNMAN (1954) An Absorber Theory of Radiation. *Revs. Mod. Phys.*, 17, 157–181. [77]
WHITROW, G. J. (1961) *The Natural Philosophy of Time.* Nelson; 2nd Edition, 1980, Oxford U.P. [77]
WHITTAKER, E. T. & G. N. WATSON (1927) *A Course of Modern Analysis.* 4th Edition; Cambridge U.P. [35, 107]
ZANDBERGEN, P. J. & D. DIJKSTRA (1977) Nonunique Solutions of the Navier-Stokes Equations for the Kármán Swirling Flow. *J. Eng. Math.*, 11, 176–188. [168]
ZEMAN, J. (Ed.) (1971) *Time in Science and Philosophy.* Elsevier. [77]
ZEMANIAN, A. H. (1954) An Approximate Means of Evaluating Integral Transforms. *J. Appl. Phys.*, 25, 262–266. [368]
ZEMANIAN, A. H. (1965) *Distribution Theory and Transform Analysis.* McGraw-Hill. [33, 34, 41, 51, 62]
ZEMANIAN, A. H. (1968) *Generalised Integral Transforms.* Interscience. [78, 96]

Subject Index

A CATALOG OF SELECTED

DOVER BOOKS

IN SCIENCE AND MATHEMATICS

THE HISTORICAL BACKGROUND OF CHEMISTRY, Henry M. Leicester. Evolution of ideas, not individual biography. Concentrates on formulation of a coherent set of chemical laws. 260pp. 5⅜ x 8½. 61053 5

A SHORT HISTORY OF CHEMISTRY, J. R. Partington. Classic exposition explores origins of chemistry, alchemy, early medical chemistry, nature of atmosphere, theory of valency, laws and structure of atomic theory, much more. 428pp. 5⅜ x 8½. (Available in U.S. only.) 65977-1

GENERAL CHEMISTRY, Linus Pauling. Revised 3rd edition of classic first-year text by Nobel laureate. Atomic and molecular structure, quantum mechanics, statistical mechanics, thermodynamics correlated with descriptive chemistry. Problems. 992pp. 5⅜ x 8½. 65622-5

Engineering

DE RE METALLICA, Georgius Agricola. The famous Hoover translation of greatest treatise on technological chemistry, engineering, geology, mining of early modern times (1556). All 289 original woodcuts. 638pp. 6¾ x 11. 60006-8

FUNDAMENTALS OF ASTRODYNAMICS, Roger Bate et al. Modern approach developed by U.S. Air Force Academy. Designed as a first course. Problems, exercises. Numerous illustrations. 455pp. 5⅜ x 8½. 60061-0

DYNAMICS OF FLUIDS IN POROUS MEDIA, Jacob Bear. For advanced students of ground water hydrology, soil mechanics and physics, drainage and irrigation engineering and more. 335 illustrations. Exercises, with answers. 784pp. 6⅛ x 9¼. 65675-6

ANALYTICAL MECHANICS OF GEARS, Earle Buckingham. Indispensable reference for modern gear manufacture covers conjugate gear-tooth action, gear-tooth profiles of various gears, many other topics. 263 figures. 102 tables. 546pp. 5⅜ x 8½. 65712-4

MECHANICS, J. P. Den Hartog. A classic introductory text or refresher. Hundreds of applications and design problems illuminate fundamentals of trusses, loaded beams and cables, etc. 334 answered problems. 462pp. 5⅜ x 8½. 60754-2

MECHANICAL VIBRATIONS, J. P. Den Hartog. Classic textbook offers lucid explanations and illustrative models, applying theories of vibrations to a variety of practical industrial engineering problems. Numerous figures. 233 problems, solutions. Appendix. Index. Preface. 436pp. 5⅜ x 8½. 64785-4

STRENGTH OF MATERIALS, J. P. Den Hartog. Full, clear treatment of basic material (tension, torsion, bending, etc.) plus advanced material on engineering methods, applications. 350 answered problems. 323pp. 5⅜ x 8½. 60755-0

A HISTORY OF MECHANICS, René Dugas. Monumental study of mechanical principles from antiquity to quantum mechanics. Contributions of ancient Greeks, Galileo, Leonardo, Kepler, Lagrange, many others. 671pp. 5⅜ x 8½. 65632-2

Math–Geometry and Topology

ELEMENTARY CONCEPTS OF TOPOLOGY, Paul Alexandroff. Elegant, intuitive approach to topology from set-theoretic topology to Betti groups; how concepts of topology are useful in math and physics. 25 figures. 57pp. 5⅜ x 8½. 60747-X

COMBINATORIAL TOPOLOGY, P. S. Alexandrov. Clearly written, well-organized, three-part text begins by dealing with certain classic problems without using the formal techniques of homology theory and advances to the central concept, the Betti groups. Numerous detailed examples. 654pp. 5⅜ x 8½. 40179-0

EXPERIMENTS IN TOPOLOGY, Stephen Barr. Classic, lively explanation of one of the byways of mathematics. Klein bottles, Moebius strips, projective planes, map coloring, problem of the Koenigsberg bridges, much more, described with clarity and wit. 43 figures. 210pp. 5⅜ x 8½. 25933-1

CONFORMAL MAPPING ON RIEMANN SURFACES, Harvey Cohn. Lucid, insightful book presents ideal coverage of subject. 334 exercises make book perfect for self-study. 55 figures. 352pp. 5⅝ x 8¼. 64025-6

THE GEOMETRY OF RENÉ DESCARTES, René Descartes. The great work founded analytical geometry. Original French text, Descartes's own diagrams, together with definitive Smith-Latham translation. 244pp. 5⅜ x 8½. 60068-8

THE THIRTEEN BOOKS OF EUCLID'S ELEMENTS, translated with introduction and commentary by Sir Thomas L. Heath. Definitive edition. Textual and linguistic notes, mathematical analysis. 2,500 years of critical commentary. Unabridged. 1,414pp. 5⅜ x 8½. Three-vol. set.
Vol. I: 60088-2 Vol. II: 60089-0 Vol. III: 60090-4

GEOMETRY OF COMPLEX NUMBERS, Hans Schwerdtfeger. Illuminating, widely praised book on analytic geometry of circles, the Moebius transformation, and two-dimensional non-Euclidean geometries. 200pp. 5⅝ x 8¼. 63830-8

DIFFERENTIAL GEOMETRY, Heinrich W. Guggenheimer. Local differential geometry as an application of advanced calculus and linear algebra. Curvature, transformation groups, surfaces, more. Exercises. 62 figures. 378pp. 5⅜ x 8½. 63433-7

CURVATURE AND HOMOLOGY: Enlarged Edition, Samuel I. Goldberg. Revised edition examines topology of differentiable manifolds; curvature, homology of Riemannian manifolds; compact Lie groups; complex manifolds; curvature, homology of Kaehler manifolds. New Preface. Four new appendixes. 416pp. 5⅜ x 8½. 40207-X

TOPOLOGY, John G. Hocking and Gail S. Young. Superb one-year course in classical topology. Topological spaces and functions, point-set topology, much more. Examples and problems. Bibliography. Index. 384pp. 5⅝ x 8¼. 65676-4

Physics

OPTICAL RESONANCE AND TWO-LEVEL ATOMS, L. Allen and J. H. Eberly. Clear, comprehensive introduction to basic principles behind all quantum optical resonance phenomena. 53 illustrations. Preface. Index. 256pp. 5⅜ x 8½. 65533-4

ULTRASONIC ABSORPTION: An Introduction to the Theory of Sound Absorption and Dispersion in Gases, Liquids and Solids, A. B. Bhatia. Standard reference in the field provides a clear, systematically organized introductory review of fundamental concepts for advanced graduate students, research workers. Numerous diagrams. Bibliography. 440pp. 5⅜ x 8½. 64917-2

QUANTUM THEORY, David Bohm. This advanced undergraduate-level text presents the quantum theory in terms of qualitative and imaginative concepts, followed by specific applications worked out in mathematical detail. Preface. Index. 655pp. 5⅜ x 8½. 65969-0

ATOMIC PHYSICS (8th edition), Max Born. Nobel laureate's lucid treatment of kinetic theory of gases, elementary particles, nuclear atom, wave-corpuscles, atomic structure and spectral lines, much more. Over 40 appendices, bibliography. 495pp. 5⅜ x 8½. 65984-4

AN INTRODUCTION TO HAMILTONIAN OPTICS, H. A. Buchdahl. Detailed account of the Hamiltonian treatment of aberration theory in geometrical optics. Many classes of optical systems defined in terms of the symmetries they possess. Problems with detailed solutions. 1970 edition. xv + 360pp. 5⅜ x 8½. 67597-1

THIRTY YEARS THAT SHOOK PHYSICS: The Story of Quantum Theory, George Gamow. Lucid, accessible introduction to influential theory of energy and matter. Careful explanations of Dirac's anti-particles, Bohr's model of the atom, much more. 12 plates. Numerous drawings. 240pp. 5⅜ x 8½. 24895-X

ELECTRONIC STRUCTURE AND THE PROPERTIES OF SOLIDS: The Physics of the Chemical Bond, Walter A. Harrison. Innovative text offers basic understanding of the electronic structure of covalent and ionic solids, simple metals, transition metals and their compounds. Problems. 1980 edition. 582pp. 6⅛ x 9¼. 66021-4

HYDRODYNAMIC AND HYDROMAGNETIC STABILITY, S. Chandrasekhar. Lucid examination of the Rayleigh-Benard problem; clear coverage of the theory of instabilities causing convection. 704pp. 5⅜ x 8¼. 64071-X

INVESTIGATIONS ON THE THEORY OF THE BROWNIAN MOVEMENT, Albert Einstein. Five papers (1905–8) investigating dynamics of Brownian motion and evolving elementary theory. Notes by R. Fürth. 122pp. 5⅜ x 8½. 60304-0

THE PHYSICS OF WAVES, William C. Elmore and Mark A. Heald. Unique overview of classical wave theory. Acoustics, optics, electromagnetic radiation, more. Ideal as classroom text or for self-study. Problems. 477pp. 5⅜ x 8½. 64926-1

METHODS OF THERMODYNAMICS, Howard Reiss. Outstanding text focuses on physical technique of thermodynamics, typical problem areas of understanding, and significance and use of thermodynamic potential. 1965 edition. 238pp. 5⅜ x 8½. 69445-3

TENSOR ANALYSIS FOR PHYSICISTS, J. A. Schouten. Concise exposition of the mathematical basis of tensor analysis, integrated with well-chosen physical examples of the theory. Exercises. Index. Bibliography. 289pp. 5⅜ x 8½. 65582-2

RELATIVITY IN ILLUSTRATIONS, Jacob T. Schwartz. Clear nontechnical treatment makes relativity more accessible than ever before. Over 60 drawings illustrate concepts more clearly than text alone. Only high school geometry needed. Bibliography. 128pp. 6⅛ x 9¼. 25965-X

THE ELECTROMAGNETIC FIELD, Albert Shadowitz. Comprehensive undergraduate text covers basics of electric and magnetic fields, builds up to electromagnetic theory. Also related topics, including relativity. Over 900 problems. 768pp. 5⅝ x 8¼. 65660-8

GREAT EXPERIMENTS IN PHYSICS: Firsthand Accounts from Galileo to Einstein, edited by Morris H. Shamos. 25 crucial discoveries: Newton's laws of motion, Chadwick's study of the neutron, Hertz on electromagnetic waves, more. Original accounts clearly annotated. 370pp. 5⅜ x 8½. 25346-5

RELATIVITY, THERMODYNAMICS AND COSMOLOGY, Richard C. Tolman. Landmark study extends thermodynamics to special, general relativity; also applications of relativistic mechanics, thermodynamics to cosmological models. 501pp. 5⅜ x 8½. 65383-8

LIGHT SCATTERING BY SMALL PARTICLES, H. C. van de Hulst. Comprehensive treatment including full range of useful approximation methods for researchers in chemistry, meteorology and astronomy. 44 illustrations. 470pp. 5⅜ x 8½. 64228-3

STATISTICAL PHYSICS, Gregory H. Wannier. Classic text combines thermodynamics, statistical mechanics and kinetic theory in one unified presentation of thermal physics. Problems with solutions. Bibliography. 532pp. 5⅜ x 8½. 65401-X